Wind Energy
Engineering

About the Author

Pramod Jain, Ph.D., is founder and president of Innovative Wind Energy, Inc., a wind energy consulting company. He is recognized as a global expert in the planning of wind projects and has worked on projects in the United States, the Caribbean, and Latin America that range from a single 100-kW turbine to a 100-plus MW wind farm. He has worked on wind projects for a variety of clients including Fortune 100 companies, the US government, universities, utilities, municipalities, and land developers. He was a cofounder and Chief Technologist at Wind Energy Consulting and Contracting, Inc. He has a Ph.D. in Mechanical Engineering from the University of California, Berkeley, an M.S. from University of Kentucky, Lexington, and a B.Tech. from the Indian Institute of Technology, Mumbai.

To my mother Manchi Jain, and late father U.M. Jain

Contents

Preface

I have been interested in writing short technical articles from my graduate school days. I was never good at it. In those days, I supposedly wrote dense stuff, and the audience I had in mind were experts in the field. This changed as I wrote for a corporate audience. When I got into the wind business, I wrote white papers and blogs regularly but never considered writing a book. The idea of writing this book came to me from a dear friend Satya Komatineni, author of books on Android. He encouraged me to send a proposal to McGraw-Hill about the book. This led me down to a nine-month long adventure. The best metaphor to describe the adventure is that writing a book is akin to the nine-month process of gestation and birthing of the first child. Although I have not personally experienced it, I have lived with someone who has. It is exciting, uncomfortable, painful, at times really painful, and in the end, the product makes you forget the pain.

The impetus for writing this book was the lack of books on the market that targeted engineers. Specifically, I wanted to write a book that would give an engineer, from any discipline, sufficient knowledge about the multidisciplinary field of wind energy. This book intends to bring to bear at least five disciplines in order to provide a reasonably comprehensive understanding of the field of wind energy. The five disciplines are meteorology, mechanical and aeronautical engineering, civil engineering, electrical engineering, and environmental engineering. In addition, to these core engineering disciplines, the book has chapters on finance and project management, two business-related disciplines that are key to wind energy.

I wrote the book with the following audiences in mind. First are engineers and scientists in the wind industry but who practice in a narrow segment of the industry that covers their specific discipline. Second are engineers and scientists who want to enter the wind industry. Third are undergraduate engineering students and technical college students who want to learn about the various disciplines in wind energy engineering. Finally, another intended audience is comprised of business people and project managers who work in the wind energy industry.

Engineers will find sufficient detail about each of the topics. I have kept the math to a level that would be comfortable for a practicing engineer. In areas that require sophisticated math, I have attempted to provide insights into the relationships.

As with any endeavor, I had to make decisions about what to include in the book and what to leave out. I chose to leave out of the book discussions and debates about climate change and energy policy. Although these are critical to understanding the big picture, I am not particularly qualified to write about these issues. Wherever appropriate, I have briefly discussed these two topics. This book is not an engineering design manual for turbines. The exposition on turbines is limited to describing the major components and their functions; it does not cover the complexity of computing forces and displacements nor design and engineering of the components.

The book starts with a brief description of the wind energy business with an emphasis on the explosive growth witnessed by the wind energy industry. Although such an explosive growth rate is difficult to sustain for long periods, I believe that the wind industry will experience sustained 15 to 20% growth over the next decade. On the basis of this conservative estimate, there will be a healthy demand for engineers, technicians, scientists, project managers, and financiers for years to come.

The second chapter of the book introduces readers to the concepts of energy and power, what kind and how much energy is contained in wind, and how much of it can be captured by a wind turbine.

The third chapter describes properties of wind from a meteorological perspective. It starts with a description of how wind is generated. Next, the statistical nature of wind speed is described, followed by the impact of height on wind speed. The chapter then concludes with dependence of wind energy on air density and dependence of air density on temperature, pressure, and humidity.

The fourth chapter describes the mechanics of how wind energy is converted into mechanical energy using aerodynamics of blades. This is important in order to understand the functioning of a wind turbine. The fifth chapter presents a more detailed exposition on the aerodynamics of blades and how power performance curves of turbines are created.

The sixth chapter switches from the science of energy and airflow to the science of measurement. Measurement of wind speed is a crucial step in a wind project because all utility scale projects require it, and in most cases, it is the longest duration task. Measurement is a key step in reducing uncertainty related to the financial performance of a wind project.

The seventh chapter deals with wind resource assessment. It is another pivotal step in the development phase of a wind project. In this chapter, different methods of assessment are covered, from methods based on publicly available wind data and no onsite measurements, to methods that extrapolate measured data along three spatial axes and the temporal axis. In the eighth chapter, advanced wind resource assessment topics such as computation of extreme wind speed, and modeling of rough terrain and wake are described. Losses and uncertainty associated with the various components of wind resource assessment are also covered in this chapter.

The ninth chapter describes the components of a wind turbine generator. The rotor system, nacelle, and tower and foundation systems are described. The components of these three systems are described for different types of utility scale turbines.

The tenth chapter deals with the electrical side of wind energy. Basic concepts of electricity and magnetism are covered followed by description of various types of generators used in wind turbines. In the eleventh chapter, the integration with an electricity grid is described. It covers how the variability of wind energy is incorporated in the grid, the grid interconnection standards, and the protection systems required in a wind farm. In addition, several topologies of wind farm from an electrical standpoint are explained.

The twelfth chapter covers the environmental impact of wind projects. It begins by setting the context for relative impact relative to fossil fuel-based generation. In the chapter, each of the environmental impacts: wildlife, noise, esthetics, shadow flicker, and others are described. In addition, impact on aviation, radar, and telecommunications are described.

The thirteenth chapter describes financial models used to evaluate wind energy projects. In this chapter, the various components of revenue, capital costs, and recurring costs are described. The impact of incentives, in particular tax incentives in the United States, on the financial performance is detailed. Finally, the financial performance measures used to evaluate wind projects are described.

The fourteenth and final chapter describes planning and execution of wind projects. This chapter will serve as a guide to project managers of wind energy projects during development, construction and commissioning, and operations.

I learned a lot while writing this book. There were quite a few things that I was certain were true but which turned out to be not so true. There were more things that I had explained with confidence to colleagues and clients, which turned out to be full of holes and superficial, at best. I hope the book serves a similar purpose in helping you to better understand wind energy.

PRAMOD JAIN

Acknowledgments

The first acknowledgment goes to the family. This book would not have been possible without the support of my wife Shobhana and two wonderful daughters Suhani and Sweta. The book took a significant toll on the family; I am grateful for their wholehearted support and backing. I also want to thank my mother and sisters Savita and Rekha for their support.

The second acknowledgment goes to my colleagues at Wind Energy Consulting and Contracting, Inc. I am grateful to Wayne Hildreth, who got me into the wind industry and Glenn Mauney and Mike Steinke for helping me to sell the products to clients and giving me the opportunity to hone my skills and to all the other colleagues. I learned a lot from Per Nielsen of EMD, who always responded to my strangest queries. Other people that helped me learn about the various facets of wind industry are Tim Printy, Kirk Heston, Mark Tippett, Craig White, and Ralph Wegner.

The third acknowledgment goes to companies that shared pictures and data for the book including Alan Henderson of P&H, Vergnet, Vensys, Bosch-Rexroth, SKF, Vestas, GE, WindPower Monthly, World Wind Energy Association, American Wind Energy Association, Lawrence Berkeley National Lab, and National Renewable Energy Lab.

Next, I would like to thank the International Electrotechnical Commission (IEC) for permission to reproduce Information from its International Standard IEC 61400-1 ed.3.0 (2005). All such extracts are copyright of IEC, Geneva, Switzerland. All rights reserved. Further information on the IEC is available from www.iec.ch. IEC has no responsibility for the placement and context in which the extracts and contents are reproduced by the author, nor is IEC in any way responsible for the other content or accuracy therein.

Finally, I want to thank McGraw-Hill for accepting my proposal for the book and helping me with the editing and publishing process.

Wind Energy
Engineering

CHAPTER 1

Overview of Wind Energy Business

First, there is the power of the Wind, constantly exerted over the globe... Here is an almost incalculable power at our disposal, yet how trifling the use we make of it.
—Henry David Thoreau, American naturalist and author (1834)

Introduction

The energy of wind has been exploited for thousands of years. The oldest applications of wind energy include extracting water from wells, making flour out of grain, and other agricultural applications. In recent times, the use of wind energy has evolved to, primarily, generation of electricity.

The field of wind energy blossomed in 1970s after the oil crisis, with a large infusion of research money in the United States, Denmark, and Germany to find alternative sources of energy. By the early 1980s, incentives for alternative sources of energy had vanished in the United States and, therefore, the wind energy field shrank significantly. Investments continued in Europe and, until recently, Europe led in terms of technology and wind capacity installations.

Worldwide Business of Wind Energy

The data presented in this section is from the World Wind Energy Report 2009 by the World Wind Energy Association.[1] According to this report, in 2009, wind energy was a 50 billion Euro business in terms of revenue and it employed about 550,000 people around the world.

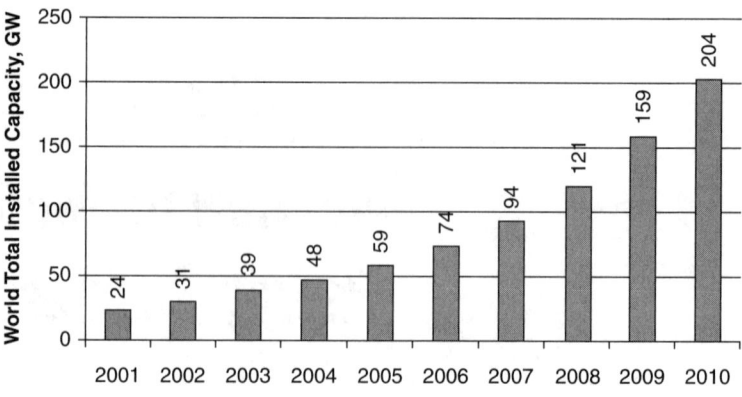

FIGURE 1-1   Total installed capacity of wind power worldwide.[1] 2010 data is a forecast.

Figure 1-1 shows the installed wind capacity in the world by year. In 2009, 159.2 GW of wind capacity was online. Figure 1-2 shows the new installed capacity by year. The pace of growth of new installed capacity has increased. In fact, the world market for wind capacity grew by 21.3% in 2004 and has steadily increased to 31.7% in 2009.

Figure 1-3 illustrates the total wind capacity by country. The United States leads in wind capacity installations with 35.1 GW, followed by China and Germany at 26 and 25.7 GW, respectively. The UK leads in offshore installations, with a total capacity of 688 MW followed by Denmark at 663 MW (see Fig. 1-4).

In terms of penetration of wind energy in the total electricity supply, Denmark leads with 20%, followed by Portugal, Spain, and Germany at 15, 14, and 9%, respectively. Penetration in the United States is slightly below 2%.[2]

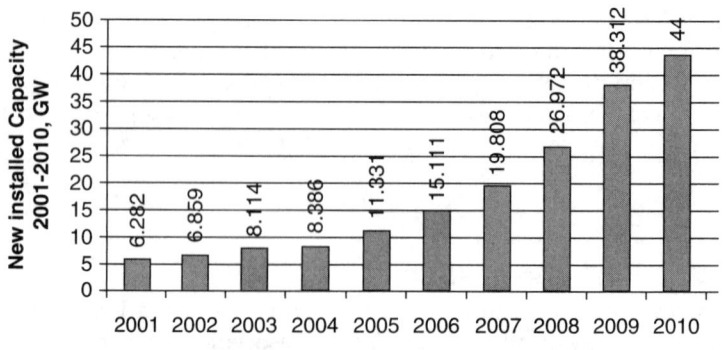

FIGURE 1-2   New installed capacity of wind power worldwide[1] 2010 data is a forecast.

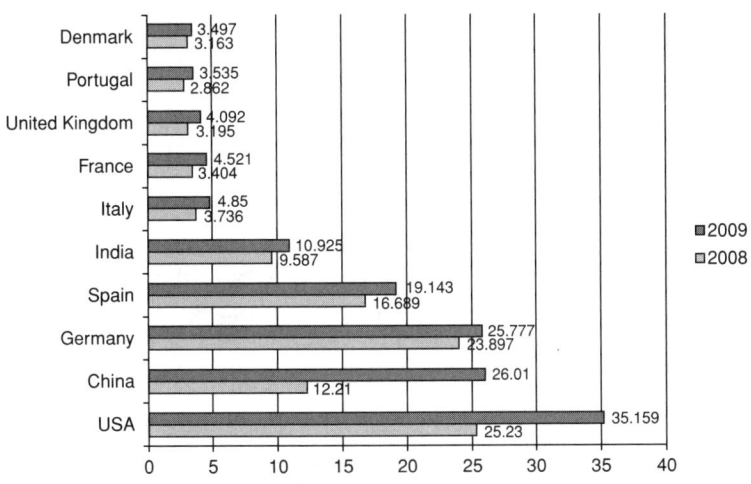

Figure 1-3 Total installed capacity of wind power (GW) by country for top ten countries.[1]

The prominence of wind in the last half of the first decade of the twenty-first century is evident in the fact that it is the leading source of newly installed electricity generation capacity in the United States. In the United States, out of a total of 20 GW of new electricity generation in 2008, 42% was from wind energy.[2] The percentage has risen steadily since 2005, when wind was 12% among generation types in annual capacity addition. From an energy standpoint, the prominence of wind is even more impressive. The Lawrence Berkeley National Laboratory (LBL) report[2] predicts, "almost 60% of the nation's projected increase in electricity generation from 2009 through 2030 would be met with wind electricity. Although future growth trends are hard to predict, it is clear that a significant portion of the country's new generation

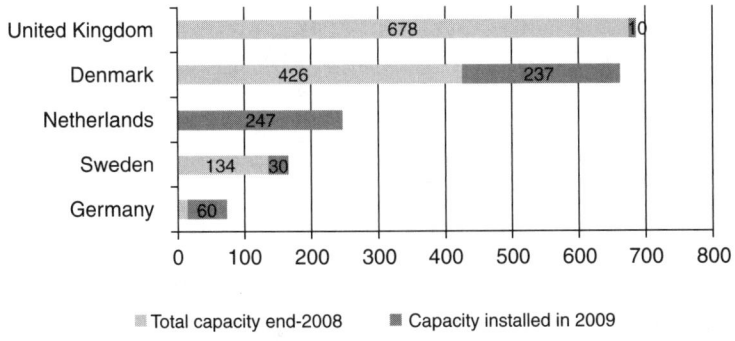

Figure 1-4 Total installed capacity of offshore wind power (MW) in the top five countries.[1]

needs is already being met by wind." The LBL report used forecast data from Energy Information Administration of the US Department of Energy (DOE).

Cost of Wind Energy

The cost of wind energy is comparable to fossil-fuel–based energy, when cost of greenhouse gas emissions is taken into account. Average cost of energy[3] from coal is about €80 per MWh, while wind energy at a site with average annual wind speed of 7 m/s is slightly less than €80 per MWh. Figure 1-5 is a plot of levelized cost of energy from coal, natural gas, nuclear, and onshore and off-shore wind for average wind speed in the range of 6 to 10 m/s.

Table 1-1 compares the components of cost of wind energy projects to other source of electricity generation. Capital cost and O&M cost for onshore wind projects are comparable to coal-fired projects. The advantage of wind is that it has no fuel cost.

According to the DOE report,[4] the amount of economically viable onshore wind power is 8000 GW that can be produced at a cost of $85 per MWh or less. Figure 1-6 is a plot of potential of wind energy and the cost of energy in the United States, as a function of class of wind resource.

Benefits of Wind Energy

The primary benefits of wind energy are environmental and cost. Wind energy production results in zero emissions. Compared to fossil

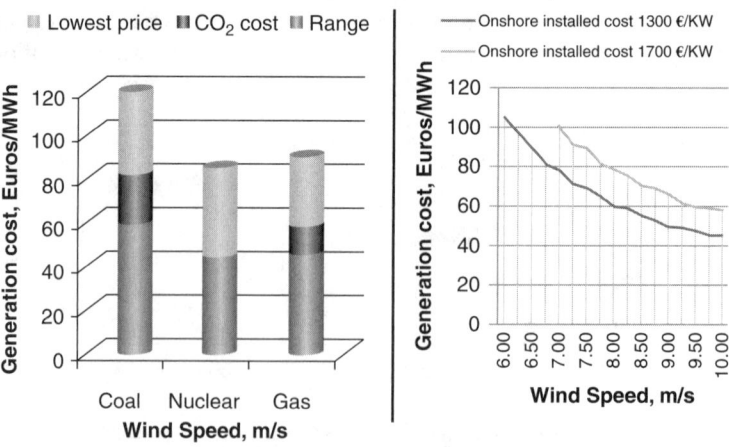

FIGURE 1-5 Levelized cost of energy from different sources. Costs are in euros per MWh. Cost of wind energy is a function of wind speed.[3]

Technology	Installed Cost, €/kW	Fuel Price, €/MWh	O&M Cost, €/kW
Gas-fired	635–875	US: 16 EU: 27	19–30
Coal-fired	1300–2325	US: 12 EU: 18	30–60
Nuclear	1950–3400	3.6–5.5	80–96
Onshore wind	1300–1500	N/A	33–50
Offshore wind	3000	N/A	70

Source: Milborrow, D. "Annual Power Costs Comparison: What a Difference a Year Can Make." *WindPower Monthly.* 2010, January.

TABLE **1-1** Total Installed Cost, Fuel Cost, and O&M Cost of Energy from Different Sources

fuel–based energy generation, no pollutants are produced. In the United States every megawatt-hour of wind energy production that is not produced by a conventional source reduces greenhouse gas emission by an equivalent of 0.558 tons of CO_2. According to the DOE's 20% Wind Energy by 2030 Technical Report,[4,5] overall 25% of CO_2 emissions from the electricity production sector can be reduced in the United States if 20% of electricity is produced by wind energy. In the United States, wind energy production in 2007 reduced CO_2 emissions by more than 28 million tons.

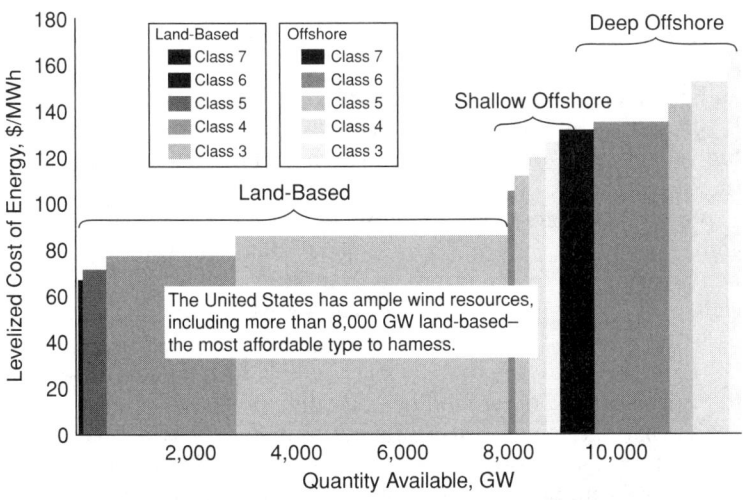

FIGURE **1-6** Estimated cost of energy production in the United States based on wind classes.[4] Cost excludes cost of transmission and integration.

Wind energy is among the cheapest sources of renewable energy. The cost of electricity production using wind is comparable to fossil fuel–based electricity production. In most cases, the cost is lower or about the same when cost of greenhouse gas emissions are taken into account. In addition, wind energy is available in abundance in most countries.

In addition to the above benefits, wind energy provides income to farmers, ranchers, and landowners that have sufficient wind resources on their property. The income is in terms of land lease payments, while majority of the land is still available for other uses.

Wind turbine generators are available in wide range of capacities, from small to utility scale. On small scale, wind energy can be used to power remote locations that do not have access to an electricity grid.

Wind Energy Is Not a Panacea

Despite the significant benefits, wind energy is not a cure-all. The primary disadvantages of wind are variability of the resource, requirement for large investment in transmission, and impact on the environment.

Wind energy production depends on wind conditions. Unlike solar energy, which is ubiquitous and can be produced in most locations, wind energy can be produced economically only in areas that have average annual wind speeds above 6.5 m/s at 50-m height. For instance, most of the southeast part of the United States has no wind resources, other than in coastal areas. Even in areas with abundant wind resources, there is a high degree of diurnal and seasonal variability. When the wind is not blowing, there is no energy production and other sources of electricity must be deployed.

People do not like to live in areas that have high wind. Therefore, high-wind areas are usually far away from population centers. This implies electricity generated from wind energy must be transported to population centers, which requires expensive transmission lines. In conventional methods of electricity generation, fuel is transported to a population center and electricity is produced close to a population center. In contrast, wind resource cannot be transported and long-distance transmission is required.

From an environmental perspective, wind farms can cause harm to birds, bats, and other wildlife, although most studies suggest that the harm is minimal. Aesthetic impact is another area of concern if the wind plant is located in an area of scenic value. Wind farms require significantly more land per kilowatt compared to fossil fuel–based electricity plants; however, continued use of the majority of the land mitigates this concern.

Other disadvantages of wind energy are reliance on government subsidies and significantly higher cost of small wind projects. Like other electricity generation, wind relies on moderate to low-level subsidies from governments. Over time, as the cost of greenhouse gas emission is built into the cost of traditional forms of electricity generation, these subsidies may not be required. Small winds projects (less than 100 kW), especially wind projects of size 15 kW or less, are expensive. The capital cost per kilowatt may be 3 to 5 times the cost per kilowatt of a large wind farm.

In conclusion, any potential negative impacts should be rigorously analyzed and strategies put in place to mitigate the impact. On balance, there is compelling evidence that wind energy delivers significant benefits to the environment and the economy.

References

1. World Wind Energy Association. *World Wind Energy Report 2009*, World Wind Energy Association, Bonn, Germany, March, 2010.
2. Wiser, R., and Bolinger, M. *2008 Wind Technologies Market Report*, Lawrence Berkeley National Laboratory, Berkeley, CA, 2009.
3. Milborrow, D. "Annual Power Costs Comparison: What a Difference a Year Can Make," *Windpower Monthly*, 2010, January.
4. Energy Efficiency and Renewable Energy, US Department of Energy. *20% Wind Energy by 2030*. US Department of Energy, Washington, DC, 2008. www.nrel.gov/docs/fy08osti/41869.pdf. DOE/GO-102008-2567.
5. American Wind Energy Association. *20% Wind Energy by 2030: Wind, Backup Power, and Emissions*, American Wind Energy Association, Washington, DC, 2009. http://www.awea.org/pubs/factsheets/Backup_Power.pdf.

CHAPTER 2

Basics of Wind Energy and Power

It should be possible to explain the laws of physics to a barmaid.
—Albert Einstein

Introduction

In this chapter, basic concepts of physics as they relate to energy and power of wind are discussed. The treatment of the concepts is at the high school or first-year college level. However, since most do not remember it anymore, this chapter will provide a quick overview. Basic laws of physics like conservation of mass, conservation of energy, and Newton's second law are used to explain concepts related to the amount of energy that is available in wind and the limits on the amount of energy that can be captured by wind turbines.

The chapter starts with kinetic energy of wind and its relationship to rotor radius and wind speed. Next, conservation of mass, energy, and momentum are described in the context of wind energy. These concepts are used to derive the Betz limit. It defines a limit on the amount of energy that can be extracted by a rotor disk turbine as a percentage of the total energy contained in wind. Finally, a comparison between water and wind turbines is presented.

Kinetic Energy of Wind

The kinetic energy contained in wind is:

$$E = \frac{1}{2}mv^2 \qquad (2\text{-}1)$$

where m is mass and v is speed; units of energy are kg m^2/s^2 = Joule.

9

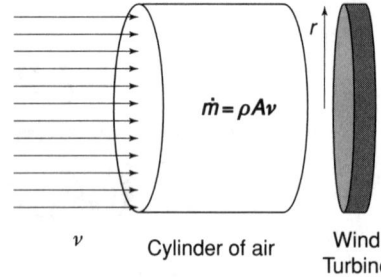

FIGURE 2-1 Cylinder of air in front of the rotor.

$\dot{m} = \rho A v$

v Cylinder of air   Wind Turbine

The mass (m) from which energy is extracted is the mass contained in the volume of air that will flow through the rotor. For a horizontal axis wind turbine (HAWT), the volume of air is cylindrical, as shown in Fig. 2-1. Approximation of a uniform cylinder will be relaxed later in the chapter.

Most are familiar with kinetic energy of a solid object of fixed mass. With air flow, it is convenient to think of mass in a cylinder of air of radius r. Since v m/s is the wind speed, the mass contained in cylinder of length v meters and radius r is the amount of mass that will pass through the rotor of turbine per second. It is, therefore, convenient to use mass per second ($\dot{m}$) in Eq. (2-1).

$$\dot{E} = \frac{1}{2}\dot{m}v^2 \qquad (2\text{-}2)$$

$$\dot{m} = \rho A v \qquad (2\text{-}3)$$

where ρ is air density and A is the cross-section area. $\dot{m}$ is the amount of matter contained in a cylinder of air of length v. $\dot{E}$ is energy per second, which is the same as power P

$$\dot{E} = P = \frac{1}{2}\rho A v v^2 = \frac{1}{2}\rho A v^3 \qquad (2\text{-}4)$$

Units of power are $(kg/m^3)\, m^2 m^3/s^3 = kg\, m^2/s^3 = J/s = $ Watts. Other units of power are kiloWatts (kW), megaWatts (MW), gigaWatts (GW), and horsepower (HP). Units of energy are Watt-seconds (=1 J), Watt-hours, kiloWatt-hours (kWh), megaWatt-hours (MWh), etc.

For a HAWT, $A = \pi r^2$, where r is the radius of the rotor, therefore:

$$P = \dot{E} = \frac{1}{2}\rho \pi r^2 v^3 \qquad (2\text{-}5)$$

The distinction between power and energy is important. If a wind turbine operates at a constant power of 10 kW for 2 h, then it will produce 20 kWh of energy, which is 72 million J (or Watt-seconds).

Sensitivity of Power to Rotor Radius and Wind Speed

The impact of change in radius by a small amount Δr, while all else is constant, can be expressed as:

$$\Delta P/P = 2\,\Delta r/r \qquad (2\text{-}6)$$

This means that if the radius is increased/decreased by 1%, power will increase/decrease by 2%. For larger changes in radius, the above formula does not apply; for instance, a 10% increase in radius will lead to increase by 21% in power. A 20% increase in radius will lead to 44% increase in power.

If speed is changed by a small amount and all else is constant, then

$$\Delta P/P = 3\Delta v/v \qquad (2\text{-}7)$$

This means that if the speed is increased/decreased by 1%, energy will increase/decrease by 3%. However, if the wind speed is increased by 20%, the power will increase by:

$$\frac{P_1}{P_2} = \frac{v_1^3}{v_2^3} = (1.2)^3 = 1.728 \qquad (2\text{-}8)$$

This is a 72.8% increase in power. The relationship between power and wind speed, and power and rotor diameter are seen in Figs. 2-2 and 2-3.

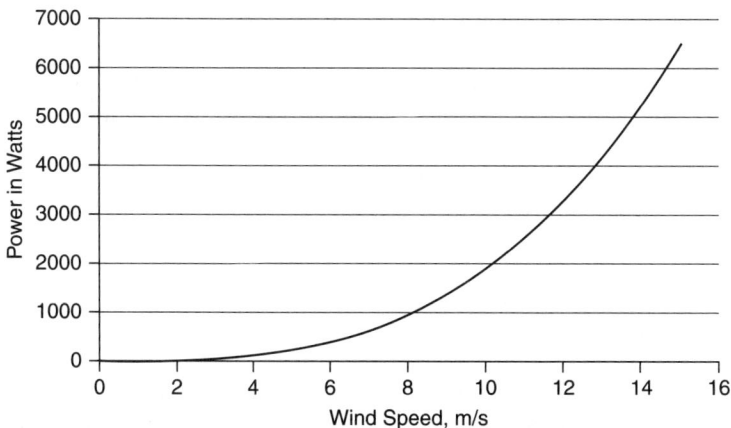

Figure 2-2 Cubic relationship between power and wind speed for a horizontal axis wind rotor with radius = 1 m.

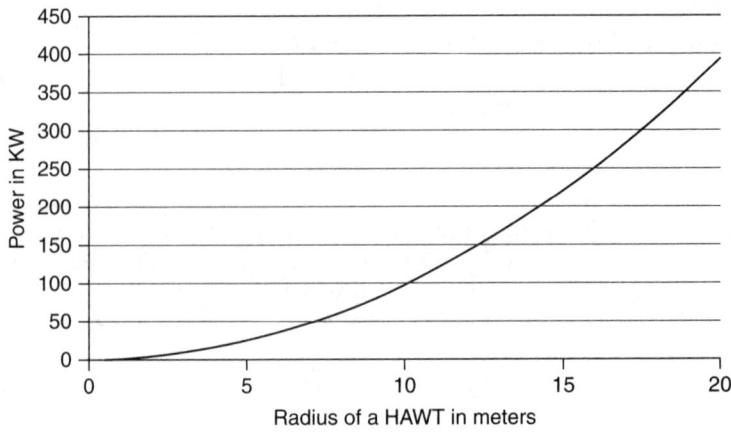

Figure 2-3   Quadratic relationship between power and rotor size. Wind speed is 8 m/s.

Basic Concepts/Equations

Three basic principles of physics are often used in this and subsequent chapters: Conservation of mass, conservation of energy, and conservation of momentum. Before these equations can be described, there is an important concept of control volume. The above principles must be applied in a defined control volume. The right and left side of the equation must be referring to the same control volume; in a derivation as one moves from one equation to another, all the equations must refer to the same control volume. This initial control volume may be of any shape; the most useful shapes are constant-radius cylinder and variable-radius streamlined cylinder, as shown in Fig. 2-4.

Streamlines can be conceptualized as infinitesimal tubes. Fluid flows in these tubes axially and not perpendicular to the tubes, so there is no exchange of matter across streamlines. This implies that in the control volume above, there is no mass gain or loss except from A_0 and A_2.

Conservation of Mass

Assumptions:

- All air that enters at A_0 leaves from A_2. Fluid flow is streamlined and so there is no loss of mass from the surface of the control volume.

- Fluid is incompressible, that is, there is no change in density.

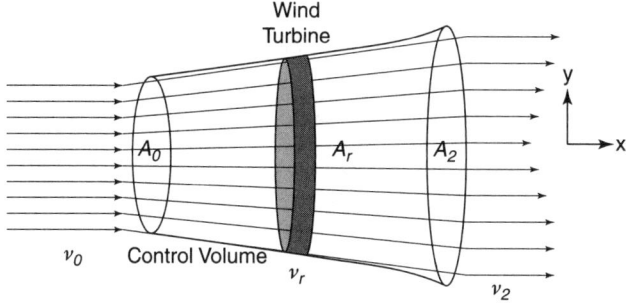

Figure 2-4 Illustration of a control volume that follows streamlines that pass through the rotor. v_0, v_r, v_2 are upstream, rotor, and downstream wind speeds. A_0, A_r, A_2 are upstream, rotor, and downstream cross-sectional areas.

Under these assumptions, conservation of mass is:

$$\dot{m} = \rho A_0 v_0 = \rho A_r v_r = \rho A_2 \bar{v}_2 \tag{2-9}$$

where $\bar{v}_2$ is the average wind speed, where the average is taken over cross-sectional A_2; v_r is assumed to be uniform over A_r, where A_r is the area of the rotor. Since the rotor of turbine is extracting energy from air, the kinetic energy of air will reduce, so, $v_0 > v_r > v_2$. Why is $v_0 > v_r$? This will be answered in the section on conservation of momentum.

Conservation of Energy

A simplified conservation of energy equation is used initially, under the assumptions listed below.

Total energy = Kinetic energy + Pressure energy + Potential energy (2-10)

The kinetic energy is because of the directed motion of the fluid; pressure energy is because of the random motion of particles in the fluid; potential energy is because of relative position of the fluid.

Assumptions:

- Fluid is incompressible, meaning the density does not change. Note that pressure can change.
- Fluid flow is inviscid, meaning the equation applies to fluid flow outside a boundary layer. The boundary layer is where the friction between a surface and fluid causes slower fluid flow.
- All the flow is along streamlines.
- There is no work done by shear forces.

- There is no heat exchange.
- There is no mass transfer.
- Relative position of fluid with respect to the earth's surface does not change, that is, the potential energy remains constant.

The first two assumptions define an ideal fluid. The above assumptions lead to Bernoulli's equation:

$$\text{Total energy per unit volume} = \rho\frac{v^2}{2} + p = \text{constant} \qquad (2\text{-}11)$$

$\rho\frac{v^2}{2}$ is the kinetic energy term, which is also called the dynamic pressure, and p is the static pressure.

Bernoulli's equation, therefore, states that along a streamline when speed increases, then pressure decreases and when speed decreases, then pressure increases. The magnitude of change in pressure is governed by the quadratic relationship.

Note that Bernoulli's law can be applied from A_0 to the left of the rotor; and then from right of the rotor to A_2 (see Fig. 2-4). Bernoulli's law cannot be applied across the device that extracts energy; the constant in Eq. (2-11) will be different for the two regions.

Conservation of Momentum

Since the wind rotor is a machine that works by extracting kinetic energy from wind, the wind speed is reduced. Since momentum is mass multiplied by speed, there is a change in momentum. According to Newton's second law, the rate of change of momentum in a control volume is equal to the sum of all the forces acting. In order to simplify the equations, the following assumptions are required:

There are no shear forces in the x-direction.
The pressure forces on edges A_0 and A_2 are equal.
There is no momentum loss or gain other than from A_0 and A_2.
The equation for Newton's second law along the x-axis becomes:

$$\dot{m}_0 v_0 - \dot{m}_2 v_2 = F \qquad (2\text{-}12)$$

Because of change in momentum in the control volume, there must be external force acting. In this case, rotor provides the external force. According to Newton's third law, there must be an equal, but opposite, force that acts on the rotor. This force is exerted by wind.

Because wind is exerting a force on the rotor, there must be a pressure difference across the rotor equal to the force divided by the area of rotor. Since the rotor hinders the flow of air, the pressure at the front of the rotor (p_r^0) is higher than the free-stream pressure (p_0); the pressure at the back surface of rotor (p_r^2) is below the free-stream pressure (see Fig. 2-5).

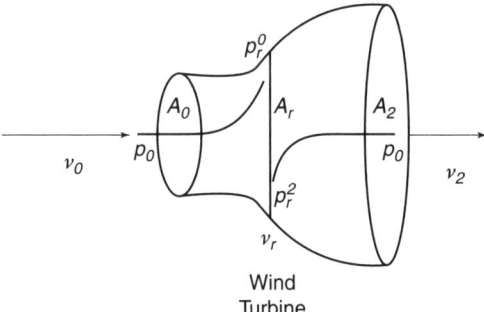

Figure 2-5 Illustration of the increase in pressure (p_0 to p_r^0) as wind approaches the rotor; drop in pressure behind the rotor (p_r^2) and then an increase in pressure in the wake to p_0.

Because the pressure is higher at the front of the rotor, according to Bernoulli's equation, the wind speed decreases from the free-stream wind speed (v_0) as it approaches the front of the rotor. Because v_r, the wind speed in front rotor, is less than v_0, conservation of mass mandates that the area increase; since $v_0 > v_r$, cross-sectional areas must have the relationship $A_0 < A_r$. Note, the wind speed does not change as it passes through the rotor; that is, the wind speed is the same immediately in front of the rotor and immediately behind the rotor. The reason is explained later in the chapter.

Because the pressure is low immediately after wind has passed through the rotor and the pressure will increase to the free-stream pressure as air moves toward A_2, the wind speed will decrease and, therefore, the area will increase from the right face of rotor to A_2, that is, $A_2 > A_r$. The volume to the right of the rotor is called the wake.

From the above exposition, two key follow-up questions arise. The first question is: What if there is a uniform cylindrical tube around the rotor and wind is forced to stay in this volume?[1]

To answer the first question, consider conservation of mass. Since density and cross-sectional area remain constant along the axis of the cylinder, the wind speed must remain constant throughout the cylinder in a streamlined flow. In Fig. 2-6, breaking the cylinder into two regions, one to the left of the rotor and another to the right of the rotor, and applying Bernoulli's equation to each region will result in the conclusion that the pressure must remain constant. If the wind

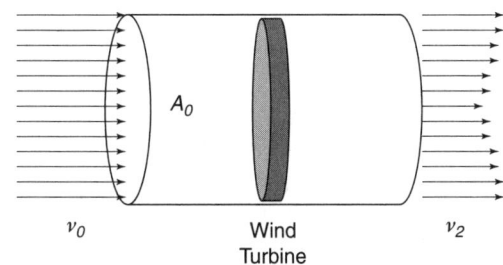

Figure 2-6 Can the flow around a rotor be a uniform cylinder?

FIGURE 2-7 Illustration of rotor acting as an impenetrable wall.

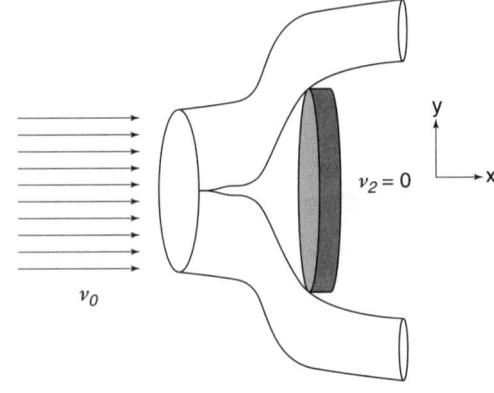

$v_2 = 0$

speed is the same and pressure is the same on both sides of the rotor, then no energy is extracted by the rotor.

The second question is: Can a rotor of a wind turbine extract all the kinetic energy from wind and make $v_2 = 0$?

To answer this question, see Fig. 2-7. If v_2 is zero, then there is no wind passing through the rotor. The rotor acts like an impenetrable wall and the wind flows around the wall. Since no wind is passing through the rotor, there is no energy extraction.

Derivation of Betz Limit

In 1919, Albert Betz postulated a theory about the efficiency of rotor-based turbines. Using simple concepts of conservation of mass, momentum, and energy, he postulated that a wind turbine with a disc-like rotor cannot capture more than 59.3% of energy contained in a mass of air that will pass through the rotor. The Betz limit is derived next.

Applying conservation of mass, Eq. (2-9), in control volume A_0, A_r, and A_2 with constant density (see Fig. 2-5):

$$A_0 v_0 = A_r v_r = A_2 v_2$$

where v_2 is the average wind speed at A_2. Applying Newton's second law from Eq. (2-12), force exerted on rotor by wind:

$$F = \dot{m}_r (v_0 - v_2) = \rho A_r v_r (v_0 - v_2) \qquad (2\text{-}13)$$

The force exerted on the rotor is also because of the pressure difference across the rotor:

$$F = A_r (p_r^0 - p_r^2) \qquad (2\text{-}14)$$

Equating (2-13) and (2-14):

$$F = A_r \left(p_r^0 - p_r^2 \right) = \rho A_r v_r \left(v_0 - v_2 \right) \tag{2-15}$$

Bernoulli's law is next applied in two volumes: (a) Flow along streamlines from A_0 to the front face of the rotor; and (b) flow from the back surface of rotor to A_2.

$$p_0 + \frac{1}{2}\rho v_0^2 = p_r^0 + \frac{1}{2}\rho v_r^2 \tag{2-16}$$

$$p_r^2 + \frac{1}{2}\rho v_r^2 = p_0 + \frac{1}{2}\rho v_2^2 \tag{2-17}$$

Subtracting (2-16) from (2-17):

$$p_r^0 - p_r^2 = \frac{1}{2}\rho \left(v_0^2 - v_2^2 \right) \tag{2-18}$$

Equating (2-15) and (2-18):

$$\frac{F}{A_r} = p_r^0 - p_r^2 = \rho v_r \left(v_0 - v_2 \right) = \frac{\rho}{2} \left(v_0^2 - v_2^2 \right) \tag{2-19}$$

$$v_r = \frac{\left(v_0 + v_2 \right)}{2} \tag{2-20}$$

Equation (2-20) implies that v_r, the wind speed at the rotor, is average of the free-stream wind speed and the wind speed in the wake. Note, the wind speed in wake (v_2) is where the pressure reaches free-stream pressure (p_0). Equation (2-20) also implies that one-half the wind speed loss occurs in front of the rotor and the other one-half occurs downstream.

This is counterintuitive because in the rotor itself—between the front and the back face of the rotor—there is no loss in wind speed, and all the speed loss happens upstream and downstream. The power is delivered (or work is done) by the force exerted because of pressure difference across the rotor. Power is defined as force multiplied by speed $= Fv_r$.

Therefore, the mechanism that delivers power to the rotor is:

- In the volume that is upstream of the rotor, some of the free-stream kinetic energy is converted into static pressure (Bernoulli's equation). Kinetic energy is reduced and pressure is increased as wind approaches the face of rotor. Since it is assumed that air is incompressible, that is, density is assumed to be constant, the reduction in wind speed is accompanied by increase in the flow area.

- The pressure difference across the rotor creates the force that performs the work and generates power. This is counterintuitive. One would expect the wind speed to drop abruptly across the rotor. In fact, the wind speed does not drop, instead, there is an abrupt drop in pressure and the pressure energy is transferred to the rotor. Note, an abrupt drop in wind speed would cause large undesirable acceleration and force.

- In the downstream volume, the static pressure rises from p_r^2 to p_0. Again assuming there is no mass transfer across the flow boundary, this pressure rise is because of transfer of kinetic energy from wind to static pressure. The wind speed, therefore, reduces from v_r to v_2.

The net effect is that the flow starts with p_0 as the upstream pressure and ends with p_0 as the downstream pressure. In the middle, the pressure rises to p_r^1 in front of the rotor, drops abruptly to p_r^2 behind the rotor, and then climbs back to p_0. During this same interval, kinetic energy of wind is extracted, converted to pressure energy, and delivered to the rotor. In fact, theoretically, 62.5% of the energy delivered to an ideal rotor is extracted from kinetic energy in front of the rotor. The rest of the 37.5% is extracted from kinetic energy in the wake of the rotor. In other words, although 100% of the pressure energy is delivered to energy extraction device at the rotor, only 62.5% of this energy came from kinetic energy to pressure energy conversion in front of the rotor. This means that at the rotor, the pressure energy goes into a "deficit" by delivering more energy to rotor than was imparted to it by kinetic energy of wind. This deficit in pressure energy is recovered in the wake of the rotor.

In all the discussions above, v_2 is the wake wind speed at a distance from the rotor where the pressure is restored to p_0. It is assumed that until this imaginary point in the wake is reached where average wake speed is v_2 and pressure is p_0, there is no mass transfer from the surrounding air. Beyond this imaginary point and further downstream, p_0 will remain the same, but the wind speed will start increasing and eventually reach the free-flow speed of v_0. This will happen because the air around slower wake will cause the wake to accelerate through either shear force or mass transfer.

Note this is an idealized rotor and no reference is made to the blades and the aerodynamics of the blades of the rotor. In the next chapter, the behavior of wind at the rotor, that is, interactions of wind with the blades (aerodynamics) is discussed.

The power delivered to the idealized rotor by the wind is:

$$P = Fv_r = (p_r^0 - p_r^2)A_r v_r \qquad (2\text{-}21)$$

Pressure difference was computed in Eq. (2-19), so power is:

$$P = (p_r^0 - p_r^2)A_r v_r = \frac{1}{2}\rho A_r v_r \left(v_0^2 - v_2^2\right) = \frac{1}{2}\rho A_r v_r \left(v_0 - v_2\right)\left(v_0 + v_2\right)$$

(2-22)

Note that

$$P = \frac{1}{2}\rho A_r v_r \left(v_0^2 - v_2^2\right) = \frac{1}{2}\dot{m}\left(v_0^2 - v_2^2\right)$$

which is change in kinetic energy applied to the flow of mass per unit time through the rotor. That is, the work done by force due to pressure difference is equal to the change in kinetic energy.

Combining Eqs. (2-20) and (2-22):

$$P = \rho A_r v_r^2 \left(v_0 - v_2\right) = 2\rho A_r v_r^2 \left(v_0 - v_r\right)$$

(2-23)

Maximum power is realized when:

$$\frac{\partial P}{\partial v_r} = 0 = 2v_r v_0 - 3v_r^2$$

(2-24)

$$v_r = \frac{2}{3}v_0$$

(2-25)

This implies:

$$v_2 = \frac{1}{3}v_0$$

(2-26)

$$P = 2\rho A_r v_r^2 \left(v_0 - v_r\right) = \rho A_r v_0^3 \left(\frac{8}{27}\right)$$

$$\frac{\text{Max power extracted}}{\text{Power available}} = P/\frac{1}{2}\rho A_r v_0^3 = \frac{16}{27} = 0.593 = C_p$$

(2-27)

C_p is called the power coefficient. A related concept is the thrust coefficient, C_T, which is

$$\frac{F}{\frac{1}{2}\rho A_r v_0^2} = \frac{8}{9} = C_T$$

(2-28)

C_p is referred to as the Betz limit and states that the maximum power an ideal rotor can extract from wind is 59.3%. See Fig. 2-8 for a graphical representation of the above function.

An ideal rotor of the type described above is called an "actuator disk." The actuator disk induces a reduction of the free-stream wind

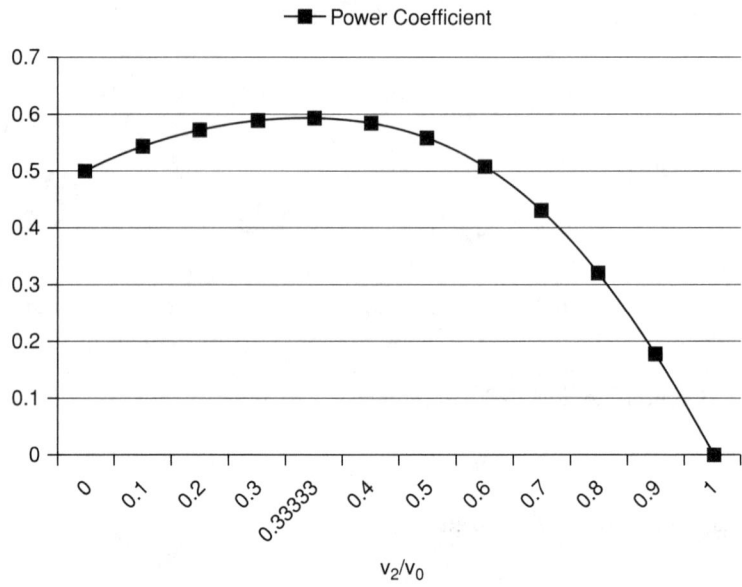

Figure 2-8 Power coefficient of a wind rotor (P/P_{ideal}) as a function of ratio of wind speed at the wake to input wind speed.

speed. If a is the induction factor, then:

$$v_r = (1 - a)v_0 \qquad (2\text{-}29)$$

In terms of a the wake wind speed, force and power are:

$$v_2 = (1 - 2a)v_0 \qquad (2\text{-}30)$$

$$F = 2\rho A_r v_0^2 a (1 - a) \qquad (2\text{-}31)$$

$$P = 2\rho A_r v_0^3 a (1 - a)^2 = \left(\frac{1}{2}\rho A_r v_0^3\right) 4a(1 - a)^2 \qquad (2\text{-}32)$$

Note, a must be less than $1/2$, otherwise, according to Eq. (2-30), $v_2 < 0$. Therefore, the above derivation does not apply when $a \geq 1/2$. Equation (2-32) is an alternate derivation of Betz limit in terms of a; $\frac{\partial P}{\partial a} = 0$ will yield $a = 1/3$.

The Meaning of Betz Limit

Wind rotors in idealized conditions can extract, at most, 59.3% of energy contained in the wind. This is an important limit because it defines the upper limit of the efficiency of any rotor disk type energy-extracting device that is placed in the flow of a fluid. A large fraction of the 59.3% of total wind energy that is extracted from wind is transferred to the turbine, but some of it is used to overcome viscous drag on blades and create vortices in the wake. Within the turbine, most of the energy is converted into useful electrical energy, while some of it is lost in gearbox, bearings, generator, power converter, transmission and others. Most practical rotors with three blades reach an overall efficiency of about 50%. From a theoretical standpoint, Okulov et al.[2] show that a rotor reaches the Betz limit with a large number blades operating at a very high tip speed ratio (the ratio of tip speed to wind speed).

Often, inventors claim to have created a rotor that achieves an efficiency that is greater than 59.3%; such claims are suspect and must be analyzed. There are turbine configurations that may violate the Betz limit. A shrouded rotor[3] that augments axial velocity is an example. No large turbine has been created, to the best of author's knowledge, with a shroud; however, there are a few small horizontal and vertical axis turbines with shroud. The following are illustrative examples of the Betz limit.

Example 1

As an example, consider 1-MW rated turbine with rotor diameter = 70 m and power curve, as shown in Fig. 2-8. Power curve is a key performance indicator of a turbine that is provided by the turbine manufacturer. It describes the relationship between power produced and wind speed. A Betz limit curve is also plotted in Fig. 2-9. It is always above the power curve, which implies that the turbine is within the Betz limit at all wind speeds.

$$P_{Betz}(v_0) = \frac{16}{27} \frac{\rho A_r v_0^3}{2} = \frac{16}{27} \frac{\rho A_r v_0^3}{2} \tag{2-33}$$

To show calculations at one point in the curve, $v_0 = 12$ m/s with air density of 1.225 kg/m³:

$$P_{Betz}(v_0) = \frac{16}{27} \frac{1.225\pi \left(\frac{70}{2}\right)^2 12^3}{2} = 2.4 \text{ MW} \tag{2-34}$$

Note the power curve indicates a power production capacity of slightly less than 1 MW, which is less than P_{Betz}.

Example 2

As a second example, consider a turbine with rotor diameter = 2 m and power rating of 2 KW at 12 m/s.

$$P_{ideal} = \frac{\rho A_r v_0^3}{2} = \frac{1.22\pi \left(\frac{2}{2}\right)^2 12^3}{2} = 3.3 \text{ kW} \tag{2-35}$$

$$P_{Betz} = 0.59 \, P_{ideal} = 1.953 \text{ kW} \tag{2-36}$$

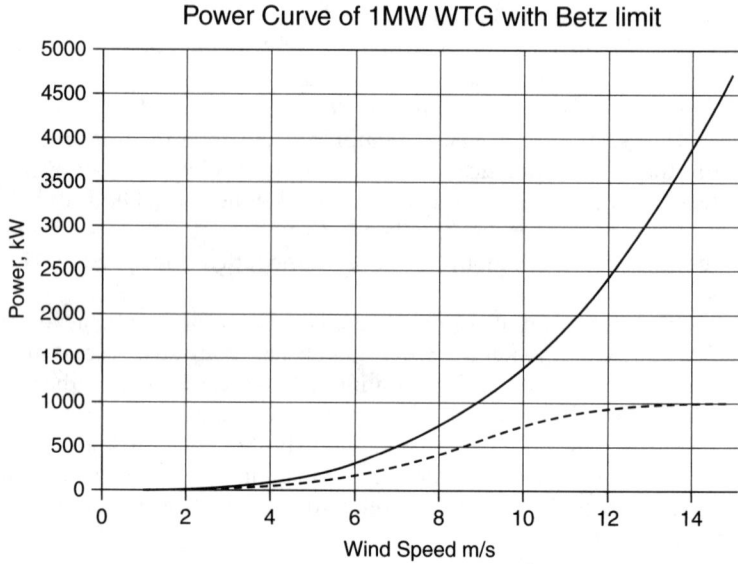

Figure 2-9 Dashed line is the power curve of 1-MW wind turbine generator (WTG) with rotor diameter = 70 m. The solid curve is the Betz limit curve for the same rotor.

Since the power rating of turbine is greater than the maximum power that can be extracted, this turbine rotor, therefore, does not pass the Betz limit test; at 12 m/s, the turbine cannot produce 2 KW of power, unless it uses a shroud or some other means to enhance axial wind speed.

Example 3

As a third example, consider a vertical axis wind turbine (VAWT) in Fig. 2-10. The height (h) = 6.1 m, diameter (d) = 1.2 m, and swept area = $h \cdot d$ = 7.43 m^2.

Power rating of rotor at 12 m/s = 1.2 kW.

$$P_{ideal} = \frac{\rho A_r v_0^3}{2} = \frac{1.22\pi \cdot 7.43 \cdot 12^3}{2} = 7.8 \text{ kW} \qquad (2\text{-}37)$$

$$P_{Betz} = 0.59 \, P_{ideal} = 4.6 \text{ kW} \qquad (2\text{-}38)$$

This VAWT passes the Betz limit test.

Wind versus Water

How do the limits on capturing energy from wind compare with limits of capturing energy from water in a hydrorotor?

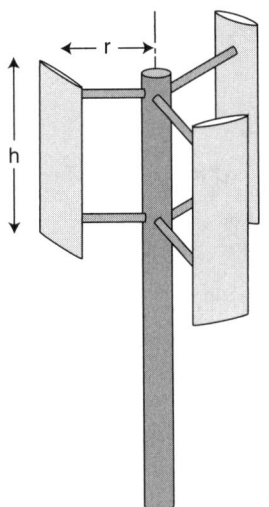

Figure 2-10 Vertical axis wind turbine. The height of blade is h and diameter of rotation is $2r = d$.

Assume a jet of water with speed v_0 strikes blades of Pelton wheel rotor as shown in Figure 2-11. Assuming a frictionless surface on the blade, the output speed of water will remain v_0.

According to Newton's second law, the force required to change the momentum is:

$$F = \dot{m}(-v_0 - (+v_0)) = -2\,\dot{m}v_0 \tag{2-39}$$

where $\dot{m}$ is the mass of water that hits a stationary blade every second. Because the blade is moving, the speed of water relative to the blade

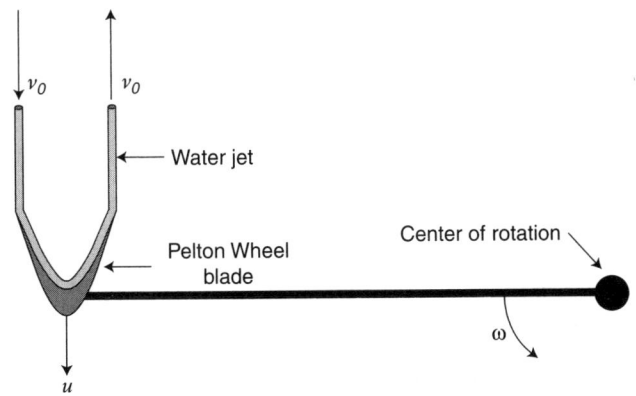

Figure 2-11   Illustration of a Pelton wheel hydrorotor.

must be used. If the blade moves at speed u, then with respect to the blade,

$$F = \dot{m}\left[-(v_0 - u) - (v_0 - u)\right] = -2\,\dot{m}(v_0 - u) \tag{2-40}$$

The power that is delivered to the blade (and, therefore, the rotor) is:

$$P = Fu = 2\dot{m}\,(v_0 - u)\,u \tag{2-41}$$

When $u = 0$, then force F is the largest, but power P is zero. When $u = v_0$, the rotor blade is moving away at the same speed as the water jet. Therefore, as expected, the force is zero and power is zero. Highest power is delivered when $u = v_0/2$.

$$P_{max} = \dot{m}v_0^2/2 \tag{2-42}$$

Therefore, the maximum power that can be extracted by a Pelton wheel rotor is equal to the kinetic energy in water, which implies that the theoretical efficiency of the Pelton wheel water rotor is 1 when it operates at $u = v_0/2$.

From a mechanical efficiency standpoint, considering friction and other losses, most water rotors operate at 90% or higher efficiency. In contrast, the maximum efficiency at which a wind rotor operates is 50%. The reason for such a large discrepancy in efficiency is the mechanics of energy transfer. With water, the direction of flow of almost all the particles is the same as the overall flow direction and, therefore, it is able to deliver most of the energy to the rotor. If air were used in a Pelton wheel, a significant amount of the energy would be used up to increase the randomness in velocity of air particles as opposed to imparting energy to the rotor.

References

1. Danish Wind Industry Association. Wind Turbines Deflect the Wind [Online] June, 2003. http://guidedtour.windpower.org/en/tour/wres/tube.htm.
2. Okulov, V. L., and Sorensen, J. N. "Optimum operating regimes for the ideal wind rotor," *Journal of Physics: Conf. Ser.*, 75: 2007.
3. Hansen, Martin O.L. *Aerodynamics of Wind Turbines*, Chapter 5: Shrouded rotors. 2nd Edition. Earthscan, Sterling, VA, 2008.
4. Patel, Mukund R. "Wind speed and energy," *Wind and Solar Power Systems*, CRC Press, Boca Raton, FL, 2006.
5. Burton, T., Sharpe, D., Jenkins, N., and Bossanyi, E. *Wind Energy Handbook*, Wiley, Hoboken, NJ, 2001.
6. Wagner, H-J., Mathur, J. *Introduction to Wind Energy Systems*, Springer, Berlin, 2009.

CHAPTER 3

Properties of Wind

Among the absolutes in man's world are the stars and the weather, and
as inescapable as the existence of weather is the presence of winds.
—Goerge M. Hidy

Introduction

In this chapter, the basic properties of wind and air are discussed.
The treatment will be from the meteorological standpoint. The chapter begins with a short exposition on how wind is created. The second
section deals with a popular classification scheme in the wind industry
that is based on annual average wind speed. A more rigorous method
to classify wind speed that is based on power density is described next.
The third section describes the statistical properties of wind speed
along with the probability density function and its impact on the average energy density. The next topic is wind shear, which describes
the variation in wind speed as a function of height. It is used to extrapolate measured wind speed to hub height and describes the variation
of wind speed in the plane of rotation of the rotor. The final section
describes air density and the impact of temperature, elevation, and humidity on air density. This chapter, therefore, describes two of the three
terms in the wind power equation, namely, wind speed and air density.

How Is Wind Generated?

Uneven heating of the surface of the earth creates wind. Solar energy absorbed by land or water is transferred to the atmosphere. A
larger amount of solar radiation is received at the tropics compared
to the poles, which causes hot air to rise at the tropics and flow toward the poles. This flow occurs 10 to 15 km above the earth's surface.
The Coriolis force causes the flow of hot air in the upper atmosphere
to turn right. This flow does not continue beyond 30° latitude. This

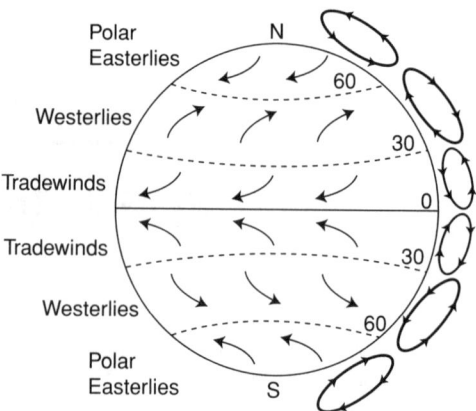

Atmospheric circulation of air. The arrows between the latitude lines indicate the direction of surface winds. The closed circulation or convection shown on the right indicates the vertical flow of air.

vertical motion of hot air causes low pressure at the tropics. Cold air from the higher latitudes flows toward the tropics; these are "surface" winds that are called trade winds. Figure 3-1 illustrates the direction of prevailing wind because of solar radiation differences and Coriolis force in each 30° latitude band. This large-scale atmospheric circulation defines the global flow of air. The effect of this circulation is to redistribute the heat.

The totalities of the atmospheric mechanisms that play a role in wind are complex and include local effects. Some of the primary local effects are:

- Landmass heats and cools faster than water body. This causes sea breeze, which is wind from the sea to land during the day, with reverse flow during the night.

- Orography and roughness. Orography is change in elevation of earth's surface; roughness is a measure of the friction on the surface of the earth. Changes in elevation can cause a mountain breeze.

Statistical Distribution of Wind Speed

Wind speed is a stochastic quantity. The most common density function used to represent wind speed is Weibull, whose probability density function pd(v) is:

$$\text{pd}(v) = (k/A)(v/A)^{k-1}e^{-(v/A)^k} \text{ for } v > 0 \tag{3-1}$$

where v is the wind speed, k is the shape factor, and A is the scale factor. As the names suggest, k determines the shape of the curve and A determines the scale of the curve (Fig. 3-2; and Fig. 3-4 [see later]). k is dimensionless and A has the units of m/s).

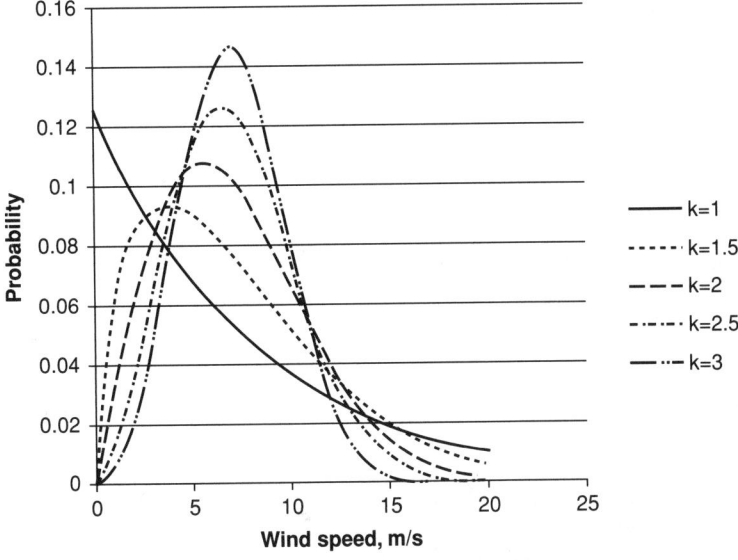

Figure 3-2 Weibull probability density function for $A = 8$ m/s.

As a convention when speaking about Weibull density function, wind speed v is the 10-min average. In a wind measurement campaign, for each 10-min interval the average wind speed and standard deviation are recorded. The Weibull probability density function is a model that represents the 10-min average wind speed. This assumes that over the 10-min interval the wind conditions are stationary. However, not all wind measurements are at 10-min intervals, therefore, it is important to mention the time interval when a reference is made to wind speed density function.

Instead of a probability density function that represents the fraction of time wind speed is at v, it is sometimes customary to speak in terms of hours in a year.[1] That is, pd(v) is multiplied by 8760 (number of hours in a year), see Figure 3-3. For instance, the area under the a curve between 5 and 10 m/s represents the total number of hours in a year the wind speed is likely to be in that wind speed range. An example of Weibull distribution in terms of number of hours is seen in Figure 3-3.

Note the Weibull distribution is defined only for positive value of wind speed. The other properties of the Weibull for different value of k are:[1]

$k = 1$, the Weibull distribution becomes an exponential distribution.

$k = 2$, the Weibull distribution becomes a Rayleigh distribution (Fig. 3-4).

$k > 3$, the Weibull distribution approaches a Gaussian distribution.

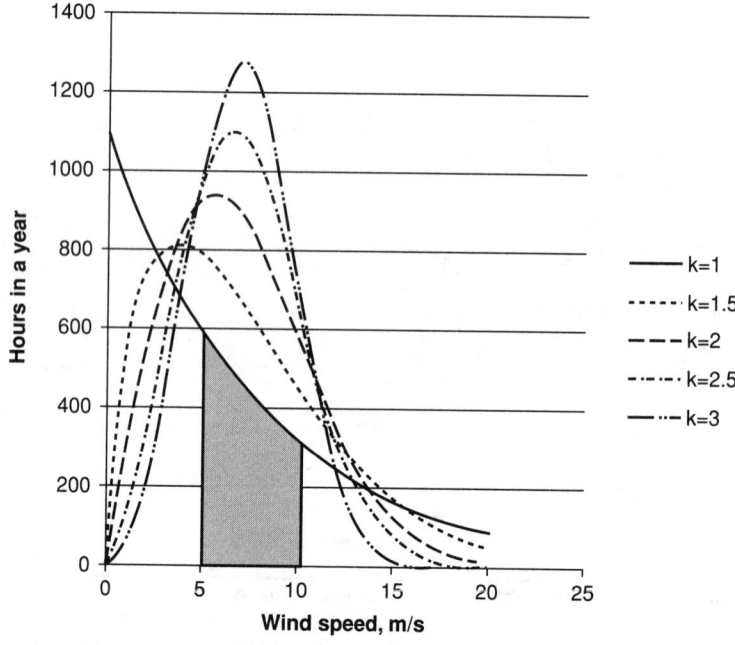

FIGURE 3-3 The Weibull probability density function expressed in hours per year. The shaded area represents the number of hours in a year that wind speed will be between 5 and 10 m/s for the $k = 1$ case.

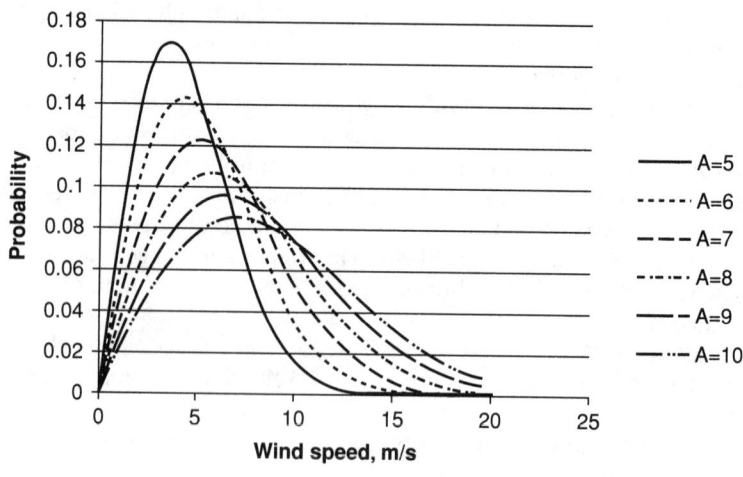

FIGURE 3-4 Weibull probability density function for $k = 2$ and different values of A.

Empirically, it has been observed that wind speed in most locations is a Weibull distribution. Furthermore, the value of k is approximately 2 for most wind profiles.

A few observations are in order for Weibull as a distribution for wind speed. As k increases, the height of the curve increases, which implies that the curve becomes narrower with a smaller tail—probability drops quickly at higher wind speeds. This can lead to lower energy at higher values of k. That is, if two wind speed distributions have the same average wind speed, but different value of k, say, $k = 2$ and $k = 3$, the latter case will yield lower energy.

Mean and Mode of Weibull Distribution for Wind Speed

The mode of a Weibull distribution is:

$$\text{Mode} = A\left(\frac{k-1}{k}\right)^{1/k} \tag{3-2}$$

$$\text{Mean} = \bar{v} = A\Gamma\left(1 + \frac{1}{k}\right) \tag{3-3}$$

$$\text{Variance} = \sigma^2 = c^2\left[\Gamma\left(1 + \frac{2}{k}\right) - \Gamma^2\left(1 + \frac{1}{k}\right)\right] = \bar{v}^2\left[\frac{\Gamma\left(1 + \frac{2}{k}\right)}{\Gamma^2\left(1 + \frac{1}{k}\right)} - 1\right] \tag{3-4}$$

where $\Gamma(x)$ is the gamma function; if x is an integer, then $\Gamma(x) = x!$. Table 3-1 contains sample values of the mean and mode.

K	A	Mean	Mode
1	8	8.000	0
1.5	8	7.222	3.846
2	8	7.090	5.657
2.5	8	7.098	6.522
3	8	7.144	6.989
1	10	10.000	0
1.5	10	9.027	4.807
2	10	8.862	7.071
2.5	10	8.873	8.152
3	10	8.930	8.736

TABLE 3-1 Mean and Mode of a Weibull Distribution for Different Shape and Scale Factors

Computation of A and k given mean and variance requires approximations.[2]

$$k = \left(\frac{\sigma}{\bar{v}}\right)^{-1.086} \tag{3-5}$$

$$A = \frac{\bar{v}}{\Gamma\left(1 + \frac{1}{k}\right)} \tag{3-6}$$

Equation (3-5) is a good approximation when $1 \leq k \leq 10$.

Power Density

In order to understand the impact on power generation of statistical distribution of wind speed, consider the impact on power density. Power density is defined as:

$$\text{PD} = \frac{\text{Power}}{\text{Area}} = \frac{1}{2}\rho v^3, \text{ units are } \frac{W}{m^2} \tag{3-7}$$

If the statistical distribution of wind is ignored and it is assumed that there is no variation in wind speed, then the power density is incorrectly computed (see column 2 in Table 3-3).

$$\text{Incorrect Power Density} = \frac{1}{2}\rho(\bar{v})^3 \tag{3-8}$$

where $\bar{v}$ is the average wind speed.

However, if the energy density is computed correctly while taking into account probability density of wind speed, then the power density numbers are very different. (See column 3 in Table 3-2.)

$$\text{Correct Power Density} = \int_0^\infty \frac{1}{2}\rho v^3 \text{pd}(v)dv \tag{3-9}$$

where $\text{pd}(v)$ is the Weibull probability density function in Eq. (3-1).

As Table 3-2 illustrates, the power density of rotor is underestimated if computed based on average wind speed in Eq. (3-8).

Comparison of the probability density of wind speed and power density is illuminating. As an illustration, a wind speed profile with $A = 8$ m/s and $k = 2$ is chosen, (see Fig. 3-5). The mean wind speed is 7.09 m/s and mode is 5.657 m/s. The power contained in this wind profile peaks at approximately 11 m/s. Reason is power delivered at a particular value of v is $\rho v^3 \text{pd}(v)\,dv/2$; although $\text{pd}(v)$ is decreasing above 5.657 m/s, $\rho v^3/2$ keeps rising; the product of the two quantities, which is power delivered at a particular wind speed, peaks at 11 m/s.

In most cases, during prospecting for wind projects, the only data available is mean wind speed. This is because actual measurements

Average Wind Speed, m/s	Incorrect Power Density, W/m²	Correct Power Density, W/m²
0	0	0
1	1	2
2	5	9
3	17	32
4	39	75
5	77	146
5.6	108	206
6	132	253
6.5	168	321
7	210	401
7.5	258	494
8	314	599
8.5	376	719
9	447	853
10	613	1170
11	815	1557

* The average wind speeds chosen in the table correspond to Weibull distribution with $k = 2$.

TABLE 3-2 Comparison of Power Density Computation*

have not been performed and, therefore, the shape of the wind speed distribution is not available. The sensitivity of the shape factor k to the power density is illustrated in Table 3-3. As the table illustrates, even if the average wind speed is correct, a difference in 0.1 in shape factor leads to about 5% changes in power density.

Wind Classes

As a convention, the strength of wind at a site is classified based on power density at an elevation of 50 m above the ground level (AGL). Table 3-4 lists the definition of wind classes in terms of power density at 10 and 50 m.[3] For sake of convenience, wind speed ranges are associated with the power density ranges. Note, this mapping of power density to wind speed is correct only if $k = 2$, that is, it is only correct if wind at the location has a Rayleigh distribution.

Although wind class definition in terms of wind speed range is widely used, it is an approximation. The nature of approximation will

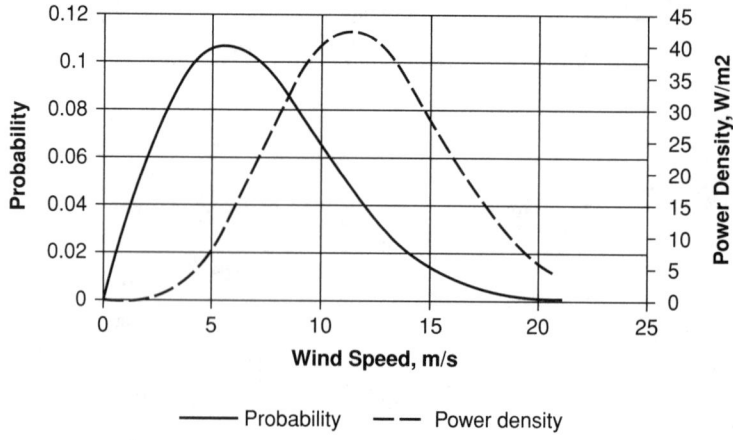

Probability ——— Power density

Figure 3-5 Illustration of the power density function for wind speed with a Weibull distribution $A = 8$ and $k = 2$. The mean wind speed is 7.09 m/s. Notice the power density curve peaks at 11 m/s.

be explained next. A popular misconception is that wind class can be determined if the annual average wind speed at 50 m is given. Although this works for certain probability density functions of wind speed, it may not always yield the correct wind class (see Table 3-3).

As an example, consider a location with wind measurements at 50-m elevation from the ground. Statistical analysis of the data yields good fit with Weibull distribution with the following parameters:

- Annual mean wind speed $= 7.9$ m/s
- $A = 8.85$ m/s
- $k = 3$

Average Wind Speed	Power Density, W/m²		
	$k = 1.9$	$k = 2$	$k = 2.1$
5	154	146	140
5.5	206	195	186
6	267	253	241
6.5	339	321	307
7	424	401	383
7.5	521	494	471

Table 3-3 Change in Power Density as Shape Factor Is Changed

	10 m		50 m	
Wind Class	**Wind Class Name**	**Power Density, W/m²**	**Average Wind Speed, m/s**	**Power Density, W/m²**	**Average Wind Speed, m/s**
1	Poor	0–100	0–4.4	0–200	0–5.6
2	Marginal	100–150	4.4–5.1	200–300	5.6–6.4
3	Fair	150–200	5.1–5.6	300–400	6.4–7.0
4	Good	200–250	5.6–6.0	400–500	7.0–7.5
5	Excellent	250–300	6.0–6.4	500–600	7.5–8.0
6	Outstanding	300–400	6.4–7.0	600–800	8.0–8.8
7	Superb	400–1000	7.0–9.4	800–2000	8.8–11.9

* Wind speed ranges and power density are at specific heights.

TABLE **3-4**   Definition of Wind Classes*

The power density of this location is 424 W/m². A look at column 5 in Table 3-4 shows that this location belongs to Class 4 wind regime. A common mistake is to look at the mean wind speed in column 6 and conclude that this location belongs to Class 5 wind regime.

Wind Shear

Wind shear describes the change in wind speed as a function of height. Assuming there is no slippage on the surface, the surface wind speed is zero. That is, wind speed is zero at an elevation of zero. There are two methods to describe shear: Power law profile and logarithm profile.

The power law is the most common method to describe the relationship of wind speed and height. This is an engineering approximation and must be used with caution.

$$\frac{v_2}{v_1} = \left(\frac{h_2}{h_1}\right)^{\gamma} \qquad (3\text{-}10)$$

where v_2 and v_1 are wind speeds at heights h_2 and h_1, and exponent γ is called wind shear.

Figure 3-6 is a plot of the wind speed ratio and height ratio for different values of shear. Figure 3-7 is a plot of height versus wind speed for different values of shear.

An alternate method to extrapolate wind speed is to use the logarithmic profile, which uses roughness of the surface.[4]

$$\frac{v_2}{v_1} = \frac{\ln(h_2/z_0)}{\ln(h_1/z_0)} \qquad (3\text{-}11)$$

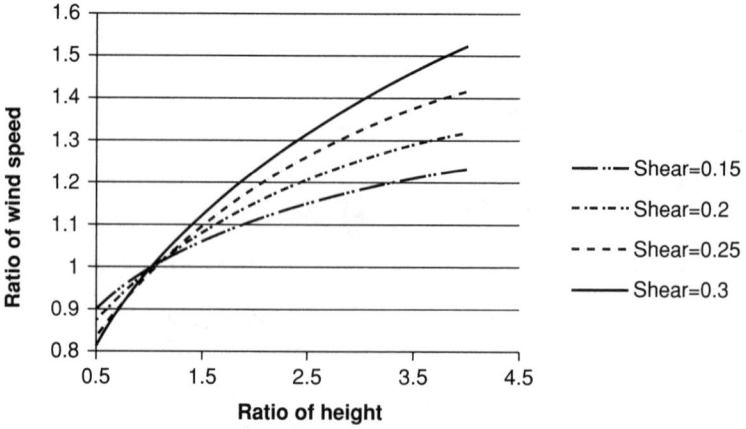

FIGURE 3-6 Plot of the ratio of wind speed to ratio of height for different values of shear.

where z_0 is called the roughness length. If wind speed v_1 is available at $h_1 = 10$ m, then Eq. (3-11) may be used to compute v_2.

The value of shear can then be derived using Eqs. (3-10) and (3-11) as:

$$\gamma = \ln\left(\ln\frac{h_2}{z_0} \Big/ \ln\frac{h_1}{z_0}\right) \Big/ \ln(h_2/h_1) \qquad (3\text{-}12)$$

Shear, therefore, depends on the heights and roughness length.

Roughness length is the extrapolated height above the surface at which the mean wind speed is zero. This assumes that the speed

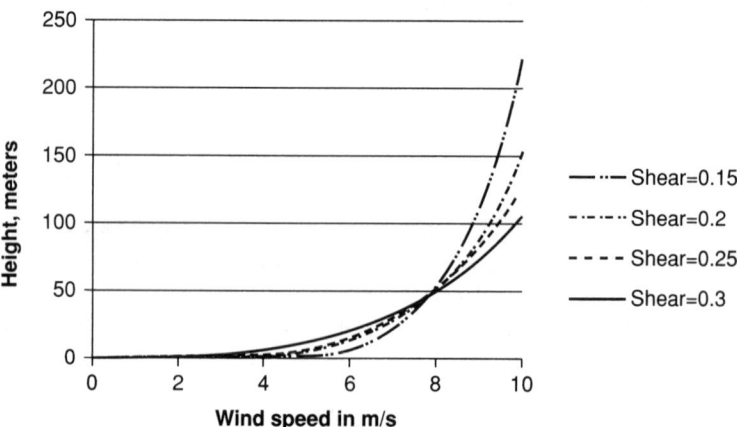

FIGURE 3-7 Plot of wind speed versus height for different values of shear, given the average wind speed is 8 m/s at 50 m.

Description	Roughness Class	Roughness Length, m	Shear
Open sea	0	0.0001–0.003	0.08
Open terrain with a smooth surface, like concrete runway, mowed grass	0.5	0.0024	0.11
Open agricultural area without fences and hedgerows and very scattered buildings. Only softly rounded hills	1	0.03	0.15
Agricultural land with some houses and 8-m-tall sheltering hedgerows with a distance of approx. 1250 m	1.5	0.055	0.17
Agricultural land with some houses and 8-m-tall sheltering hedgerows with a distance of approx. 500 m	2	0.1	0.19
Agricultural land with many houses, shrubs and plants, or 8-m tall sheltering hedgerows with a distance of approx. 250 m	2.5	0.2	0.21
Villages, small towns, agricultural land with many or tall sheltering hedgerows, forests, and very rough and uneven terrain	3	0.4	0.25
Larger cities with tall buildings	3.5	0.8	0.31
Very large cities with tall buildings and skyscrapers	4	1.6	0.39

Source: Nielsen, Per. WindPRO 2.5 Users Guide. EMD International, Aalborg, Denmark, 2006.

TABLE 3-5   Description of Roughness Classes, Roughness Length, and Wind Shear

has a logarithmic variation. Table 3-5 describes classes of roughness, roughness length, and shear.

Often, the following approximations are used for shear. However, these approximations can lead to significant error in energy prediction and, therefore, must be used with caution.

1. *Approximation 1:* Shear $= 1/7 = 0.14$.[4] This is the most widely used value when wind speed is available at single height. If extrapolating 10-m data, this value may underestimate wind

speed at hub height in stable laminar flow situations and over-estimate when there is mixing and convection.

2. *Approximation 2:* Extrapolating 10-m wind speed data to 50 m or higher using a constant shear value. The shear formula in Eq. (2-10) is most accurate when it is used to extrapolate wind speeds at heights that satisfy:

$$0.5 < h_2/h_1 < 2$$

3. *Approximation 3:* Constant value of shear in all seasons and all hours of a day. It is not uncommon to observe negative shear during the daytime hours when there is significant thermal mixing of air because of convection. Therefore, constant shear based on annual wind speed averages can lead to inaccuracies in energy computations.

Section "Uncertainty in Wind Speed Measurement with Anemometers" in Chapter 6 contains guidelines on wind speed measurement heights for accurate shear computation. Also, see section "Computed Quantities: Wind Shear" in Chapter 6 for an example of wind shear computation.

Understanding Wind Shear

In order to illustrate the impact of wind shear, consider the following: Which location is preferable for a wind project?

   a. Location with 4 m/s wind speed at 10 m in the desert with low roughness

   b. Location with 4 m/s wind speed at 10 m in a forested area with high roughness

Location (*a*) has low roughness, which implies low friction on the surface and low shear. For illustration purposes, choose shear = 0.15, a constant value regardless of h_1 and h_2. Using Eq. (3-10), at hub height of, say, 80 m, the wind speed is 5.46 m/s.

Location (*b*) has high roughness, which implies high friction on the surface and high shear. In this example, choose shear = 0.25. Using Eq. (3-10), at hub height of 80 m, the wind speed is 6.72 m/s.

A location with higher roughness has higher wind speed at 80 m height; this is counterintuitive. The reason is that both locations have the same wind speed at 10 m. Therefore, at location (*b*) the "driving energy" of wind at upper elevations is much higher and that energy is able to overcome high level of friction near the surface.

An alternate way to visualize shear is to consider wind speed of 10 m/s at 200-m elevation and examine two locations with two different

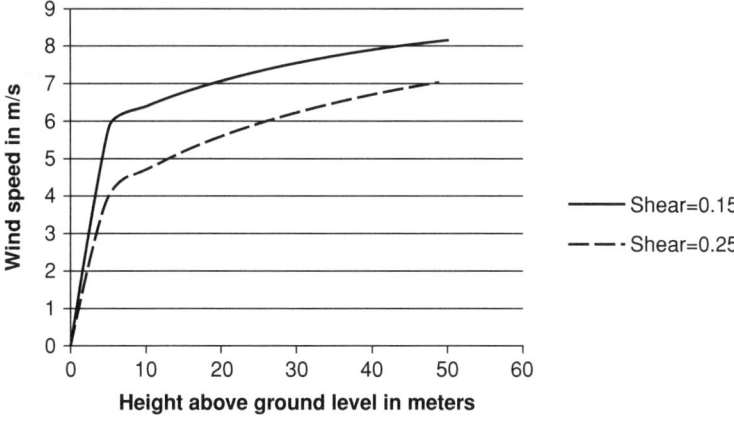

FIGURE 3-8   Plot of wind speed as a function of height with wind speed = 10 m/s at 200-m elevation.

roughness conditions on the surface (see Fig. 3-8). A location with high roughness will have lower wind speed at 10 m because of higher friction compared to a location with lower roughness. However, if the higher roughness location has the same wind speed as the lower roughness location at 10-m elevation, then the higher roughness location (*b*) must have higher wind speed at higher elevations compared to location (*a*).

Density of Air as a Function of Elevation

The other parameter that influences power is air density, $P = \rho A v^3/2$. The relationship between P and ρ is linear. If ρ, the air density is lower by 10%, then the power will be lower by 10%. Air density depends on pressure, temperature, and relative humidity. As elevation increases, both pressure and temperature decrease. To derive the relationship, begin with the ideal gas law:

$$pV = nRT \qquad (3\text{-}13)$$

where p is pressure in Pascals, V is volume in m^3, n = number of moles, R is the gas constant which is 287.05 J/(kg · K), and T is the temperature in Kelvin (K = °C + 273.15).
    Density = $\frac{n}{V}$, number of moles per unit volume.

$$\rho = \frac{p}{RT} \qquad (3\text{-}14)$$

Density expressed in kg/m^3 is

$$\rho = \frac{p}{RT}\frac{M}{1000} \tag{3-15}$$

where M is the molecular weight of dry air in grams $= 28.9644$ g/mole and R expressed in moles is 8.31432 J/(mol K)

In the troposphere (elevation up to 11 km from sea level), the variations in temperature are described by:

$$T = T_0 - Lh \tag{3-16}$$

where $T_0 = 288.15$ K and L, the temperature lapse rate is 6.5 K/km, and h is the elevation from sea level in kilometers.

$$p = p_0 \left(1 - \frac{Lh}{T_0}\right)^{\frac{gM}{RL}} \tag{3-17}$$

where p_0 is 101,325 Pa, g is gravitational constant, 9.80665 m/s^2.

$$\rho = p_0 \left(1 - \frac{Lh}{T_0}\right)^{\frac{gM}{RL}} \frac{1}{R\,(T_0 - Lh)}\frac{M}{1000} \tag{3-18}$$

Table 3-6 contains values of density using Eq. (3-18). Density is lower by 2% at 200 m and 17.8% lower at 2000 m. Since power is directly

Height, m	Density, kg/m^3
0	1.224999
5	1.224411
10	1.223824
50	1.21913
100	1.213282
150	1.207456
200	1.201651
250	1.195867
500	1.167268
1000	1.111642
1500	1.058067
2000	1.00649

TABLE 3-6 Density of Air as Function of Elevation for Dry Air

Molecule	Percent in Dry Air, %	Molecular Weight
N_2	78.1	28
O_2	20.95	32
Ar	0.934	18
CO_2	0.038	44

TABLE 3-7 Composition of Air and the Molecular Weight of the Components

proportional to air density, there will be a 2 or 17.8% drop in power depending on elevation of the site. The change in density measured at the ground level versus density at a 100 m rotor hub is less than 1%.

Density of Air as a Function of Humidity

The composition of dry air is in Table 3-7. The average molecular weight of dry air is 28.9644 g/mole. When moisture is added to air, H_2O of molecular weight 18 displaces heavier molecules in dry air. Therefore, moist air is lighter and has lower density—a counterintuitive conclusion. So, moist air delivers less power.

The density of mixture of dry air and water vapor is:[5,6]

$$\rho = \frac{p_d}{R_d T} + \frac{p_v}{R_v T} = \left(\frac{p}{R_d T}\right)\left(1 - \frac{0.378 p_v}{p}\right) \tag{3-19}$$

Relative Humidity, %	Density, kg/m^3
0	1.225012
10	1.224233
20	1.223454
30	1.222674
40	1.221895
50	1.221116
60	1.220337
70	1.219557
80	1.218778
90	1.217999
100	1.217219

TABLE 3-8 Air Density as a Function of Relative Humidity for $p = 101,325$ Pa, $T_c = 15°C$

where p_d is the partial pressure of dry air, p_v is the partial pressure of water vapor, R_d is gas constant for dry air = 287.05, R_v is gas constant for water vapor = 461.495, and pressure $p = p_d + p_v$. An approximation for p_v is:[5]

$$p_v = RH * 610.78 * 10^{\frac{7.5T_C}{237.3+T_C}}$$ (3-20)

where RH is the relative humidity, T_C is the temperature in degrees Celsius.

Table 3-8 contains value of density for different values of relative humidity. At 100% relative humidity, the density is lower by 0.6%.

References

1. Patel, M. R. *Wind and Solar Power Systems*, CRC Press, Boca Raton, FL, 2006.
2. Johnson, G. L. *Wind Energy Systems*, electronic edition, Manhattan, KS, 2001.
3. Elliott, D. L., Holladay, C. G., Barchet, W. R., Foote, H. P., and Sandusky, W. F. Wind power classes. *Wind Energy Resource Atlas of the United States* [Online] 1986. http://rredc.nrel.gov/wind/pubs/atlas/tables/A-8T.html.
4. Gipe, P. *Wind Energy Comes of Age*, Wiley, New York, 1995.
5. Air Density and Density Altitude Calculations [Online] http://wahiduddin.net/calc/density_altitude.htm#15.
6. Density of Air. *Wikipedia.* [Online] http://en.wikipedia.org/wiki/Density_of_air.

CHAPTER **4**

Aerodynamics of Wind Turbine Blades

We had taken up aeronautics merely as a sport. We reluctantly entered upon the scientific side of it. But we soon found the work so fascinating that we were drawn into it deeper and deeper.

—Orville and Wilbur Wright

Introduction

Chapter 2 used the actuator disk theory to describe the Betz limit, a limit on how much energy can be extracted by a rotor-based wind turbine. In this chapter, a more realistic look at flow of air over blades will be undertaken. The chapter starts with description of airfoils, which is the shape of the cross-section of turbine blade. The second section describes frame of reference for velocity and force vectors on a rotating blade. Rotor disk theory is described next, which describes the flow of wind through the rotor, unlike the actuator disk theory. The following section describes the basic mechanisms associated with creation of lift and drag forces on blades and creation of torque on the rotor. The chapter ends with a contrast between a drag- and lift-based machine.

Airfoils

The cross-section of a wind turbine blade is an airfoil. Figure 4-1 is a schematic of a symmetrical airfoil. Chord line connects the leading to the trailing edge. Most airfoils used in wind turbines have a larger area above compared to below the chord line. A line connecting the leading and trailing edge that bisects the area of an airfoil is called a camber line.

Figure 4-1 Airfoil terminology.

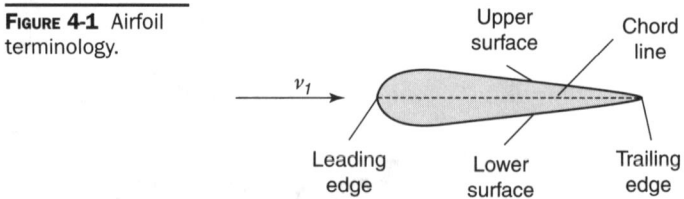

Upper surface — Chord line — Leading edge — Lower surface — Trailing edge — v_1

Figure 4-2 contains a symmetrical airfoil in an ideal fluid. The airfoil is placed in airflow with uniform wind speed of v_0, ambient pressure of p_0, and the direction of wind speed is parallel to the chord. The changes in speed and pressure are plotted in the two graphs in Fig. 4-2. At the leading edge, the wind speed is zero, so the fluid decelerates from free stream speed v_0 to zero. The static pressure, therefore, increases and reaches the maximum value at the leading edge (according to Bernoulli's principle, Eq. 2-11). As the fluid flows over the upper edge of the airfoil, it accelerates, reaching a speed of v_1 at the shoulder of the airfoil. $v_1 > v_0$, therefore, $p_1 < p_0$. The shoulder of the airfoil is the point of highest thickness. Between the shoulder and the trailing edge of the airfoil, the fluid decelerates and reaches zero speed at the trailing edge; the reason for zero speed will be discussed later in the chapter. The pressure, therefore, increases and reaches a maximum at the trailing edge. Beyond the trailing edge, the speed

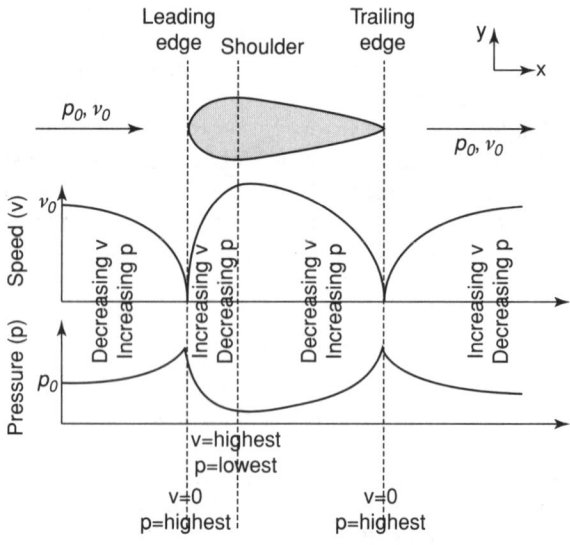

Figure 4-2 Plot of speed and pressure along a symmetrical airfoil.

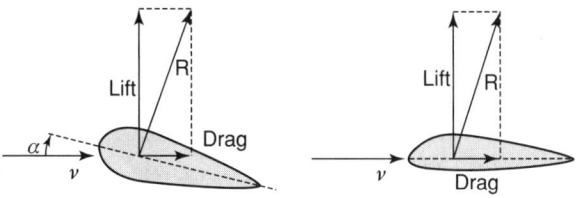

Figure 4-3 Symmetrical airfoil with positive angle of attack and a nonsymmetrical airfoil with zero angle of attack. Air flow around both types of airfoil result in lift and drag. R is the resultant aerodynamic force.

increases to the free-stream speed and pressure decreases to free-stream static pressure.

Since the airfoil is symmetric, the speed and pressure are identical on the upper and lower surface of the airfoil. There is no imbalance of force along the y-axis. Since the fluid is inviscid, there is no friction force and the static pressure along the x-axis is equal and opposite.

When the airfoil is tilted at an angle to the fluid flow, as shown in Fig. 4-3, then there is an imbalance in the pressure along the y-axis resulting in a lift force. In an ideal fluid, the pressure remains balanced along the x-axis and, therefore, there is no net force along the x-axis. α is called the angle of attack.

When the airfoil is not symmetrical and the upper surface is curved more than the lower surface, then a lift force occurs because the pressure decrease and speed increase in the upper surface is larger than the pressure decrease and speed increase in the lower surface. As a convention, lift force is perpendicular to the direction of wind and drag force is parallel to the direction of wind. Positive α or nonsymmetrical airfoils cause airplanes to fly and wind turbines to produce energy. Each type of blade has an optimal value of α that produces maximum lift and minimal drag (see Fig. 4-4). This is discussed in greater detail later in the chapter.

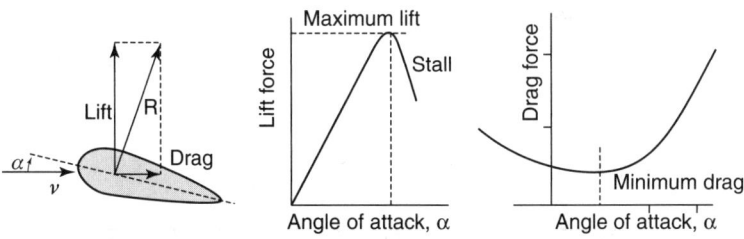

Figure 4-4 Relationship between lift and drag force and angle of attack for an airfoil.

Relative Velocity of Wind

The flow of air over an airfoil in the case of airplane is different compared to a wind turbine. In the former, the angle of attack of wind is constant along most of the length of the wing of an aircraft; in some aircraft, the end of wing turns up, where the angle of attack changes. In the case of wind turbines, the angle of attack changes along the length of a blade. The angle of attack is with respect to the blade, meaning, it is the angle at which wind strikes a blade as seen by an observer on the blade. This relative velocity is at an angle of γ, as shown in Fig. 4-5. In this figure, the axis of rotation is parallel to the x-axis and the blades move in the y-z plane. Consider point labeled A in the schematic, which is a point at a distance r from the center when the longitudinal axis of the blade is parallel to the z-axis. The velocity diagram is drawn for point A and is shown in the schematic to the right. The velocity diagram is with respect to an observer on the blade at point A. Since the observer itself is moving with tangential velocity of $v_t = +\omega r$, the observer will experience wind velocity of v_{rel}.

$$v_{rel} = \sqrt{v^2 + (\omega r)^2}$$
$$\gamma = \tan^{-1} \frac{\omega r}{v} \tag{4-1}$$

The magnitude and direction of the relative velocity of wind, as experienced by an observer sitting on the blade, changes with radius r, the distance from the hub.

Consider a turbine turning at 20 revolutions per minute (rpm) or 1/3 revolutions per second (rps), wind speed of 10 m/s and blade of length 50 m. Figure 4-6 illustrates the relative wind speed and γ at different values of r . v_{rel} and γ increase with distance from the hub of the rotor.

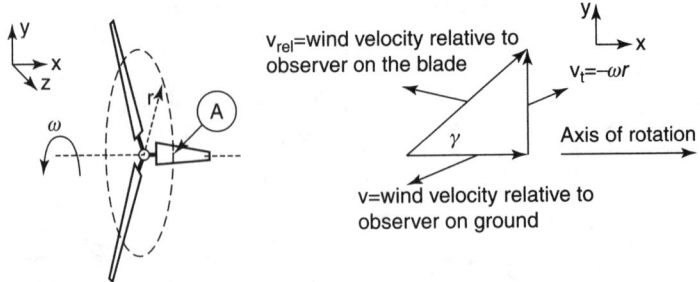

Figure 4-5   Schematic of a wind turbine rotating in the y-z plane. The relative velocity of wind (v_{rel}) as seen by an observer on the blade when the blade is at point A.

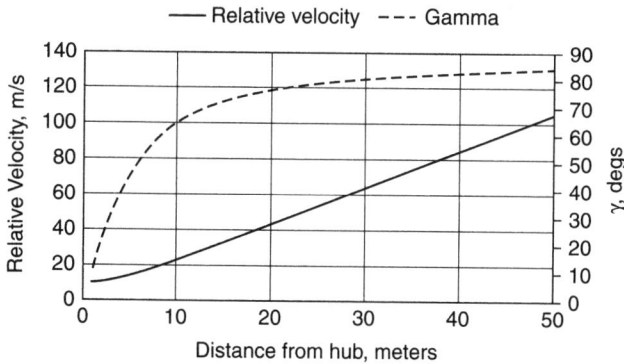

FIGURE 4-6 Plot of relative velocity and angle between wind velocity vector and the relative velocity vector.

For purposes of illustration, the impact of changing γ on blade design, the relationship between γ and angle of attack is discussed next. As stated in the previous section, there is an optimal angle of attack, which is the angle between the chord of the airfoil and the relative velocity vector v_{rel}. This optimal angle of attack will yield high lift and low drag forces. In order to maintain an optimal angle of attack α along the entire length of the blade while γ changes as a function of radius, the orientation of chord has to change along the length of blade. This orientation is called the pitch, ϕ. Pitch is the angle between the chord and the direction of motion, as shown in Fig. 4-7.

If the angle of attack is held constant, then the pitch of the blade has to decrease from the root of the blade to the tip of the blade. Close to the root of the blade, the pitch (ϕ) is approximately 90-α. As the distance from root, r, increases, the value of ϕ decreases (see Fig. 4-9).

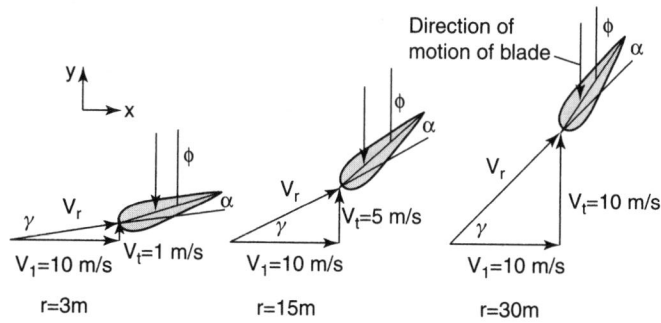

FIGURE 4-7 Velocity of wind relative to the blade at three locations on the blade, $r = R/10$, $R/2$, and R. The wind speed is constant at 10 m/s along the x-axis. Angular velocity $\Omega = 20$ rpm, $R = 30$ m. Tangential velocity of blade is in –y direction. Angle of attack α is constant.

Figure 4-8 Relation-
ship between angle
of attack, pitch
angle, and angle
between relative
velocity and
tangential velocity.

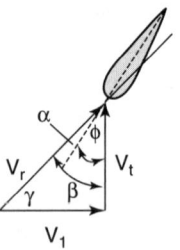

This is the reason why blades of most large turbines have a twist. Close to the hub, the blade airfoil chord is almost perpendicular to the plane of rotation. At the farthest point from the hub, which is the tip of the blade, the chord is at a small angle to the plane of rotation. Figure 4-7 shows the cross-section of blade at three different distances from the hub. The blades are moving in the negative y direction.

Figure 4-8 explains the angles in more detail. The angle of attack (α) plus the pitch angle (ϕ) is equal to the angle of relative velocity with the direction of motion of the blade (β).

$$90 - \gamma = \beta = \alpha + \phi \tag{4-2}$$

Therefore, $\phi = 90 - \gamma - \alpha$. Figure 4-9 is a plot of pitch as a function of radius for a blade with optimal angle of attack, $\alpha = 6°$.

Smaller turbine blades usually do not have a twist; this is to reduce the cost of manufacturing blades. Figure 4-10 shows such a blade in two positions. When the chord of airfoil is perpendicular to the plane of rotation (which is same as parallel to the direction of wind), the turbine is not in operational mode; it is in the feather position. The feather position is used to stop/brake the turbine or it is used after the brake has been applied to allow the wind to flow around

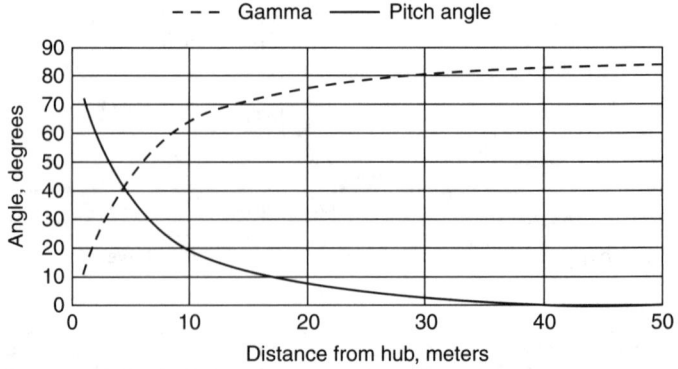

Figure 4-9   Relationship between gamma and pitch angle as a function of radius for $\alpha = 6°$.

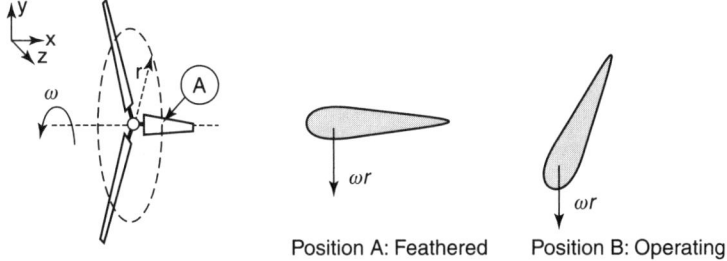

Position A: Feathered   Position B: Operating

Figure 4-10 Illustration of three-blade wind turbine with no twist. For a viewer looking in the –z direction, two blade positions are shown. In position A, the chord of airfoil is parallel to the wind direction.

the blades without generating lift. This may seem counterintuitive that when the chord is parallel to the wind direction, no energy is produced. The fallacy in the intuition is to expect lift at 90° pitch in both cases—stationary and moving blade. The correct way to examine this is to consider a rotating blade and the relative velocity of wind from the perspective of the blade. That is, to an observer on the blade that is rotating, the angle of attack should be positive for lift to occur.

With the basic airfoil terminology and flow of air over an airfoil covered, a more sophisticated theory of how energy is transferred from wind to a turbine rotor is presented next.

Rotor Disk Theory

In the exposition of Betz limit in Chapter 2, the actuator disk theory was used. It was postulated that the pressure difference across the turbine rotor leads to thrust, which performs work on the turbine. Performing work is the same as delivering energy. This energy delivered to the turbine comes from the loss of kinetic energy in the control volume that contains the upstream and downstream volumes of the turbine rotor. The loss of kinetic energy of wind does not happen at the turbine rotor; rather it happens upstream and downstream.

The rotor disk theory[1] will work with torque, thereby moving from Betz's completely abstract turbine rotor to a more realistic turbine rotor that delivers energy to a generator using torque. It will also introduce a more realistic model of airflow through the turbine rotor. However, this theory assumes infinite blades.

Let v_1 be the wind speed at the face of the rotor and ω be the rotational speed of the rotor. v_1 is referred to as the axial speed. As the wind passes through the blades of the rotor, it will acquire a tangential component of velocity. This is due to the rotation of the rotor, as shown in Fig. 4-11. The tangential component of wind velocity is opposite in direction to the tangential blade speed.

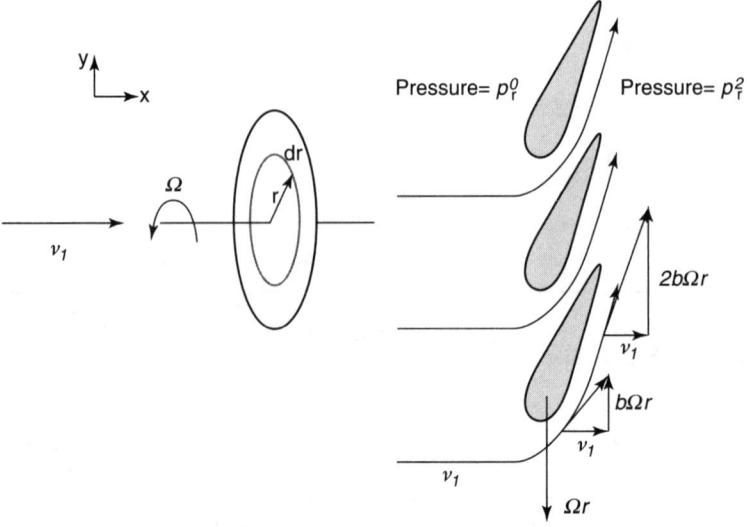

Figure 4-11 Schematic of rotor disk theory. Axis of rotation of the rotor is the x-axis; the wind direction is along the x-axis. The cross section of the rotor at a distance r from the center is shown in the right. Blades move down with tangential velocity; wind acquires a tangential velocity in the opposite direction. (*Adapted from Burton et al.[1].*)

According to Newton's second law, this change in angular momentum of wind must be accompanied by a torque. Stated differently, the geometry of the flow is such that wind is forced to acquire tangential component of velocity (forced to acquire angular momentum). Using Newton's second law, this forced increase in angular momentum will demand that the blade/rotor deliver a torque to wind. According to Newton's third law, wind will then deliver an equal but opposite torque to the blade/rotor. Angular momentum is the cross-product of the radius vector and the tangential momentum vector. The power that is delivered to the rotor is the torque multiplied by the angular velocity.

$$\text{Angular momentum} = \vec{r} \times m\vec{v} \tag{4-3}$$

$$Q = \text{Torque} = \text{Rate of change of angular momentum} = \dot{m}v_t r \tag{4-4}$$

$$P = \text{Power} = \text{Torque . angular velocity} = Q.\omega \tag{4-5}$$

The rate of mass flow through an annulus ring of the rotor is:

$$\delta\dot{m} = \rho v_1 \delta A = \rho v_1 2\pi r \delta r = \rho v_0 (1 - a) 2\pi r \delta r \tag{4-6}$$

where a is the axial flow induction factor, v_1 is the wind speed at the rotor, and v_0 is the wind speed upstream (see Eq. 2-29).

The final tangential speed of air is expressed in terms of the tangential flow induction factor b:

$$v_t = 2\omega r b \qquad (4\text{-}7)$$

The tangential speed of air at the middle of the rotor is one-half of the final tangential speed. The torque and power because of wind blowing through the annulus ring is:

$$\delta Q = \delta \dot{m} v_t r = \delta \dot{m}(2\omega r b)r \qquad (4\text{-}8)$$

$$\delta P = \delta Q\omega = 2\delta \dot{m}(\omega r)^2 b \qquad (4\text{-}9)$$

$\lambda_r = \left(\frac{\omega r}{v_0}\right)$ is called the local speed ratio, and $\lambda = \left(\frac{\omega R}{v_0}\right)$ is called the tip speed ratio, where R is the length of the blade.

$$\delta P = 2\left(\rho v_0 (1 - a) 2\pi r \delta r\right) v_0^2 \lambda_r^2 b = \left[\frac{1}{2}\rho v_0^3 2\pi r \delta r\right] 4(1 - a)\lambda_r^2 b \qquad (4\text{-}10)$$

The term in the square parenthesis is the power contained in an annulus of air upstream. Note that in the above equations a and b may be functions of r. The efficiency of power capture in this annulus is defined as:

$$\varepsilon = \frac{\delta P}{\left(\frac{1}{2}\rho v_0^3 2\pi r \delta r\right)} = 4(1 - a)\lambda_r^2 b \qquad (4\text{-}11)$$

Maximum efficiency will require:

$$\frac{d\varepsilon}{db} = 0 = -\frac{da}{db}\lambda_r^2 b + (1 - a)\lambda_r^2$$

$$\frac{da}{db} = \frac{1 - a}{b} \qquad (4\text{-}12)$$

The relationship between variables a and b in Eq. (4-12) will hold true at maximum efficiency. Power coefficient function (C_P) can be written using Eq. (4-10) and dividing by the total power contained in disk of radius R.

$$\delta C_P = \frac{\delta P}{\left(\frac{1}{2}\rho v_0^3 \pi R^2\right)} = \frac{8r(1 - a)\lambda_r^2 b \delta r}{R^2}$$

$$\frac{\delta C_P}{\delta r} = \frac{8r(1 - a)\lambda_r^2 b}{R^2} \qquad (4\text{-}13)$$

According to the actuator disk theory from Chapter 2, Eq. (2-32), the power extracted by the turbine in an annulus of area δA_r at distance r from the center is:

$$\delta P = \left(\frac{1}{2}\rho\delta A_r v_0^3\right) 4a(1-a)^2 \tag{4-14}$$

Equating (4-10) and (4-14) gives:

$$4(1-a)\lambda_r^2 b = 4a(1-a)^2$$
$$\lambda_r^2 = \frac{a(1-a)}{b} \tag{4-15}$$

λ_r is a function of angular speed, radius, and upstream wind speed, so it is not a function of a and b. Differentiating by b gives:

$$\frac{da}{db} = \frac{\lambda_r^2}{1-2a} = \frac{a(1-a)}{b(1-2a)} \tag{4-16}$$

From Eqs. (4-12) and (4-16):

$$a = \frac{1}{3}$$

This is the same axial flow induction factor that was derived from the actuator disk theory in Chapter 2.

Substituting Eq. (4-15) in (4-13) yields:

$$\frac{\delta C_P}{\delta r} = \frac{8ra(1-a)^2}{R^2}$$

$$C_P = \int_0^R \frac{8ra(1-a)^2}{R^2} dr = 4a(1-a)^2 = \frac{16}{27} \tag{4-17}$$

The maximum power coefficient as computed using the rotor disk theory is the same as the actuator disk theory. The rotor disk theory does not alter the axial component of velocity, but introduces a tangential component in the wake (after the wind passes through the turbine).

Question: How can the maximum power coefficient be the same even though additional kinetic energy (in the form of tangential speed) has been imparted to air at the rotor?

The explanation lies in the observation that the tangential speed causes additional pressure drops, which is in addition to the pressure drop because of the actuator disk theory.

$$\text{Additional pressure drop} = \frac{1}{2}\rho v_t^2 = \frac{1}{2}\rho(2\omega r b)^2 \qquad (4\text{-}18)$$

Similar to the pressure drop from actuator disk theory, this additional pressure drop will result in expansion of the wake as the tangential speed slows in the wake. In addition, the tangential component of velocity will cause the wind to follow a helical path in the wake.

In summary, the actuator disk theory of Chapter 2 stated that the energy is delivered by the work done by the pressure drop across the rotor. The exact mechanism within the rotor disk was not required in the actuator disk theory. The rotor disk theory, on the other hand, specifies a mechanism—impart of tangential momentum by the rotor to the wind as it passes through the blades of the rotor. When power from the two mechanisms is compared, the same value of power coefficient is obtained.

In the remainder of the chapter, basic theory of airfoils is developed.

Lift Force

This section will first begin with an explanation of lift. The fundamental mechanism for generation of lift forces in an airfoil is still debated. In this section, the commonly stated mechanisms will be discussed—some have been proved to be wrong.

Equal Transit Time Fallacy

This theory states that particles of air that separate at the leading edge of the airfoil will meet at the trailing edge, hence, the equal transit time theory. This requires that the particles at the top surface travel faster than the lower surface. Bernoulli's principle can then be applied: the pressure at the top surface will be lower than the lower surface and this causes lift.

The fallacy in this theory is the equal transit time. This has been shown to be not true; that is, particles that separate at the leading edge do not meet at the trailing edge. A more accurate method of explaining lift is presented below.

Rotation Fluid Flow, Circulation, and Vortices

Generation of lift can be explained by circulation. For a circulation of strength Γ, the lift force is given by:

$$L = \rho v_0 \Gamma \qquad (4\text{-}19)$$

where ρ is the density and v_0 is the free stream wind speed. If there is no circulation, there is no lift. What is circulation? Mathematically, circulation around a body is defined as the sum of velocity along a closed path.

$$\Gamma = \oint \vec{v} \cdot d\vec{s} \qquad (4\text{-}20)$$

where the closed path is around the object that is subject to lift. The integral sums the dot product of the velocity along the path. The circulation around any closed path around a cylinder spinning with angular speed ω is:

$$\Gamma = \omega 2\pi R^2 \qquad (4\text{-}21)$$

where R is the radius of the cylinder. The lift generated by the spinning cylinder, also called the Magnus effect, is shown in Fig. 4-12 and is expressed as:

$$L = \rho v_0 \omega 2\pi R^2 \qquad (4\text{-}22)$$

Intuitively, circulation creates higher flow speeds above the cylinder and lower flow speed below the cylinder. Using Bernoulli's principle, pressure will be lower above the cylinder and higher below the cylinder. This pressure difference causes lift.

According to this theory, the flow around an airfoil is viewed as a sum of two flows: Flow with no circulation and flow with circulation (see Fig. 4-13)[2]. The circulation around the airfoil causes lift. If there is no circulation, then there is no lift. Intuitively, however, there is no circulation around an airfoil; it is not spinning. Martin Wilhelm Kutta, a German aerodynamicist, conceptualized that a body with a sharp

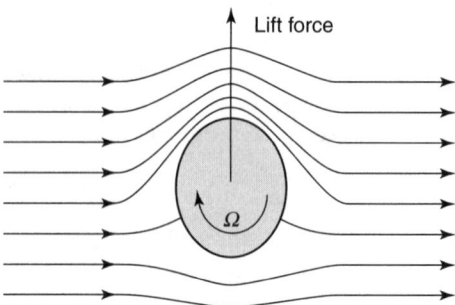

Figure 4-12 Lift force on a rotating cylinder in a fluid flow due to Magnus effect.

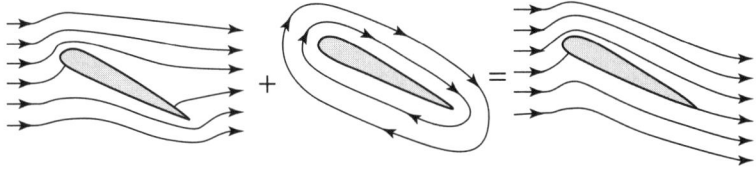

FIGURE 4-13   Illustration of how flow over an airfoil is considered as consisting of two components, inviscid flow plus circulation.

trailing edge creates a circulation similar to a spinning cylinder. The reasoning is as follows:

- In an inviscid (frictionless) flow with an airfoil, as shown in Fig. 4-14, two stagnation points (where the speed is zero) are created. The streamlines separate at the stagnation point near the leading edge and then rejoin at the stagnation point near the trailing edge. The flow at the bottom surface has to turn around the trailing edge.
- At the pointed trailing edge where the flow turns, the velocity will become infinite.
- In real fluids with viscosity, infinite velocity is not possible. Therefore, the area between the rear stagnation point and the trailing edge generates a "starting vortex." The starting vortex is clockwise and combines with the upper flow to accelerate the flow over the upper surface and then sheds into the wake. As the upper flow accelerates, the stagnation point moves toward the trailing edge.

The Kutta condition specifies that the real fluid flow will leave tangentially at the trailing edge (see Fig. 4-15). This requires that a clockwise circulation exist that moves the trailing stagnation point to coincide with the trailing edge. This Kutta condition defines the amount of circulation that must exist.

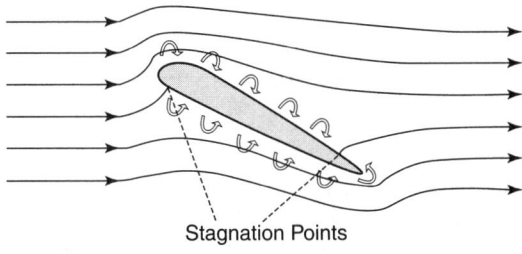

Stagnation Points

FIGURE 4-14   Fluid flow around an airfoil with two stagnation points.

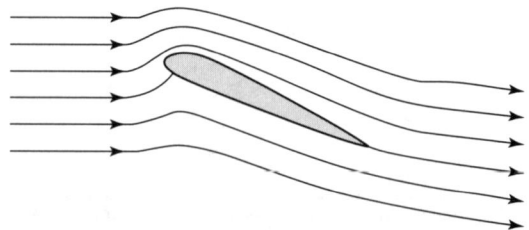

Figure 4-15 Fluid flow around an airfoil with the Kutta condition. The second stagnation point is at the trailing edge. Note the downstream airflow is not parallel to the upstream airflow.

The following equation defines the circulation:

$$\Gamma = \pi v_o c \sin \alpha \tag{4-23}$$

where v_o is the free stream speed, c is the chord length of the airfoil, and α is the angle of attack. The lift force per unit length of blade is:

$$L = \rho v_0 \left(\pi v_o c \sin \alpha \right) = \rho v_0^2 \pi c \sin \alpha \tag{4-24}$$

Lift force is traditionally expressed in terms of a lift coefficient (C_L) as:

$$L = \frac{1}{2} \rho S v_0^2 C_L \tag{4-25}$$

where S is the area of the blade, which is equal to chord length (c) multiplied by the length of the blade (l).

$$C_L = \frac{\rho v_0^2 \pi c l \sin \alpha}{\frac{1}{2} \rho S v_0^2} = 2\pi \sin \alpha$$

This is a theoretical relationship between C_L and α, the attack angle. Empirically, for small values of α, the relationship is linear. How small depends on the airfoil design; typical values for alpha are in the range of -15 to $15°$. Outside this range, the linear relationship between C_L and α ceases to exist. As α increases, the lift drops off resulting in a stall condition (see Fig. 4-16).

Salient features of the coefficient of lift curve in Fig. 4-16 are:

- Curves are obtained empirically by conducting experiments on specific airfoil shapes
- In the linear region, the slope is 2π
- For nonsymmetrical airfoils with camber line (line that is equidistant from upper and lower surface of the airfoil) that is higher than chord line, the entire lift curve shifts up, resulting in positive lift for zero angle of attack

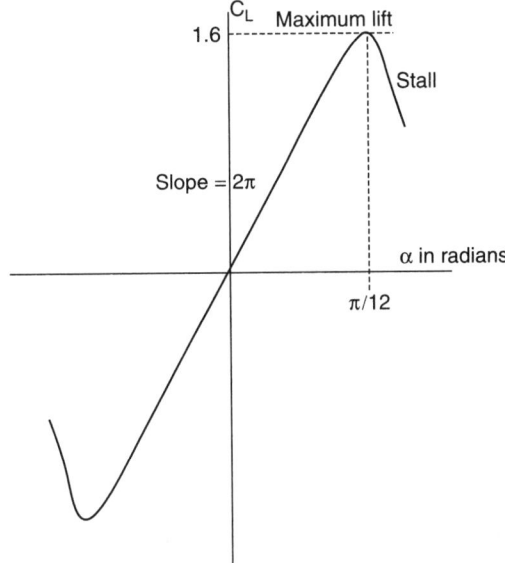

FIGURE 4-16 General form of the coefficient of lift as a function of attack angle for a symmetric airfoil.

Real Fluids

To explain the drag forces acting on an airfoil, this section starts with a short introduction to real fluids. Real fluids, as opposed to ideal fluids are not inviscid, that is, real fluids possess viscosity. Viscosity is a measure of the force required to induce shearing in materials. Consider a fluid flow on a plate with flow velocity along the x-axis, and fluid thickness along the y-axis. Newton's definition of viscosity is:

$$\mu = \tau \bigg/ \frac{\partial v}{\partial y} \qquad (4\text{-}26)$$

where μ is viscosity, τ is the shear force parallel to the fluid flow, and $\frac{\partial v}{\partial y}$ is the gradient of speed.

Viscosity creates boundary layers and skin-friction drag, a force that is parallel to the fluid flow.

There are two types of boundary layers: Laminar and turbulent. This is illustrated in Fig. 4-17 with fluid flow over an aerofoil. In a laminar boundary layer, the streamlines are maintained, whereas in a turbulent boundary layer there is exchange of energy between the layers.

When speaking of viscous flow, Reynolds number is a key concept

$$R = \frac{\rho v}{\mu / l} = \frac{\text{Inertial forces}}{\text{Viscous forces}} \qquad (4\text{-}27)$$

where l is the characteristic length; it is a dimension that is chosen by convention. For flow inside a tube, characteristic length is the diameter

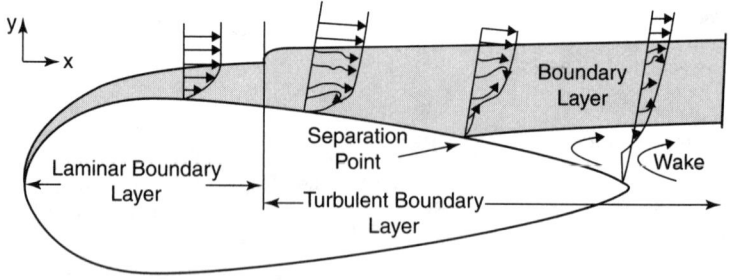

Figure 4-17 Formation of boundary layer on an airfoil.

of the tube; for an open rectangular channel, l is the ratio of cross-sectional area of fluid and perimeter of the channel in contact with the fluid; for a flat plate, l is the distance from the edge; for an airfoil, l is the chord length.

Reynolds number is a dimensionless quantity that is used to describe when a flow transitions from laminar to turbulent. Consider two examples below:

Flow in a pipe of diameter d,

$$\text{Laminar flow when } R < 2{,}300$$
$$\text{Transient when } 2{,}300 < R < 4{,}000 \tag{4-28}$$
$$\text{Turbulent when } R > 4{,}000$$

Irrespective of the type of fluid, size of pipe, and free-stream speed v, the flow will satisfy the conditions in Eq. (4-28).

Flow of Fluid over an Airfoil

On the surface of the airfoil, there is no slip, so the speed is zero. Away from the surface, the speed increases and reaches the free-stream speed. This region is the boundary layer. Skin-friction drag acts on the airfoil because there is friction on the surface and shear force that opposes the change in speed between layers.

In a laminar boundary layer, all the fluid flow is parallel to the surface. In turbulent boundary layer:

- The speed near the surface of the aerofoil is higher; that is, there is a rapid change in speed from zero on the surface to nonzero a short distance away from the surface.
- The thickness of the boundary layer is larger.
- Energy is exchanged from the faster moving particles in the boundary layer to the slower moving particles near the surface.

- In addition to the predominant component of velocity that is parallel to the surface, there is a random component of velocity perpendicular to the surface that causes mixing.

As the fluid moves over the shoulder of the airfoil, pressure reaches its lowest point and the viscous forces become large as the change in velocity becomes large in the boundary layer close to the surface. As the fluid moves away from the shoulder and toward the trailing edge, pressure increases. Higher pressure leads to amplification of disturbances leading to turbulence. A combination of higher viscous force and higher pressure causes the boundary layer to separate from surface at the separation point. This separation causes an imbalance in the pressure forces along the x-axis resulting in another drag force called pressure drag.

How can pressure drag be reduced? The answer is, by moving the boundary layer separation point closer to the trailing edge. The two primary factors that influence the location of boundary layer separation point are angle of attack and surface roughness. As the angle of attack increases, the separation point moves closer to the leading edge. The pressure drag is proportional to the square of the attack angle.

The drag force is traditionally expressed in terms of a drag coefficient (C_D) as:

$$D = \frac{1}{2}\rho S v_0^2 C_D \tag{4-29}$$

$$C_D = C_{fD} + C_{pD} \tag{4-30}$$

where C_{fD}, C_{pD} are coefficients of skin-friction drag and pressure drag. For normal wind speeds (<< speed of sound), skin-friction drag is small and the pressure drag is larger. Since C_L is a linear function of α, and C_{pD} is a quadratic function of α, see Fig. 4-18. The relationship between drag and lift at normal wind speeds and small values of α

FIGURE 4-18 Drag as a function of attack angle.

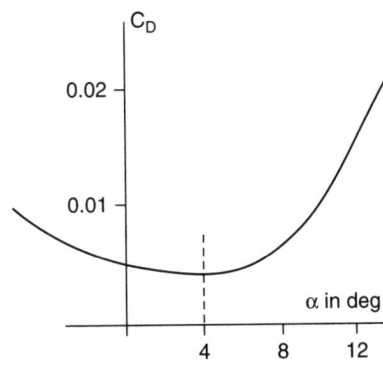

(less than 15 degrees) is:[2]

$$C_{pD} = \frac{C_L^2}{\pi A_r \epsilon} \qquad (4\text{-}31)$$

where ϵ is the spanwise efficiency factor, A_r is the aspect ratio, defined as:

$$A_r = \frac{l^2}{S} \qquad (4\text{-}32)$$

l is the length of the blade, and S is the blade area. For a rectangular blade with constant chord length, $A_r = l/c$, where c is the chord length. As the aspect ratio increases, the pressure drag decreases. Therefore, a long blade with short chord is preferable.

Effect of Reynolds Number on Lift and Drag Coefficients

Reynolds number of blades of wind turbines are in the range of one to ten million. Reynolds number has a significant impact on lift and drag coefficients. Drag coefficient falls and the lift coefficient increases as the Reynolds number is increased, as shown in Figs. 4-19 and 4-20. At low values of α, the coefficient of drag is almost constant, and then it rises rapidly. Reynolds number dictates the location on the aerofoil where the flow becomes turbulent. As the Reynolds number increases, the transition point from laminar to turbulent moves closer to the leading edge. For low values of α, there is no separation of boundary layer, as α approaches 8° in the Fig. 4-20 example, separation occurs and the pressure drag rapidly increases. This is the stall condition. A wind turbine therefore avoids the stall condition during operational mode by keeping the angle of attack low and by increasing the Reynolds number.

FIGURE 4-19 Coefficient of lift for different values of Reynolds number.

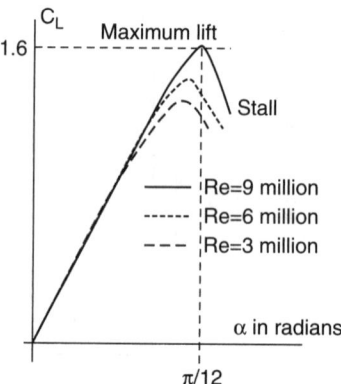

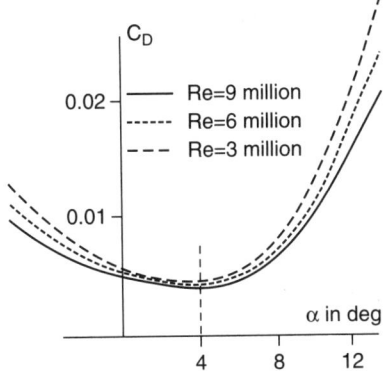

FIGURE 4-20 Coefficient of drag for different values of Reynolds number.

The relationship between Reynolds number and drag is counter-intuitive: Higher Reynolds number leading to lower drag. A higher Reynolds number leads to displacement of transition point (from laminar to turbulent) toward the leading edge. This should cause the separation of boundary layer closer to the leading edge and therefore higher drag. In reality, the opposite happens, when the transition point is closer to the leading edge, turbulence starts early. The turbulence causes energy to be exchanged between layers, which energizes the layer close to the aerofoil surface. This reenergized layer is better able to withstand the viscous force at the surface and higher pressure, which leads to occurrence of separation closer to the trailing edge.[2]

The shape of aerofoil[3] determines the shape of the lift curve during stall in Fig. 4-19. Thicker aerofoil with smooth curvature exhibit a graceful decline in lift during stall, that is, the curve after stall has a small negative slope. Thinner aerofoil has a larger negative slope, similar to the one in Fig. 4-19. The reason for the difference is the location of the start of separation. If separation starts at the trailing edge, then stall is graceful. In contrast, if separation starts in the middle or closer to the leading edge, then there is a sudden rise in pressure drag.

Drag-Based Turbines

In this section, a lift-based turbine is compared with a drag-based turbine. As described above, lift force is perpendicular to the relative velocity of wind as observed by the blade and the drag force is parallel to the direction of relative velocity. Drag force is essentially the force that pushes an object.

$$\text{Drag force } F_d = \frac{1}{2}\rho A v^2 C_D \qquad (4\text{-}33)$$

Shape	C_D
Sphere	0.47
Cone	0.5
Cube	1.05
Flat plate with surface perpendicular to air flow	1.17
Cup with air flow that fill the cavity	1.3
Flat plate with surface parallel to air flow	0.001
Streamlined body in the shape of an airfoil	0.04

TABLE 4-1 Drag Coefficient for Different Shape Objects

where C_D is the drag coefficient, a dimensionless constant. C_D depends on the shape of an object and on the Reynolds number. Table 4-1 contains drag coefficient of different shapes. In the extreme case, consider a flat plate with its surface perpendicular to the direction of wind, as in Fig. 4-21.

Old vertical axis wind turbines (VAWT) were drag-based machines. An example of VAWT is seen in Fig. 4-22. In the figure, top view of a four-blade turbine is shown. It has barrier with two openings. The first opening allows wind to push the blades in the direction of motion and the second opening allows wind to escape. The barrier prevents wind from pushing the blades when the blade is traveling against the wind direction.

The drag force per unit area on a stationary flat plate with wind speed of 10 m/s is:

$$\frac{F_d}{A} = \frac{1}{2}1.225 \cdot 10^2 \cdot 1.17 = 71.7\frac{N}{m^2} \tag{4-34}$$

Applying the drag force to a VAWT with a flat plate:

$$F_d = \frac{1}{2}\rho A(v - u)^2 C_D \tag{4-35}$$

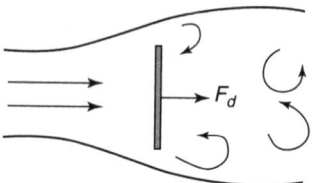

FIGURE 4-21 Drag experienced by a flat plate that is placed perpendicular to the direction of wind. F_d is the drag force.

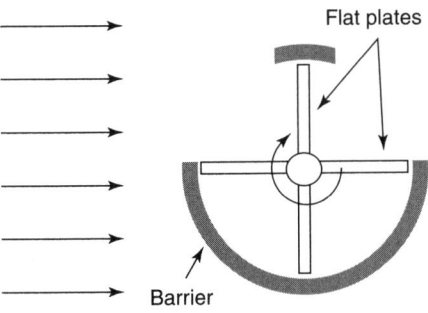

FIGURE 4-22 Top view of a vertical axis drag-based turbine. This turbine has four plates. Wind in the top-left quadrant drags or pushes the plates; wind exits from the top-right quadrant.

where u is the speed of the flat plate and $(v - u)$ is the relative speed of wind as experienced by the plate. As expected, when u is zero, F_d is maximum; when $v = u$, $F_d = 0$. Several simplifying assumptions will be made to compare the efficiency between lift- and drag-based machines. The power delivered to a drag-based machines is (see Fig. 4-23):

$$P = F_d u = \frac{1}{2} \rho A (v - u)^2 u C_D \qquad (4\text{-}36)$$

As expected, the power delivered is zero when $u = 0$ and $v = u$. The maximum power is delivered when $u = \frac{v}{3}$.

$$\text{Power coefficient} = P \left/ \frac{1}{2} \rho A v^3 \right. = C_D \frac{4}{27} \qquad (4\text{-}37)$$

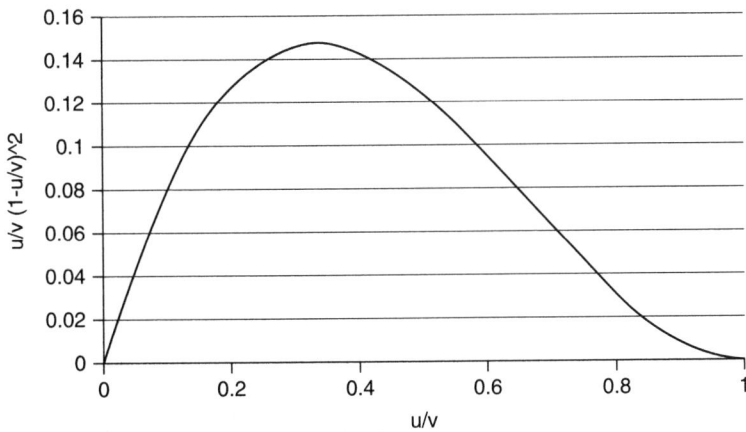

FIGURE 4-23 Plot of efficiency of a drag turbine as a function of ratio of plate speed to wind speed. Maximum efficiency is reached when $u/v = 1/3$.

Betz Law in Fig. 2-8 indicated that the maximum power coefficient is 16/27. Comparison of power coefficient of this drag turbine with Betz limit indicates that the maximum efficiency of drag-based machines is substantially lower than the maximum theoretical efficiency. Here, efficiency is used to mean efficiency in converting wind energy to rotor energy. Even if a shape with highest drag coefficient is chosen—a hemispherical cup in which air fills the cup—the value of C_D is 1.3, which makes the maximum efficiency of a drag-based device = 0.193. Therefore, drag-based machines are rarely used because of low efficiency. In addition, the amount of rotor area used in a drag-based machine is, at most, one-half of area of the rotor in other configurations. For these reasons, most efficient modern turbines are lift-based that have a small component of drag and a large component of lift.

References

1. Burton, T., Sharpe, D., Jenkins, N., and Bossanyi, E. *Wind Energy Handbook*, Wiley, Hoboken, NJ, 2001.
2. Talay, T. A. *Introduction to Aerodynamics of Flight*, NASA, Langley, 1975 (NASA SP-367).
3. Hansen, M. O. L. *Aerodynamics of Wind Turbines*, Earthscan, Sterling, VA, 2008.
4. Shankar, L. N. *Course Notes Low Speed Aerodynamics*, Georgia Institute of Technology, Guggenheim School of Aerospace Engineering [Online] 2008. http://www.ae.gatech.edu/people/lsankar/AE2020/.

CHAPTER 5

Advanced Aerodynamics of Wind Turbine Blades

Aeronautics was neither an industry nor a science. It was a miracle.
—Igor Sikorsky

Introduction

In this chapter, advanced topics related to aerodynamics of wind turbines are covered. The previous chapter was an introduction to the basic concepts of airfoil, lift, and drag. In this chapter, the first section will describe the blade element theory, which is an enhancement over the rotor disk theory. In the rest of the chapter, the manifestations of the aerodynamic theory on the performance of turbines will be discussed. Specifically, power performance curves of different configurations of wind turbines are presented. The configurations discussed include constant versus variable speed rotor and stall versus pitch regulation. The chapter ends with a brief description of the vertical axis wind turbine.

Blade Element Model

The blade element model (BEM) is more rigorous compared to the rotor disk model. This model is used to derive theoretical performance curves of wind turbines. BEM computes the axial force and the torque (tangential force multiplied with the radius along the blade). The axial force is then equated to the rate of change of momentum along

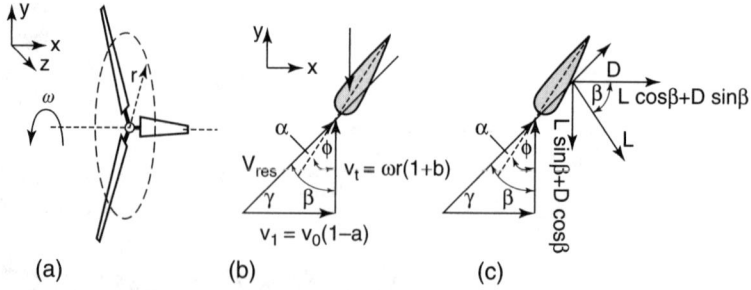

FIGURE 5-1   Velocity of wind relative to blade and lift and drag forces. Wind is in the x direction, axis of rotation is parallel to x-axis, and plane of rotation is the y-x plane. Parts (b) and (c) are views of blade when the radial orientation of the blade is parallel to the z-axis; the cross section of the blade is from a plane that is parallel to the x-y plane. The cross section of the blade is moving in the −y direction.

the x-axis, while torque is equated to the rate of change of angular momentum. Using these two equations, equations for axial and radial induction factors (a and b) are derived. Next, a relationship between the power coefficient (C_p) and tip speed ratio (λ) is derived. This relationship provides the theoretical basis for the most often used power performance curves of turbines (power output versus wind speed).

The velocity vector and force vector are described in Fig. 5-1. The velocity as seen by the blade is:

$$v_{res} = \sqrt{v_0^2(1-a)^2 + \omega^2 r^2(1+b)^2} \tag{5-1}$$

$$v_{res} = \frac{v_0(1-a)}{\sin\beta} = \frac{\omega r(1+b)}{\cos\beta} \tag{5-2}$$

where v_{res} is the resultant velocity vector. In Fig. 5-1, α is the angle of attack—angle between the resultant velocity vector and the chord of the airfoil; β is the angle between the resultant velocity vector and the tangential direction; $\gamma = 90 - \beta$; ϕ is the pitch—the angle between chord and the tangential direction; ω is the angular speed of the rotor; r is the radial distance from the axis of rotation. All the analysis is at a distance of r ($< R$; where R is the total length of the blade). Note the tangential velocity of wind with respect to an observer on the blade has a magnitude that is equal to sum of the tangential velocity of the blade (ωr) and the tangential velocity of the wake ($\omega r b$), and has direction that is opposite to the motion of the blade. Tangential velocity of wake is because of the tangential momentum imparted to wind by the rotating blades.

The lift and drag forces on a blade element δr of chord length c and turbine with N blades are:

$$\delta L = \frac{1}{2}\rho v_{res}^2 C_L c N \delta r \tag{5-3}$$

$$\delta D = \frac{1}{2}\rho v_{res}^2 C_D c N \delta r \tag{5-4}$$

In the above equations, $cN\delta r$ is the effective area or blade solidity in an annulus of thickness δr. The direction of δL is perpendicular to vector v_{res}; δD is parallel to v_{res}. Force along the axial direction is:

$$\delta L \cos \beta + \delta D \sin \beta = \frac{1}{2}\rho v_{res}^2 N c (C_L \cos \beta + C_D \sin \beta) \delta r \tag{5-5}$$

Torque is the tangential force multiplied by radius:

$$\delta Q = r(\delta L \sin -\delta D \cos \beta) = \frac{1}{2}\rho v_{res}^2 N c (C_L \sin \beta - C_D \cos \beta) r \delta r \tag{5-6}$$

Assuming that the axial force is due to change in axial momentum of air, that is, there is no radial interaction,[1] then:

$$\frac{1}{2}\rho v_{res}^2 N c (C_L \cos \beta + C_D \sin \beta) \delta r = \delta \dot{m} 2 a v_0 + \frac{1}{2}\rho (2 b \omega r)^2 2 \pi r \delta r \tag{5-7}$$

In the above equation, axial force caused by pressure drop due to wake rotation is added (last term on the right-hand side of Eq. 5-7). The force because of pressure drop is in Eq. (4-18). The value of $\delta \dot{m}$ is in Eq. (4-6). The following substitutions are made to simplify the equations:

$$C_x = (C_L \cos \beta + C_D \sin \beta) \tag{5-8}$$

$$C_y = (C_L \sin \beta - C_D \cos \beta) \tag{5-9}$$

$$\text{Tip speed ratio, } \lambda = \omega R/v_0 \tag{5-10}$$

$$\mu = r/R \tag{5-11}$$

$$\text{Blade solidity, } \sigma = \frac{Nc}{2\pi r} \tag{5-12}$$

The axial force Eq. (5-7) becomes:

$$\left(\frac{v_{res}}{v_0}\right)^2 \sigma C_x = 4(a(1-a) + (b\mu\lambda)^2) \tag{5-13}$$

$$\left[\frac{1-a}{\sin \beta}\right]^2 \sigma C_x = 4(a(1-a) + (b\mu\lambda)^2) \tag{5-14}$$

If the pressure drop term above containing b is ignored, then:

$$a = 1 \Big/ \left[\frac{4 \sin^2 \beta}{\sigma C_x} + 1 \right] \qquad (5\text{-}15)$$

Torque is equal to the rate of change of angular momentum, therefore:

$$\delta Q = \frac{1}{2} \rho v_{res}^2 Nc(C_L \sin \beta - C_D \cos \beta) r \, \delta r = \delta \dot{m}(2\omega r b)r$$
$$= \rho v_0 (1 - a) 2\pi r \, \delta r (2\omega r b) r \qquad (5\text{-}16)$$

Using Eqs. (5-9) and (5-12), the above equation becomes:

$$\frac{v_{res}^2}{v_0 \omega r} \sigma C_y = 4b(1 - a) \qquad (5\text{-}18)$$

$$\frac{(1 - a)(1 + b)}{\sin \beta \cos \beta} \sigma C_y = 4b(1 - a) \qquad (5\text{-}19)$$

$$b = 1 \Big/ \left[\frac{4 \sin \beta \cos \beta}{\sigma C_y} - 1 \right] \qquad (5\text{-}20)$$

Equations (5-15) and (5-20) for a and b require an explanation:

- In the two equations, β is a function of a and b, as indicated in Eq. (5-2).
- σ, blade solidity is a function of radius r
- C_x and C_y are functions of β, v_{res}

The values of a and b are computed iteratively; an algorithm for solving this is presented by Hansen.[2] Since pitch changes along the length of the blade, the solution algorithm involves solving the above equations for smaller slices along the length of the blade. Two corrections are required for the equations that solve for a and b, because of the assumptions made thus far:

1. Prandtl's Tip Loss Factor, which corrects for the assumption of infinite number of blades
2. Glauert correction when a is high. The momentum theory used above does not apply when $a \geq 0.5$, as discussed in Eq. (2-25). In fact, correction factors are applied for $a > \frac{1}{3}$.

The corrections are described by Hansen.[2] Having computed a and b, expressions for torque, power and the power coefficient may be derived.

$$\text{Torque } Q = \frac{1}{2} \rho v_0^2 \pi R^3 \lambda \int_0^1 8b (1 - a) \mu^3 d\mu \qquad (5\text{-}21)$$

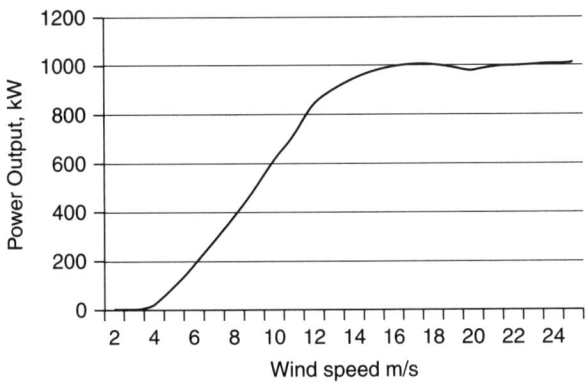

FIGURE 5-2   Power output of a turbine with Prandtl tip loss factor and Glauert correction.

$$\text{Power P} = Q\omega = \frac{1}{2}\rho v_0^3 \pi R^2 \lambda^2 \int_0^1 8b\,(1-a)\,\mu^3 d\mu \qquad (5\text{-}22)$$

$$\text{Coefficient of power,}\, C_p = Q\omega \Big/ \left[\frac{1}{2}\rho\pi R^2 v_0^3\right] = \lambda^2 \int_0^1 8b(1-a)\mu^3 d\mu \qquad (5\text{-}23)$$

A sample theoretical power curve is in Fig. 5-2. The relationship above between C_p and $\lambda = \omega R/v_0$ is not simple quadratic. For most blade designs, the peak value of C_p of approximately 0.5 occurs for values of λ between about 8 and 10 (see Fig. 5-3). Note that λ determines β, which influences the values of a and b.

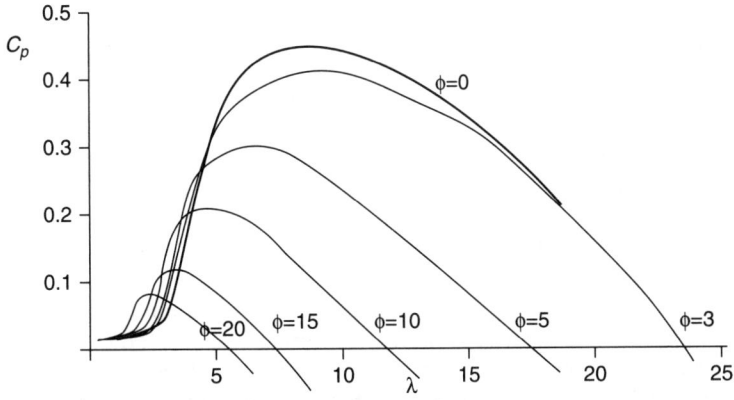

FIGURE 5-3   Typical power coefficient versus tip speed ratio ($\lambda = \omega R/v_0$) for different values of pitch (ϕ, in degrees).

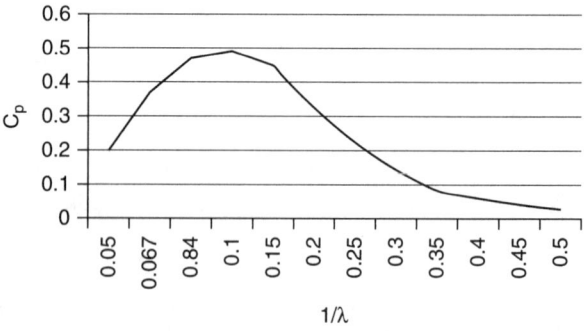

Figure 5-4   Plot of power coefficient versus reciprocal of tip speed ratio ($1/\lambda = v_0/\omega R$).

For small values of λ, which is high value of wind speed, the value of β is large and value of a is low (Eq. 5-15). This leads to stall conditions.

For larger values of λ, which is low wind speed, the value of β is low, which leads to high drag. Note the value of a is high.

In summary, for small and large values of λ, the value of a is nonoptimum and the value of C_p is low (see Fig. 5-4). It is often convenient to conceptualize the relationship between C_p and $1/\lambda$, because $1/\lambda$ is proportional to wind speed. This relationship is seen in Fig. 5-4.

For a given blade design, the theoretical relationships between power and wind speed, power coefficient and tip speed ratio, and others are calibrated with measurements. An empirical correction factor is applied to compensate for the assumptions made in the theoretical computations.

Constant-Speed Turbines, Stall-versus Pitch-Regulated

The nature of operation of a wind turbine with respect to rotor speed is determined by the type of electric generator attached to the turbine. Constant rotor speed operation is typical when a configuration is used in which a generator is directly connected to the grid. The speed of rotation of the generator is derived from the frequency of the grid. On a 50-Hz grid, the rotor speed will be 50*60/(*number of poles*/2). Number of poles is an even number that represents the number of north and south magnetic poles in the generator. Most common configuration of asynchronous generator is two or four poles. Thus, the generator operates at 3,000 or 1,500 rpm. Similarly, a 60-Hz generator operates at 3,600 or 1,800 rpm. A grid-connected asynchronous generator is the simplest and cheapest type of generator because the output energy

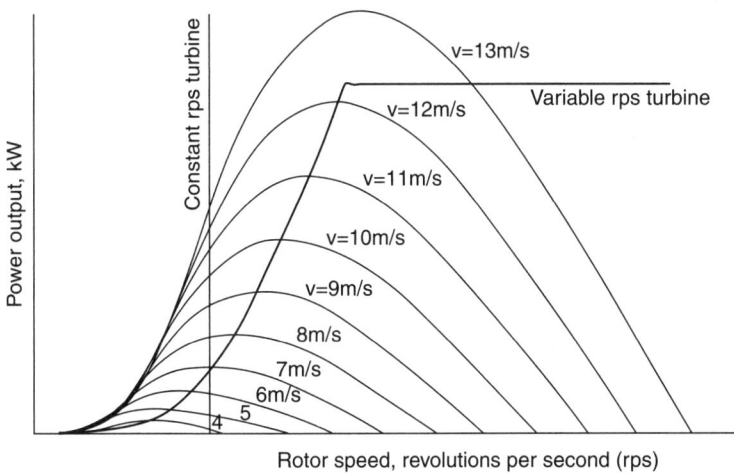

FIGURE 5-5 Power output versus rotor speed for different wind speeds. Power curves for fixed speed rotor and variable speed rotor are illustrated. *x*-axis is in rps (= rpm/60).

is at the grid frequency and, therefore, requires no power electronics to rectify or invert (AC–DC, DC–AC). However, constant-rotor speed turbines are unable to deliver the optimal power output at different wind speed (see Fig. 5-5). Generators are discussed in more detail in Chapter 9.

A stall-regulated turbine is the simplest type of turbine, in which the pitch of the blades is fixed. Since the power output depends on the pitch, as shown in Fig. 5-3, an optimum pitch angle is fixed for these turbines, based on the average wind velocity. At higher wind speed (low value of λ), the angle of attack increases to a point at which the drag force increases rapidly resulting in stall, as shown earlier in Fig. 4-19. A stall-regulated turbine is characterized by a hump in power at close to nominal wind speed and a rather sharp fall off in power above the nominal wind speed (see Fig. 5-6). Stall-regulated machines are simpler and inexpensive because there are no motors and no controls to change the pitch of the blades.

A pitch-regulated turbine, on the other hand, manages the power output at wind speeds higher than the nominal wind speed by increasing the pitch. Increasing the pitch leads to reduction in the angle of attack, which leads to reduction of drag forces. A pitch-regulated turbine is characterized by a constant power output at rated wind speed and higher (see Fig. 5-6). A higher value of pitch is also used during start up of turbine when ω is low and, therefore, λ is low. This condition yields higher C_p (see Fig. 5-3).

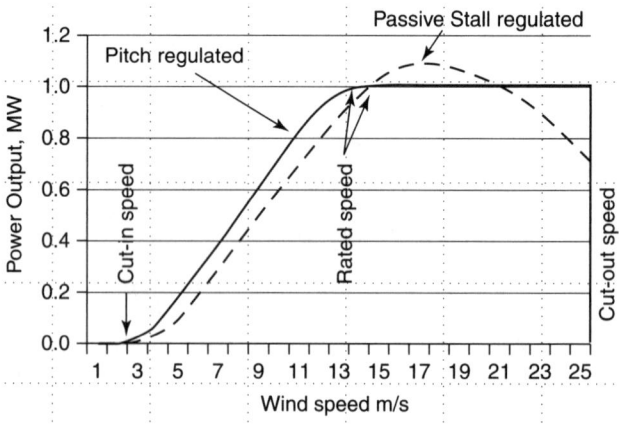

Figure 5-6   Power curve for pitch- and active stall-regulated turbines.

Variable-Speed Turbines

Wind industry is moving to variable speed turbines, where the rotational speed of rotor depends on wind speed. The advantage of variable-speed turbine is the ability to follow the optimal power curve, as shown in Fig. 5-5. Notice the difference between a constant-versus variable-speed turbine. The constant-speed turbine is at optimal performance only at single wind speed. In a variable-speed turbine, the goal is to operate the turbine at a constant tip speed ratio, λ_{opt}, in order to maintain the turbine performance at the highest value of C_p (see Fig. 5-3). Figure 5-3 is an example of power coefficient versus TSR curves for different values of pitch and a specific aerofoil profile of the blade. In this case, optimal value of pitch is zero.

A variable rotor speed turbine has a variable-speed generator. Synchronous generators with permanent magnet or DC excited rotor windings are used as variable-speed generators, which are described in Chapter 9. These generators produce energy at variable frequencies (variable revolution per minute results in variable frequency); therefore, the energy is converted to DC and then back to AC at grid frequency. This requires significant amount of power electronics.

Power Curves

The theoretical power performance curves are validated by field testing. IEC 61400-12 specifies detailed procedures for creating empirical power curves based on field testing. Three wind speed-related parameters describe a power curve (see Fig. 5-6):

1. *Cut-in speed, v_{ci}.* A turbine starts producing energy at this wind speed. At low wind speeds, the torque is small and not enough to overcome the inertia of the entire system. v_{ci} for most turbines is in the range of 3 to 5 m/s. Although turbine manufacturers like to trumpet lower cut-in speed, it must be remembered that very little amount of energy is produced at low wind speed.

2. *Rated speed, v_{rs}.* This is the wind speed at which the rated power is produced. For most turbines, v_{rs} is in the range of 11.5 to 15 m/s. This is a crucial number because it defines the shape of the power curve. Power curves with lower v_{rs}, will produce more energy overall because it will produce more energy at wind speeds between cut-in and rated. When evaluating the power production capabilities of turbines for a given wind condition, it is common to compare the capacity factor, which is the average annual energy production divided by the annual energy production at the rated power. Power curves with lower v_{rs} will yield a higher capacity factor. Larger and more efficient blades lead to lowering of v_{rs}. A deceptive way to lower v_{rs} is to take a turbine designed for, say, 1.65 MW and rate it at 1.5 MW.

3. *Cut-out speed, v_{co}.* A turbine stops operating at this wind speed. v_{co} for most turbines is 25 m/s. The primary reason for stopping is safety. Components of a turbine are not designed to handle the loads created by wind speeds above the cut-out speed. There are various methods to stop a turbine.

   • Modern pitch-controlled turbines increase the pitch to complete stall position. The control algorithm has a delay period that defines the wait time for restart of the turbine after wind speed has dropped below v_{co}.

   • The second method is a spring-loaded mechanism in the blade that turns the tip section of the blade in a feather position.

   • The third method is to turn the turbine 90° about the vertical axis and thereby change the axis of rotation. This is done in a few small turbines.

   Note, most stall-regulated turbines will start to slow down because of the sharp rise in drag force. This is a result of higher angle of attack, which leads to separation of flow and high drag.

4. *Survival wind speed.* Although this is not part of the power curve, it is an important wind speed, which specifies the design wind speed of the entire turbine structure including the tower. This is in the range of 50 to 60 m/s.

There are three primary methods of regulating power in a turbine.

- Pitch-regulated turbines control the output of power by changing the pitch of the blade in order to maintain the output power at a constant value at wind speeds higher than v_{rs}. The blade is turned along its longitudinal axis to increase the pitch and decrease the angle of attack (see Fig. 5-1).

- Active stall-regulated turbines are similar to pitch-regulated in that the blade is turned along the longitudinal axis, but instead of turning against the wind, the blades are turned to decrease the pitch and increase the angle of attack (see Fig. 5-1). The power curve looks similar to pitch-regulated turbines.

- Passive stall-regulated turbines do not control pitch; the pitch remains constant and as the wind speed increases, the angle of attack increases leading to higher drag. Stall-regulated turbines are characterized by a hump in power production and sharp fall in power output after the hump (see Fig. 5-6).

Vertical Axis Wind Turbine (VAWT)

In the early days of wind turbine commercialization (1970s and 1980s), large VAWT were popular. As technology evolved, HAWT became popular and VAWT were phased out. In recent years, there are no large VAWTs in production, to the best of the author's knowledge. Several small wind turbines are commercially available in the VAWT category. There are three types of VAWT turbines:

- *Savonius VAWT*. Savonius is a drag-based machine. An example of this was presented at the end of Chapter 4. With cup-type blades, the maximum power coefficient is below 0.2. These turbines have rotors that are typically made of simple materials and are, therefore, inexpensive. However, the relative inefficiencies do not warrant their use.

- *Darrieus VAWT*. The blades of these turbines are aerodynamic and the forces generated are based on aerodynamic lift. From an efficiency standpoint, the theoretical efficiency of Darrieus VAWT is 0.554[3] compared to 0.593 for HAWT. The blades are typically in the shape of an eggbeater. The advantages are that the generator and other components are on the ground for easy maintenance. The disadvantages are that these turbines cannot be installed high off the ground, so the wind speeds experienced by the rotor are much lower. Since the blades are rotating about the vertical axis, the angle of attack

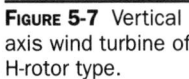

Figure 5-7 Vertical axis wind turbine of H-rotor type.

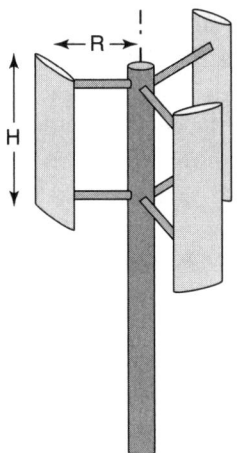

changes continuously, which causes the torque to pulsate. The pulsating torque and high centrifugal force causes structural failures.

- *H rotor or Giromill VAWT.* This is similar to a Darrieus VAWT, but the blades are parallel to the axis of rotation. Among small VAWTs, this has become a popular configuration (see Fig. 5-7). The blades are in the shape of an airfoil. The angle of attack changes as the blades rotate around the axis, which causes the lift and drag to be unfavorable in certain sectors of rotation.

The overall advantages are that the VAWT are:

- Simpler because they do not require a yaw or pitch mechanism.
- All the components that would traditionally be in a nacelle are on the ground, making maintenance easier.
- Rapid changes in the direction of the velocity vector in the horizontal plane can cause a horizontal axis turbine to "seek the wind" by changing the axis of rotation of the blades. This seeking can cause a significant loss of energy. Vertical axis turbines do not have this problem.

Despite the advantages, VAWT have not become popular because:

- Lower theoretical efficiency and lower practical efficiency.
- Lower installed heights leading to lower wind speeds and lower energy production.

- High rotational speed leading to higher centrifugal forces on structures that support blades.
- Uneven torque leads to wobbling to rotor shaft, which leads to high stresses on the bearings and the shaft.

References

1. Burton, T., Sharpe, D., Jenkins, N., and Bossanyi, E., *Wind Energy Handbook*, Wiley, Hoboken, NJ, 2001.
2. Hansen, M. O. L. *Aerodynamics of Wind Turbines*, Earthscan, Sterling, VA, 2008.
3. Wilson, R. E., and Lissaman, P. B. S., *Applied Aerodynamics of Wind Power Machines*, Oregon State University, Corvallis, 1974. http://ir.library.oregonstate.edu/jspui/bitstream/1957/8140/1/Applied_Aerodynamics.pdf.

CHAPTER 6

Wind Measurement

What we observe is not nature itself, but nature exposed to our method of questioning.
—Heisenberg, *Physics and Philosophy*, 1963

Introduction

As described in Chapter 2, wind energy production is a cubic function of wind speed. This implies that small changes in wind speed estimates can result in large changes in wind energy estimates. Therefore, a lot of emphasis is placed on wind speed measurement because this translates to higher accuracy in estimation of a project's financial returns. In the beginning of this chapter, the various properties of wind speed are discussed followed by detailed description of how wind speed and direction are measured. The following section presents the uncertainty associated with instruments used for measurement of wind speed. The next section describes design of a measurement campaign, followed by data management and reporting. The last section briefly describes use of remote sensing to measure wind condition.

Definition of Wind Speed

Although definition of wind speed may seem trivial, in reality, it is not. This exposition on wind speed will cover facets of wind speed that are relevant from the perspective of a wind energy project.

- *Wind velocity in the horizontal plane.* From the perspective of a wind energy project, the two quantities of importance are the resultant wind speed in the horizontal plane and the direction in the same plane. Vertical axis wind turbine (VAWT) captures energy irrespective of the direction of wind in the horizontal plane. Horizontal axis wind turbine (HAWT) has a yaw mechanism that aligns the normal to the plane of rotation such that it is parallel to the direction of wind. For most turbines, energy is derived from the wind velocity vector in the horizontal plane.

- *Wind velocity in the vertical direction.* The vertical component of wind velocity is caused by convection, topography of land, or other factors. This component of wind velocity can modify the flow of wind over blades and degrade the performance of a turbine.

- *Wind velocity at a point versus a volume.* Because of the stochastic nature of airflow, wind velocity vector at the particle level has a significant random component. Therefore, the wind velocity vector is spatially averaged. A cup anemometer measures wind speed in a small volume, whereas remote sensing devices measure wind speed over larger volumes.

- *Wind velocity at a point of time versus over an interval of time.* Again, because of the stochastic nature of wind velocity, there is a significant component of randomness to wind velocity at a point in space and at a point in time. Temporal averaging is, therefore, done to report wind speed. For wind energy applications, wind speed is typically measured every 1–3 s and 10-min statistics like average, minimum, maximum, and standard deviation are computed and recorded, and the 1–3-s sample data is discarded. In some meteorological applications, wind speed is measured only once an hour and the quantity that is recorded is the wind speed at the end of the hour. These single measurements are subject to significantly large error compared to averaging of, say, 200 measurements over 10 min.

As various methods of measurement and configurations of instruments are described in sequel, references will be made to the above facets of wind speed.

Configurations to Measure Wind

Broadly speaking, there are two types of measurements: *In situ* measurement and remote sensing. *In situ* measurements are done with a meteorological tower (met-tower) and remote sensing is done with SODAR (based on sound waves) or LIDAR (based on light waves).

A met-tower is used to measure wind speed, wind direction, temperature, barometric pressure, relative humidity, and few other atmospheric conditions. Met-towers may be classified as:

- *Temporary or permanent.* A met-tower installed for a period of 1 to 3 years is considered a temporary structure, while a met-tower with a life of about 20 years is considered permanent. Most temporary met-towers are tilt-up with no foundation, while permanent met-towers have a concrete foundation. However, some states in the United States, like Florida and Ohio, require even a temporary met-tower to be Rev-G compliant, which may require a foundation.

- *Maximum height of met-tower.* Some of the more common heights of met-towers are 30, 50, 60, 80, and 100 –meters.The 60 m is the most popular met-tower height and the trend is toward taller met-towers.

Anemometer

Anemometers are used to measure wind speed. There are different types including:

- *Cup anemometer*—this is the most widely used. Most modern anemometers contain three cups with a vertical axis of rotation. The rotation speed of the cups is proportional to the wind speed. The output signal of an anemometer is a low-level AC sine wave; the frequency of the sine wave is proportional to the wind speed. Figures 6-1a and 6-2 show cup anemometers.

- *Propeller anemometer.* As the name suggests, a propeller is used to measure wind speed. The axis of rotation of propeller anemometer is horizontal. In order to align the axis of rotation with the direction of wind, this type of anemometer also contains a wind vane. This instrument serves two purposes: Wind speed and wind direction measurement. A form of propeller anemometer is used to measure the vertical component of wind speed; in this case, the axis of rotation is fixed to be vertical. Figure 6-1b shows a propeller anemometer.

- *Sonic anemometer.* Ultrasound waves are used to measure wind speed and direction. Three-dimensional velocity vectors are computed by measuring travel time of sonic pulses between

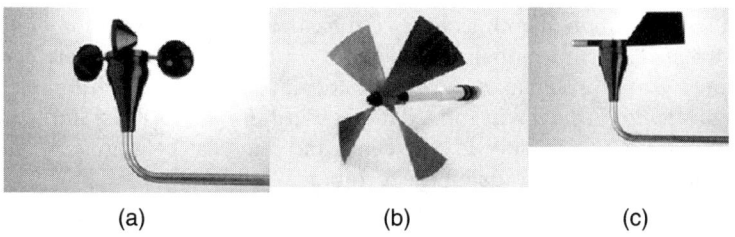

| (a) | (b) | (c) |

Figure 6-1 Anemometer and wind vane. (a) NRG#40 cup anemometer; (b) RM Young propeller anemometer; and (c) NRG200P wind vane. (With permission of NRG Systems.)

three pairs of transducers. The temporal resolution of these instruments is of the order of 20 Hz, which is appropriate for measuring very rapid changes in wind speed. Sonic anemometers have no moving parts like bearings or cups and, therefore, are more robust and appropriate for long-term use. Figure 6-3 shows a sonic anemometer.

Anemometers are classified according to IEC 61400-12-1[1] standard based on two parameters: Accuracy of measurement and terrain of measurement. The accuracy of anemometer is indicated with a class index, k, where k takes on values between 0 and 3. The terrain of measurement is indicated with a letter A, B, or S. Examples of anemometer classifications are Class 0.5B, Class 1A, Class 2B, etc.

Class 0 is the highest accuracy (unachievable) and Class 3 is a lower accuracy anemometer. To explain the classification scheme, a quick exposition of how accuracy is measured is useful. Anemometers are tested in wind tunnels in which the wind speed is varied between 0 and 16 m/s. The maximum systematic deviation of the anemometer is measured in each wind speed bin of width 1 m/s. As an example of bin, consider wind speeds between 4 and 5 m/s as belonging to bin 4. Classes are defined using the formula:

$$k = 100 \max_{1 \le i \le 15} \left| \frac{\varepsilon_i}{\frac{u_i}{2} + 5\,\text{m/s}} \right| \tag{6-1}$$

Figure 6-3 Sonic anemometer to measure all three components of wind velocity. (With permission of Campbell Scientific, CSAT3 3-D sonic anemometer.)

where k is the class index, ε_i is the maximum systematic deviation, and U_i is the wind speed in bin i. Figure 6-4 is a plot of the accuracy curves for different classes. For convenience, instead of curves describing classes, often a single point is used, $U_i = 10$ m/s, and accuracy is then defined as percentage deviation in anemometer measurement in

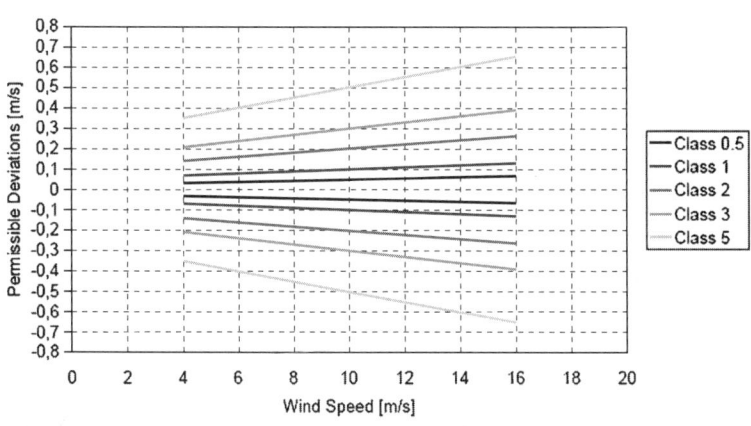

Figure 6-4 Deviations chart used to classify anemometers per IEC 61400-12-1. [From Dahlberg, J. A., Pedersen, T. F., Busche, P. *ACCUWIND-Methods for Classification of Cup Anemometers*, Riso National Laboratory, Roskilde, 2006. Riso-R-1555(EN).]

	Class A	Class B
Average flow inclination angle in degrees	−3 to +3	−15 to +15
Wind speed range (m/s)	4 to 16	4 to 16
Turbulence intensity	0.03 to 0.12+0.48/V	0.03 to 0.12+0.96/V
Turbulence structure	Nonisotropic	Isotropic
Air temperature, °C	0 to 40	−10 to 40
Air density, kg/m³	0.9 to 1.35	0.9 to 1.35

Source: From International Electrotechnical Commission (IEC). Wind Turbines – Part 12-1: Power Performance Measurements of Electricity Producing Wind Turbines. 2005. IEC 61400-12-1. With permission from IEC.

TABLE 6-1   Description of Class A and B Influence Parameter Ranges, per IEC 61400-12-1

a wind tunnel when the correct wind speed is 10 m/s. At this wind speed, a Class 0.5 anemometer has a maximum deviation of ± 0.5%; a Class 1 anemometer has a maximum deviation of ± 1%.

The second part of the anemometer classification is the influence of terrain. The inflow angle or the flow inclination angle has a significant impact on accuracy. Classes A and B are used to describe the terrain (see Table 6-1). Class S (S for special) is associated with influence parameters that are not in the ranges specified for Class A or B. The classification of most commonly used anemometers is shown in Table 6-2.

Class 2 or better sensors are used for power curve measurements of wind turbines, permanent met-towers, and temporary met-towers that are used to prove wind resources at a site. Class 1.5 or better

Anemometer	Class A	Class B
Thiess First Class Advanced	0.9	3.0
Risoe Windsensor P2546A	1.4	5.1
Thiess First Class	1.8	3.8
Vector A100LK	1.8	4.5
NRG #40	2.4	8.3

Source: From Young, M., *Met Towers & Sensors–Science & Equipment Considerations,* DNV Global Energy Concepts, AWEA Resource and Project Energy Assessment Workshop, Minneapolis, MN, 2009.

TABLE 6-2   Deviation of the Most Commonly Used Anemometers

anemometers are higher cost and, therefore, are not routinely used for prospecting. The uncertainty associated with the anemometers is one of the components in the computation of the overall uncertainty of the wind resources and eventually the uncertainty of the wind project itself. Higher uncertainty leads to lower valuation of the project. (See section "Uncertainty in Wind Speed Measurement with Anemometers" in this chapter and Chapter 7 for a more detailed exposition of uncertainty.)

Calibration of Anemometers

A cup anemometer measures wind speed by using the following process: Rotation of cups is proportional to wind speed, angular speed of the rotating cups is proportional to the frequency of the AC current that is produced, and the frequency of the AC current is measured and converted to wind speed using a transfer function. As an example, the mean consensus transfer function for a NRG #40 is:[4]

$$\text{Wind speed in m/s} = (Hz.0.765) + 0.35$$
$$\text{Wind speed in miles per hour} = (Hz.1.711) + 0.78$$
(6-2)

where Hz is the measured frequency of the AC current. In the above equation, the slope and intercept are the two parameters of the transfer function, with values of 0.765 and 0.35. In order to reduce the impact of variability between anemometers, most manufacturers calibrate individual anemometers and provide instrument-specific transfer function parameters.

For wind resource assessments, only calibrated anemometers should be used. It removes any bias introduced because of the manufacturing process. There are several agencies that calibrate anemometers, the most prominent is MEASNET (www.measnet.org), which is a collection of international institutes that calibrate anemometers.

Wind Vane

Wind vane measures the direction of wind in the horizontal plane. The output of the sensor is an analog voltage that is proportional to the direction. Depending on the type of wind vane, there is a "north" marking on the vane. During installation, the north marking must be in the true north direction. The difference between magnetic and true north must be considered when aligning the vane. In most cases, this compensation can be set in the data logger, if the vane's north marker is not pointing toward true north.

Wind vanes have a dead band around the zero degree direction. For instance, if the dead band is 8°, then measurements between 356 (−4°) and 4° are not accurate. There are two methods to deal with this: (i) Dead band compensation, which converts all readings in the range

$\alpha < 4, \alpha > 356$ to $\alpha = 0$; (ii) if accurate direction readings are desired around zero degrees, then the wind vane may be installed with an offset. In the second case, if the most frequent wind direction is north and the least frequent wind direction is east, then the vane may be oriented in the east direction with appropriate compensation in the data logger. Figure 6-1c is a picture of a wind vane.

Placement of Sensors

A met-tower has several instruments mounted on it at different heights. A frequently used configuration of met-tower is:

- 60 m monopole tower
- two anemometers at 60 m
- one wind vane at about 56 to 58 m
- two anemometers at 40 m
- two anemometers at 30 m
- one wind vane at about 26 to 28 m
- Temperature sensor at 3 m
- If present, barometric pressure and relative humidity sensors are placed at 3-m level
- Datalogger and communication device in an enclosed box at 3-m level

All the instruments are mounted on booms of length 60 to 95 in. to minimize the impact of tower in the flow. Figure 6-5 illustrates the orientation of two anemometers and wind vane at the top of the tower.

FIGURE 6-5 Monopole tower with two anemometers and one wind vane.

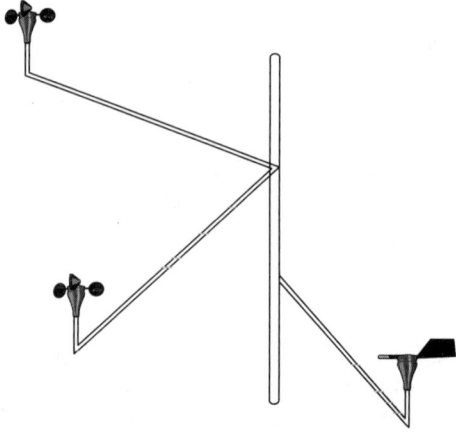

Redundant anemometers are used at each height for two reasons:

1. To remove the shadow effect of the tower. If one of the anemometers is in the shadow of the tower, then the other anemometer can measure the correct wind speed.

2. If one anemometer fails, it does not have to be replaced immediately. Lowering the tower and raising it back up is an expensive operation in remote locations, therefore, in order to minimize this expense, redundant sensors are used.

Wind vanes are usually positioned about 2 to 4 m below the anemometers. For instance, the top wind vane may be at 58 or 56 m, and the second wind vane may be at 28 or 26 m. This is done to minimize the wake effect on the anemometer because of the wind vane. The orientation of the wind vane should be same as the first anemometer. Instruments at all other levels should have the same orientation as the top-level instruments.

In this plot, airflow is along the x direction. From the center of the monopole, notice the deficit in wind speed along the negative x direction, the increase in wind speed along the negative and positive y

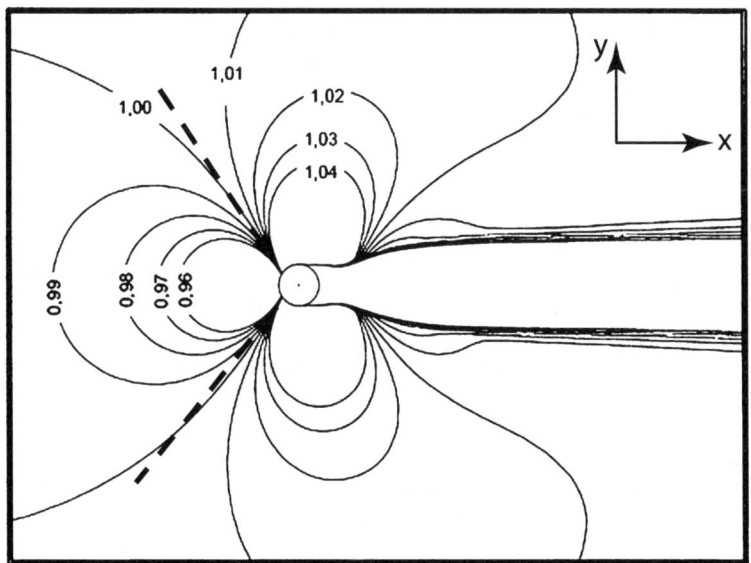

Figure 6-6 Isospeed lines around a monopole. Top view of the monopole with flow in the horizontal plane from left to right. The values are normalized to the upstream free flow wind speed. (From International Electrotechnical Commission (IEC). Wind Turbines – Part 12-1: Power Performance Measurements of Electricity Producing Wind Turbines, 2005. IEC 61400-12-1, with permission.)

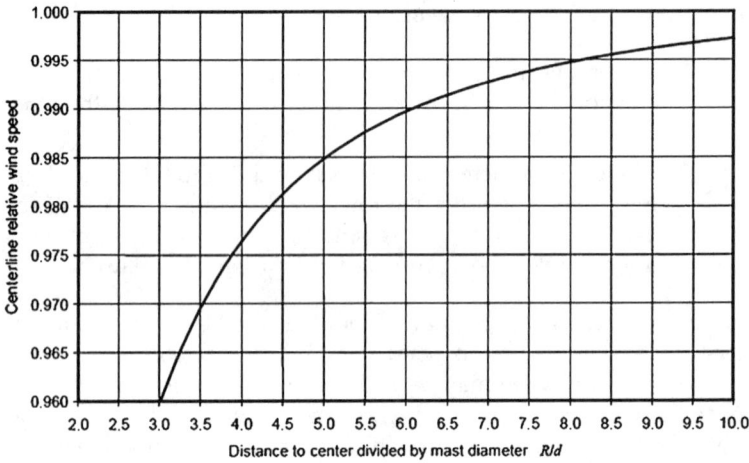

Figure 6-7 Plot of the wind speed change versus distance from the center of met-tower along the negative x direction (see Fig. 6-6). (From International Electrotechnical Commission (IEC). Wind Turbines – Part 12-1: Power Performance Measurements of Electricity Producing Wind Turbines, 2005. IEC 61400-12-1, with permission.)

directions, and significant deficit along the positive *x* direction. Given this plot, the best location to put anemometers is on booms that are at an angle of about 60° to the oncoming wind, which is at the end of the dotted lines in Fig. 6-6.[5] This would be the ideal location of anemometer if the wind direction were always along the positive *x* direction.

Two approaches are adopted by the industry to minimize error because of shadow effect of the pole of a met-tower: Longer booms and two anemometers. The impact of length of boom is shown in Fig. 6-7. The standard in the industry is boom lengths of six times the diameter of the monopole. The industry standard on orientation of the booms is loose. The complexity arises when there are multiple directions of airflow that are of significance. Most common orientation is first boom at 60° to the predominant direction of wind, and second boom at an angle of between 90 to 180° of the first boom. In situations with a large range of wind direction, a strategy for orienting the anemometers based on avoiding the wake of the tower is adopted.

The data obtained from the two anemometers at a height must be processed in order to obtain wind speed. As illustration, (see Fig. 6-8), a plot of wind speed measurement from an anemometer that is placed on a boom facing east. When the direction of the wind is from close to west (270°) direction, then the data must be ignored and data from the second anemometer should be used. When the wind direction is from the east, then measurement from the second anemometer may be used, depending on its orientation.

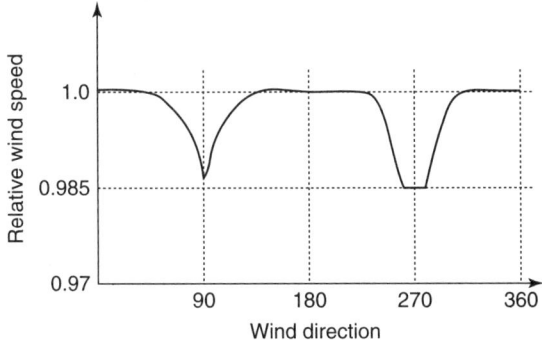

F‍ɪɢᴜʀᴇ **6-8** Conceptual graph of wind speed measurements using an anemometer on a monopole with a boom length of five times the diameter of the pole. The anemometer is mounted on a boom pointing in the east direction (90°).

Impact of Inflow Angle

Airflow is not always in the horizontal plane. The component of velocity in the horizontal plane is:

$$v_{horizontal} = v_{total} \cdot \cos \alpha$$

where α is the inflow angle, the angle of the velocity vector with respect to the horizontal plane. The actual wind measurement compared to theoretical wind speed is compared in Fig. 6-9.

When the inflow angle is greater than a few degrees, then significant deviations occur as shown in the Class B column of Table 6-2. At higher tilt angles, the wind speed is usually overestimated.

Impact of Temperature

The other parameter that can cause deviation in measurement is temperature. The response depends on the type of bearing. In greased bearings, friction increases because of higher viscosity at lower temperatures (below 0°C). In sliding bearing, the friction increases as temperature increases (above 40°C). The additional friction slows down the rotor of the anemometer.

Uncertainty in Wind Speed Measurement with Anemometers

In any measurement endeavor, it is important to understand the accuracy of measurement and reasons for uncertainties. There are two types of uncertainties: Random and systematic (or bias). Random errors are because of the variability in the quantity that is measured or variability in the measurement process. Random errors are assumed to be normally distributed around the "true" value and repeated

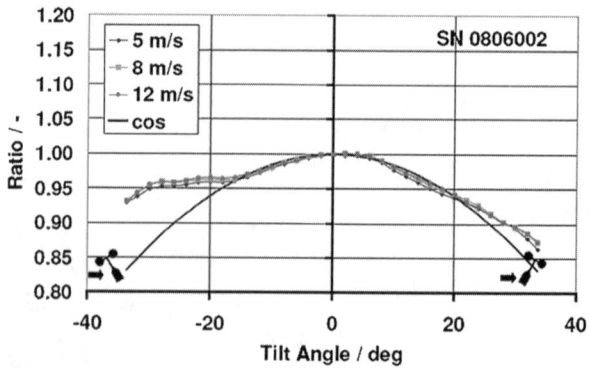

Figure 6-9　Comparison of performance of a Theis First Class Advanced for different inflow angles and different wind speed from a calibration service. cos is the theoretical cosine response. (From WindGuard, Deutsche. Summary of Cup Anemometer Classification [Online] 12 2008. http://www.skypowerinter national.com/pdf/News%20and%20Media/new/First_Class_Advanced_ Testing_Results._pdf.)

measurements provide a better estimate. Systematic error or bias, on the other hand, is a repeatable flaw and repeated measurements do not reduce the systematic error. Calibration error is an example of systematic error.

The most common uncertainties in measurement of wind speed and estimation of related quantities (wind speed at hub height) are:[5]

1. *Sensor calibration uncertainty.* This refers to uncertainty in the calibration process and differences between the test anemometer and production anemometer. Depending on anemometer, this uncertainty can be 0.1–2%

2. *Dynamic overspeeding.* Cup anemometers are susceptible to overspeeding in the presence of turbulence. Overspeeding is a phenomenon by which the anemometer speeds up more rapidly when faced with higher wind speed; it does not slow down as rapidly when faced with lower wind speed. The uncertainty has been determined to be about 0.3%.

3. *Tower shadow, boom, and mounting effects.* As explained in the section on Placement of Sensors above, tower shadow causes a negative bias; an estimate of the bias is −1.5%. Longer booms can reduce this bias; however, the booms themselves disturb the airflow. Long booms can also cause the anemometer to deviate from vertical position resulting in measurement error; an estimate of uncertainty is 0.5%.

4. Wind shear can be a large component of uncertainty in prediction of wind speed at hub height. There are several components to this uncertainty: Anemometer quality, measure-

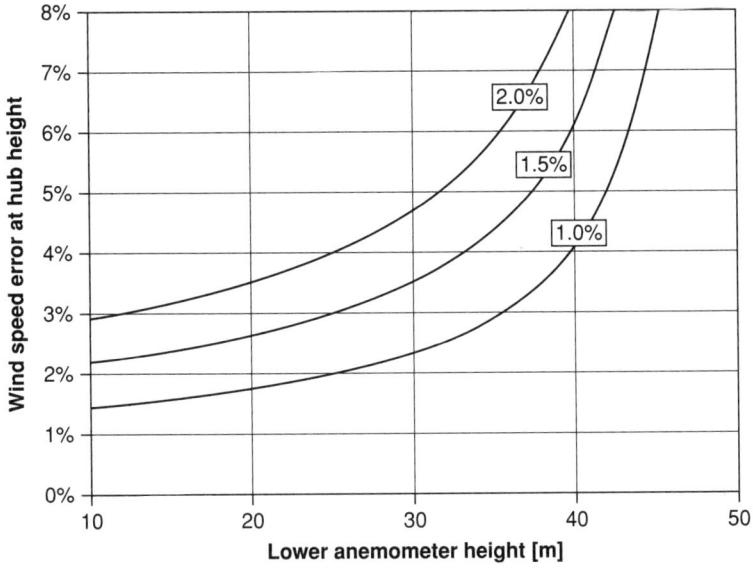

FIGURE 6-10   Error in wind speed at 80 m that is computed using wind shear based on different quality anemometers at an upper height of 50 m and different lower heights. (From Taylor, M., Mackiewicz, P., Brower, M. C., Markus, M., *An Analysis of Wind Resource Uncertainty in Energy Production Estimates*, EWEC, London, 2004.)

ment heights, and terrain. Figure 6-9 indicates the impact of measurement uncertainty of the anemometer,[6] and Fig. 6-10 indicates the impact of measurement heights. In this case, the upper anemometer is at 60 m and the hub height is 80 m. The conclusion is that a larger difference in height yields higher accuracy of wind speed extrapolation. As indicated by the points annotated in Fig. 6-10, error decreases from about 4.7 to about 2.9% as the height of lower anemometer is changed from 50 to 40 m. The error is further decreased to 2.4% when lower anemometer is at a height of 30 m. The reason for higher accuracy is explained by considering the power law for shear in the following format:

$$\gamma = \left(\log v_1 - \log v_2\right) / \left(\log h_1 - \log h_2\right) \qquad (6\text{-}3)$$

Since γ is estimated from 10-min wind speeds, it is instructive to observe that γ is the slope in the log v, log h space. A better estimate of the slope is achieved when the two measurement points are separated. However, there is a limit to the separation in terms of height of the lowest anemometer. At some point, the uncertainty in readings of the lowest height anemometer starts increasing because of higher turbulence caused by surface roughness and changes in contour. Therefore, there is a balance between increasing the distance between top and lowest height anemometers to get a better estimate of shear and decreasing the accuracy the lowest height anemometer. As a guideline, the lowest anemometer height may be set to 30 m for a 60 m met-tower.

The overall error in estimation of wind speed in a measurement campaign can be computed using the independence assumption, which assumes that each source of uncertainty is independent of their other. Under this assumption, the overall error is:

$$\varepsilon = \sqrt{\sum_{i=1}^{N} \varepsilon_i^2} \tag{6-4}$$

where ε is the overall error, ε_i is the error because of each component of uncertainty, and N is the number of components of uncertainty.

Example of Error Estimate

Consider NRG #40 calibrated anemometers installed at 60, 40, and 30 m in a complex terrain with maximum flow angle of 6°. What is the uncertainty associated with estimating wind speed at 80 m?

The purpose of this illustrative example is to expose the reader to the lack of complete data to compute uncertainties. Several assumptions based on judgment are made below regarding the source and amount of uncertainties:

- Anemometer accuracy in complex terrain. According to Table 6-2, uncertainty with 3° of inflow angle is 2.4% and up to 15° is 8.3%. As an approximation, a linear relationship will yield an uncertainty of 3.9%.

- Calibration, dynamic overspeeding, and inflow angle are included in the anemometer accuracy above. Tower shadow, boom, and mounting effects are not included and are independent of anemometer accuracy. Since information about predominant wind direction is not available, the nature of bias (positive or negative) because of tower shadow is not known. Errors of 1.5% because of tower shadow and 0.5% because of boom and mounting effects will be assumed.

- Error because of shear can be estimated by examining Figs. 6-10 and 6-11. Although complete data is not available, the error is likely to be in the range of 4.7 and 2.4%. A 4.7% error is for a case with 50-m tower and Class 2 anemometer; 2.4% is with 60-m tower and Class 1.5 anemometer; this example is of a 60-m tower and Class 2.4 anemometer. In this example,

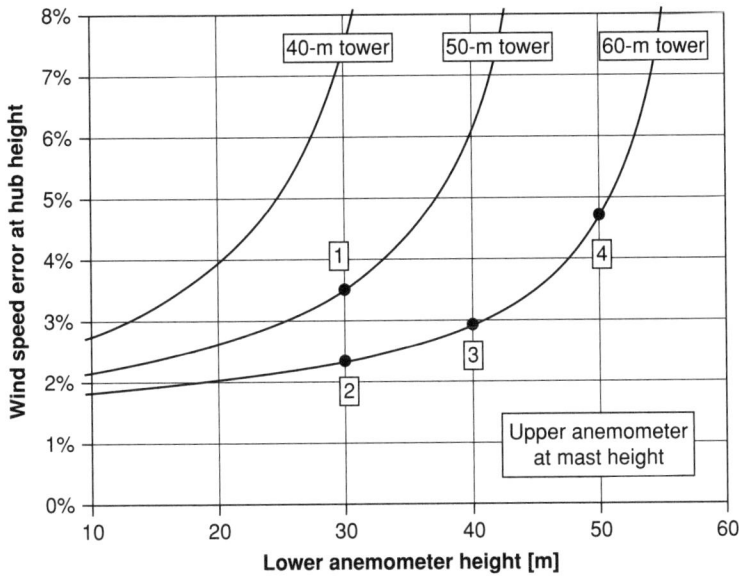

Figure 6-11 Error in wind speed at 80 m that is computed using wind shear based on wind speed measurements at two heights. The measurements were with anemometer of Class 1.5. (From Taylor, M., Mackiewicz, P., Brower, M. C., Markus, M., *An Analysis of Wind Resource Uncertainty in Energy Production Estimates,* EWEC, London, 2004.)

an uncertainty of 4% in the computation of shear and wind speed at 80 m height will be assumed.

Total uncertainty because of the three independent sources is:

$$\varepsilon = \sqrt{0.039^2 + (0.015^2 + 0.005^2) + 0.04^2} = 5.8\%$$

Other Sensors

A temperature sensor, barometric pressure sensor, and solar radiation sensor may be added to the met-tower. The temperature and pressure sensors help to compute air density accurately. The temperature sensor is also used to detect icing conditions that can cause the sensors to stop or produce errors in measurement.

Data Logger and Communication Device

Data logger is a device to which all the sensors are connected. Typical data loggers have 12–15 channels to which, for example, six anemometers, two wind vanes, a temperature sensor, a solar radiation sensor, a barometric pressure sensor, and others are connected. Data loggers also have:

Measured Quantity	Computed and Recorded Values
Wind speed	Average, standard deviation, maximum, and minimum
Wind direction	Average, standard deviation, and maximum gust direction
Temperature	Average, maximum, and minimum

TABLE 6-3   Statistical Values Computed for Measured Quantities

- Internal memory with capabilities to store data for several months.

- Interface for retrieving data through: Retrievable data storage device, computer connection to download data, or communication devices like wireless modem

- Ability to program calibration parameters of instruments, sampling intervals, and recording interval.

- Ability to process sampled data to compute average, standard deviation, minimum, and maximum values of each channel.

Average and standard deviation of sampled data requires explanation. As an example, consider a sampling interval of 2 s and a recording interval of 10 min. Anemometer, vane, and other sensors will measure wind speed and direction and send to the data logger every 2 s. The data logger collects values for 10 min. That is, it collects 300 values (30 values per minute and over duration of 10 min) and then computes average, standard deviation, maximum, and minimum from the 300 values. The average, standard deviation, maximum, and minimum values are then recorded along with a timestamp in the data logger. If the communication device is programmed to retrieve data once a day, then the 144 records (six records per hour and 24 h) are retrieved for transmission to the data center.

Table 6-3 contains a typical set of parameters that are recorded. It is common to use 3 s as the sampling interval. The maximum wind speed over a sample of 200 points or 10 min is also called the gust. A 3-s gust is used by turbine manufacturers to estimate the extreme load on a turbine. Along with the 3-s gust, the gust direction may be recorded.

Designing a Wind Measurement Campaign

In this section, three design elements of a wind measurement campaign are covered: Number of met-towers, placement of met-towers, and duration of the measurement campaign. Although the primary drivers for a measurement campaign are to improve accuracy of prediction of energy output of the wind project, other factors like economics and project schedule can play an important role. It is not unusual for a wind project to start with a small number of met-towers to prospect and to subsequently add additional met-towers as financiers demand higher accuracy in wind resources over the planned wind farm area.

In 2009, the cost (in USD) of a wind measurement campaign is:

A fully instrumented 60-m met-tower costs about $18,000 to $25,000.

Installation cost for a site with good access is $8,000–$10,000.

Annual data management with final measure-correlate-predict (MCP) analysis and estimated energy computation can cost about $16,000 to $20,000.

Decommissioning cost is $5,000 to $7,000.

Since wind is highly dependent on local conditions like contour, roughness, and obstacles, in an ideal case, wind measurement should be performed at the planned location of a turbine. This would yield the highest accuracy in predicting energy output. There are two issues with this:

1. One of the purposes of a wind measurement campaign is to determine the best location of turbines, therefore, turbine locations are not known *a priori*. Unless wind measurements are done, micrositing of turbines cannot be done.

2. When designing a wind farm consisting of multiple turbines, it is very expensive to measure wind at each turbine location.

Therefore, a wind measurement campaign must start with what is known and design a process that can economically generate reasonably accurate energy predictions. A process for designing a wind measurement campaign is described below. It may be adapted to meet a project's specific needs.

1. Conduct a preliminary wind resource assessment of the area under consideration. This applies to both single turbine and wind-farm installations. The outcome of the preliminary wind assessment will be a wind resource map of suitable granularity. For the purposes of this discussion, it is assumed that wind resources are available for a 200 m × 200 m grid. The resource map may be based on computer simulations using numerical weather prediction models or wind resource-prediction model like WAsP or others. The source of the wind data for these models may be 10-m airport data and/or reanalysis data from National Center for Atmospheric Research (NCAR). Preliminary wind assessment is described in Chapter 7.

2. The preliminary location of Wind Turbine Generators (WTGs) may be computed by running a layout optimization model

(example Optimize in WindPRO) with wind resource map from step (1) and a variety of constraints. Constraints include maximum wind-farm capacity, minimum capacity factor, turbine tower height and rotor diameter, distances between turbines, and other setback criteria like distance between turbine and property boundary, public roads, transmission, and inhabited areas. In this phase, the constraints are guidelines rather than precise. For a single turbine case, a location with the highest wind resource that satisfies all the constraints is the location of the met-tower. For a wind farm, go to step (3).

3. In the preliminary WTG layout, form clusters of WTGs. If the wind farm is in a complex terrain, then 5 to 7 WTGs may be grouped into one cluster. If it is a simpler terrain with very little changes in elevation and roughness, then 10 to 12 WTGs may be grouped into one cluster. The clusters will be based on distance. Clusters are best formed visually; the borders of clusters may be drawn manually on a Computer Aided Design (CAD) drawing or on paper. The ratio between WTG and met-towers of 5 to 7, or 10 to 12 are normal guidelines for determining number of fixed met-towers for wind measurement. Rarely will all clusters have the same number of WTG.

4. In each cluster, find the median wind speed WTG. This WTG location or a location in the vicinity would be a location for placement of met-tower. Measuring at the best or worst wind resource location in the cluster would yield wind measurements that have to be either extrapolated down or extrapolated up to all points in the cluster, which would lead to higher inaccuracies. Normally, a set of two or three locations is chosen in each cluster. For instance, the three locations are: Location with median wind speed and the two locations with the smallest difference with the median wind speed.

5. The goal of this step is to pick one location in each cluster such that, for the wind farm as a whole, the met-tower locations are sufficiently spread out geographically. This step is best done visually, starting with the median wind speed location in each cluster and then examining the proximity of these locations. If the median locations of two clusters are geographically close, then alternate locations are chosen from the set.

The duration of measurement is a minimum of 1 year with a preference of 2 to 3 years. One year is sufficient to capture the seasonal variations in wind speed, direction, turbulence, and shear. However, year-to-year variations are significant. In order to predict the energy output of a wind project accurately over its lifetime of 20 or more years, statistical methods like MCP are used. This method is described in detail in Chapter 7. One of the key steps in this process is the computation of correlation of the measured data with concurrent long-term dataset. If this correlation is low and the project economics is borderline, then there is no option but to continue measurement for 2 –3 years. Even if the project economics is favorable, but the degree of uncertainty is high because of measurement for shorter duration, investors may perceive the project as high risk and, therefore, demand a higher return. Longer measurement period will reduce the degree of uncertainty associated with wind speed and, thus reduce project risk. For large wind farms, it is normal for financiers to demand 2–3 years of met-tower data.

Installation of Met-Towers

Detailed guidelines for installing met-towers are available in the documentation supplied by the vendor of the met-tower. Proper planning and appropriate safety precautions can mitigate the risk of cost overruns and failures. Before installing a met-tower, the following should be done:

- *Examine the wind rose at the location.* This will describe the predominant direction of wind and, therefore, is the primary input into orienting the instruments on the tower.

- *Understand the terrain.* Ensure that there is sufficient lay down area to assemble the tower. For example, a 60-m tower may need a lay down area of at least 80 m × 6 m, if the wind direction is consistent.

- *Procure gin pole and winch.* The winch must have the right size drum that can hold sufficient length of wire. For example, a 60-m tower will need about 100 m of wire on the winch drum.

- A crew size of four or more personnel, or as specified by the tower vendor, is required to erect the tower.

- Obtain local permits for the type of met-tower that is being installed. Check the soil conditions and procure the right type of anchors for the guy wires. Some states require Rev-G towers, which means procuring the right kind of tower and base. In some situations, a concrete foundation for the tower may be needed.

- Procure and install proper electrical grounding rods.

- Check the weather conditions, specifically rain and wind, before scheduling an install. Met-towers cannot be installed in moderate to high winds and during thunderstorms.

- Plan the orientation of instruments based on "Placement of Sensors" section. In addition, ensure that guy wires do not disturb the airflow over the instruments.

- Program the data logger and communication equipment and prepare documentation and forms before going to site.

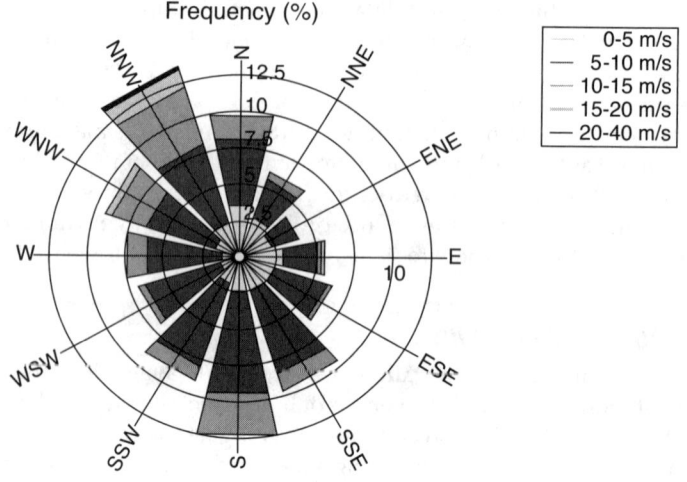

FIGURE 6-12 Wind rose of an illustrative location.

Example of Met-Tower Installation

To illustrate the design of orientation of anemometers on a tubular and guy-wire met-tower, consider a situation with a wind rose in Fig. 6-12. The predominant direction of wind is NNW at 330°, the next most frequent and high wind speed direction is S at 180°. In this example, an approach to orientation of booms will be described. These boom positions are used at each height.

To reduce the shadow effect, the first anemometer may be placed at 330 ± 60° Both directions are in low-wind frequency areas. To decide on which one, it is helpful to analyze the orientation of the redundant anemometer. Since the wind rose does not have a single wind direction that is overwhelmingly predominant, the redundant anemometer may be placed 180° from the first anemometer. The question is which of the two pairs is a better orientation (30, 210) or (270, 90)? The numbers in parenthesis are the angle in degrees of the orientation of the first and second anemometer. The second pair is a superior choice because sum of frequencies are lower than around the 270, 90° sector.

A wind vane may be installed at 270°. The wind vane may be installed two or more meters below the anemometer.

Data Management

After the data is measured by sensors, recorded in data logger, and transmitted out of the met-tower, the next steps are to manage the data in an offsite server. The steps include, see Fig. 6-13:[8]

Data Validation Flowchart

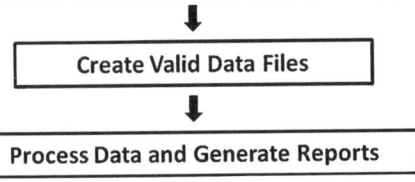

Raw Data Files

Develop Data Validation Routines
General System and Measured Parameter Checks
- Range tests
- Relational tests
- Trend tests

Fine-tune Routines with Experience

Validate Data
- Subject all data to validation
- Print validation report of suspect values
- Manually reconcile suspect values
- Insert validation codes
- Alert site operator to suspected measurement problems

Create Valid Data Files

Process Data and Generate Reports

Figure **6-13** Flowchart of data management. (*Source: Wind Resource Assessment Handbook: Fundamentals for Conducting a Successful Monitoring Program*, April 1997, created by AWS Scientific, Inc. on behalf of the National Renewable Energy Laboratory, http://www.nrel.gov/wind/pdfs/22223.pdf.)

- Import the data
- Validate the data to create valid data files
- Process the data to compute derived quantities like shear, energy density, and others
- Generate reports.

The first data management step arises during setup of the tower and data server. The first step is to check the sensor parameters and units of measure of data arriving at the server against parameters supplied by the sensor manufacturer. This will check for any data entry errors when entering parameters into the data logger. The next check is to compare the first few hours of measured data against wind data from neighboring airport or weather station. This will check for defective sensors, improper mounting of sensors, and incorrect orientation of wind vane sensor.

The next data management is a recurring task that is performed daily or with a frequency that coincides with data transfer from tower to server. This validation of data is a significant task in data management. Typical sources of errors and validation issues with measured data include:

- Wear and tear of sensors, booms and mounting, guy wires, and met-tower
- Icing of sensors
- Lightning strike
- Loose or faulty wiring

In validation check, three tests are performed on the data:[8]

- Range test: This test identifies values that are outside the validation criteria. Table 6-4 contains a sample of the validation criteria. The offset is the lowest value recorded by sensor when the wind speed is below the sensor's threshold. A reading of wind speed above 25 m/s does not necessarily mean that the data is suspect, but it requires a review.
- Relational test. This test compares concurrent measured values of redundant sensors, and sensors at different heights. If the differences do not meet the validation criteria, then a review is required. Sample test criteria are in Table 6-5.
- Trend test. This test examines temporal changes in measured values. Sample test criteria are presented in Table 6-6.

After the data issues have been identified, the next step in Fig. 6-13 titled "Validate Data" is executed.

Data Processing

In order to perform statistical analysis of the measured data and to visualize the data, a variety of data processing methods are used. For illustration purposes, publicly available wind speed data from a National Renewable Energy Laboratory (NREL) project in Valentine, Nebraska, is used in this section. The data is available at heights of 10, 25, and 40 m above the ground level, and for a period of 4 years.

For statistical analysis, the data is organized into bins. An example of two-dimensional binning of data is seen in Table 6-7. The two dimensions for the bins are wind speed and direction. For wind direction, it is normal to use 16 bins with bin size of 22.5°; some programs use 12 bins with bin size of 30°. The wind direction bins are called sectors. Figure 6-14 shows the frequency plot of the North-North-West

Sample Parameter*	Validation Criteria
Wind speed: horizontal	
• Average	offset < Avg. < 25 m/s
• Standard deviation	0 < Std. Dev. < 3 m/s
• Maximum gust	offset < Max. < 30 m/s
Wind direction	
• Average	0° < Avg. ≤ 360°
• Standard deviation	3° < Std. Dev. < 75°
• Maximun gust	0° < Max. ≤ 360°
Temperature	(Summer shown)
• Seasonal variability	5°C < Avg. < 40°C
Solar radiation	(Optional: summer shown)
• Average	offset ≤ Avg. < 1100 W/m²
Wind speed: vertical	(Optional)
• Average (F/C)†	offset < Avg. < ± (2/4) m/s
• Standard deviation	offset < Std. Dev. < ± (1/2) m/s
• Maximum gust	offset < Max. < ± (3/6) m/s
Barometric pressure	(Optional: sea level)
• Average	94 kPa < Avg. < 106 kPa
ΔT	(Optional)
• Average difference	> 1.0°C (1000 hrs to 1700 hrs)
• Average difference	< −1.0°C (1800 hrs to 0500 hrs)

*All monitoring levels except where noted.
†(F/C): Flat/Complex Terrain.
Source: Wind Resource Assessment Handbook: Fundamentals for Conducting a Successful Monitoring Program, April 1997, created by AWS Scientific, Inc. on behalf of the National Renewable Energy Laboratory, http://www.nrel.gov/wind/pdfs/22223.pdf.

TABLE 6-4   Illustration of Range Test Criteria for the Measured Quantities

(NNW) sector; it displays the frequency of occurrence at each wind speed bin. Weibull parameters are then computed from the frequency data for each sector. Table 6-8 contains the Weibull parameters for the 12 directions.

The measured data is presented sector-by-sector and as an aggregate for all sectors. Wind rose diagram is a favorite method to represent wind speed as a function of direction. Three diagrams are popular: Sector-wise frequency, sector-wise average speed, and

Sample Parameter*	Validation Criteria
Wind speed: horizontal	
• Max gust vs. average	Max gust ≤2.5 m/s avg.*
• 40 m/25 m average Δ[†]	≤2.0 m/s
• 40 m/25 m daily max Δ	≤5 m/s
• 40 m/10 m average Δ	≤4 m/s
• 40 m/10 m daily max Δ	≤7.5 m/s
Wind speed: redundant	(Optional)
• Average Δ	≤0.5 m/s
• Maximum Δ	≤2.0 m/s
Wind direction	
• 40 m/25 m average Δ	≤20°

*All monitoring levels except where noted.
[†]Δ = Difference.
[‡]In this sample, anemometer heights are 40, 25, and 10 m.
Source: Wind Resource Assessment Handbook: Fundamentals for Conducting a Successful Monitoring Program, April 1997, created by AWS Scientific, Inc. on behalf of the National Renewable Energy Laboratory, http://www.nrel.gov/wind/pdfs/22223.pdf.

TABLE 6-5 Illustration of Relational Test Criteria for the Measured Quantities[‡]

Sample Parameter*	Validation Criteria
Wind speed average	(All sensor types)
• 1 h change	<5.0 m/s
Temperature average	
• 1 h change	≤5°C
Barometric pressure average	(Optional)
• 3 h change	≤1 kPa
Δ Temperature	(Optional)
• 3 h change	Changes sign twice

*All monitoring levels except where noted.
Source: Wind Resource Assessment Handbook: Fundamentals for Conducting a Successful Monitoring Program, April 1997, created by AWS Scientific, Inc. on behalf of the National Renewable Energy Laboratory, http://www.nrel.gov/wind/pdfs/22223.pdf.

TABLE 6-6 Illustration of Trend Test Criteria for the Measured Quantities

Bin	Start	End	Sum	N	NNE	ENE	E	ESE	SSE	S	SSW	WSW	W	WNW	NNW
Mean			7.05	6.44	5.67	4.81	5.01	5.53	6.78	7.5	6.86	6.55	7.53	8.29	9.34
0		0.49	912	311	149	18	154	64	17	136	20	12	10	11	10
1	0.5	1.49	611	82	48	48	44	50	51	56	43	40	54	45	50
2	1.5	2.49	1336	155	123	135	120	112	128	116	96	103	57	92	99
3	2.5	3.49	2208	224	238	208	209	209	208	198	177	150	121	114	152
4	3.5	4.49	3152	308	292	262	345	281	306	291	247	193	184	200	243
5	4.5	5.49	3621	352	284	247	319	348	350	376	313	253	211	260	308
6	5.5	6.49	4434	423	288	225	279	422	479	458	458	344	303	347	408
7	6.5	7.49	4387	393	232	137	192	317	493	492	556	412	337	367	459
8	7.5	8.49	3898	295	150	75	112	170	449	556	480	409	353	372	477
9	8.5	9.49	3333	288	115	37	62	102	360	525	342	271	361	393	477
10	9.5	10.49	2433	195	87	13	45	53	252	421	175	142	291	334	425
11	10.5	11.49	1593	126	53	12	13	31	123	276	136	54	198	224	347
12	11.5	12.49	1079	108	41	3	8	16	70	180	68	16	105	159	305
13	12.5	13.49	714	80	32	3	4	14	33	101	39	10	40	117	241
14	13.5	14.49	489	59	28	0	4	2	20	65	21	3	24	70	193
15	14.5	15.49	304	35	11	1	3	1	11	46	4	2	7	41	142

*This has 12 direction bins.

TABLE 6-7 Binning of Four Years of 40-m Wind Speed Data at Valentine, Nebraska*

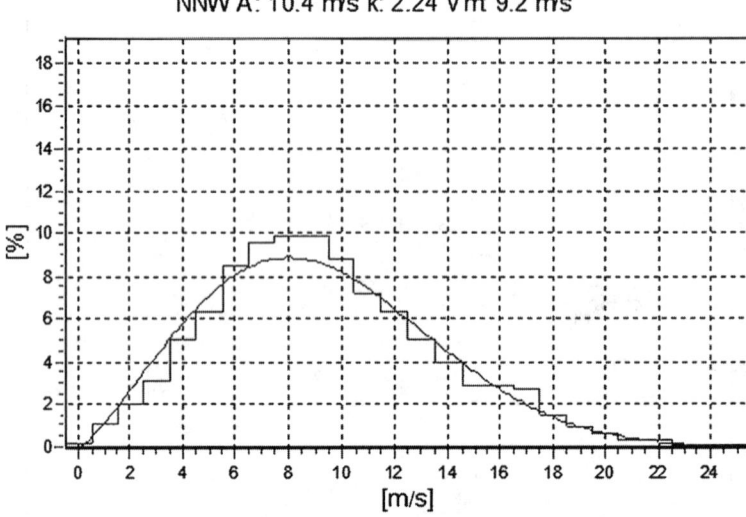

FIGURE 6-14 Frequency plot for the NNW direction. The Weibull parameters are derived by WindPRO. The jagged lines are the actual frequencies; the smooth curve is the computed Weibull distribution with $A = 10.4$ m/s and $k = 2.24$.

Sector	A (m/s)	k	Frequency (%)	Mean Wind Speed (m/s)
Mean	8.1	2.2	100.0	7.2
0-N	7.6	1.9	9.9	6.7
1-NNE	6.5	1.9	6.2	5.8
2-ENE	5.5	2.3	4.0	4.9
3-E	5.9	1.8	5.5	5.2
4-ESE	6.4	2.7	6.2	5.7
5-SSE	7.7	2.8	9.5	6.9
6-S	8.7	2.7	12.3	7.7
7-SSW	7.8	3.1	9.0	7.0
8-WSW	7.5	3.6	6.8	6.7
9-W	8.6	3.2	7.6	7.7
10-WNW	9.4	2.4	9.3	8.3
11-NNW	10.4	2.2	13.6	9.2

TABLE 6-8   Weibull Parameters for Valentine, Nebraska Binned Data

sector-wise wind energy density (see Fig. 6-15). Wind energy rose in Fig. 6-15c is a sector-wise plot of average wind energy density (proportional to the cube of wind speed) multiplied by the relative frequency of wind in each sector.

Computed Quantities

After the data has been validated, the following quantities are computed. These derived quantities are used to determine, among other items, the turbine class and uncertainty in resource.

Turbulence

Turbulence has a very specific meaning when describing wind conditions. It is the standard deviation of horizontal wind speed, vertical wind speed, and wind direction around the 10-min average. Turbulence intensity (TI) is defined as the ratio of standard deviation to the average. In wind projects, the horizontal wind speed turbulence intensity is most commonly used. For horizontal wind speed, the TI is:

$$TI = \frac{\sigma}{v_{avg}} \qquad (6\text{-}5)$$

where σ is the standard deviation of wind speed and v_{avg} is the average wind speed. In most wind applications, the standard deviation and average are based on 10-min observations. For example, with 3-s measurement interval and 10-min recording interval, the standard deviation and average used in the above formula are for a sample size of 200 values.

Turbulence intensity is a measure of the atmospheric stability; specifically, it measures rapid changes in wind speed over short intervals. A value of TI that is 0.1 or less is considered low turbulence; TI in the range of 0.1 to 0.25 is considered moderate turbulence; 0.25 or higher is considered high turbulence. TI is used to determine the turbine category appropriate for the site. High turbulence causes energy output to diminish and affects the loading, durability, and operation of a turbine.

According to IEC 61400-1, there are three categories of turbines A, B, and C with values of $I_{ref} = 0.16, 0.14$, and 0.12, respectively. I_{ref} is the expected turbulence intensity at 15 m/s wind speed. In the normal turbulence model (NTM), the standard deviation of wind speed at hub is described as:

$$\sigma = I_{ref}(0.75v_{hub} + 5.6) \qquad (6\text{-}6)$$

$$TI = \sigma/v_{hub} = I_{ref}(0.75 + 5.6/v_{hub}) \qquad (6\text{-}7)$$

The three smooth curves in Fig. 6-16 are graphs of the NTM, represented by Eq. (6-7), for different values of I_{ref}. Statistics of the measured

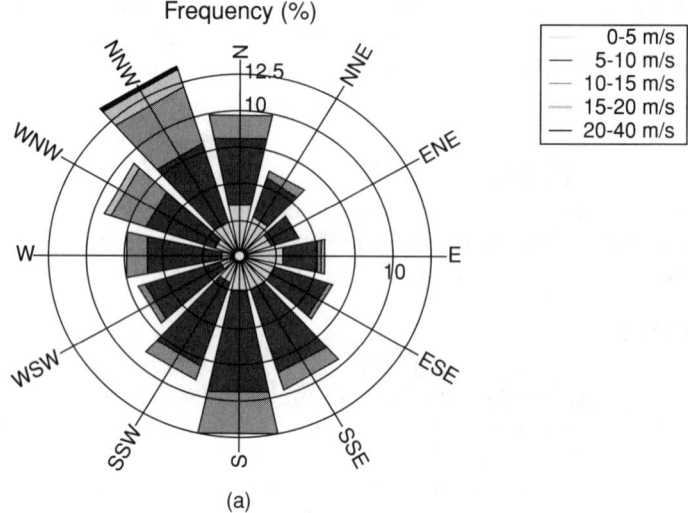

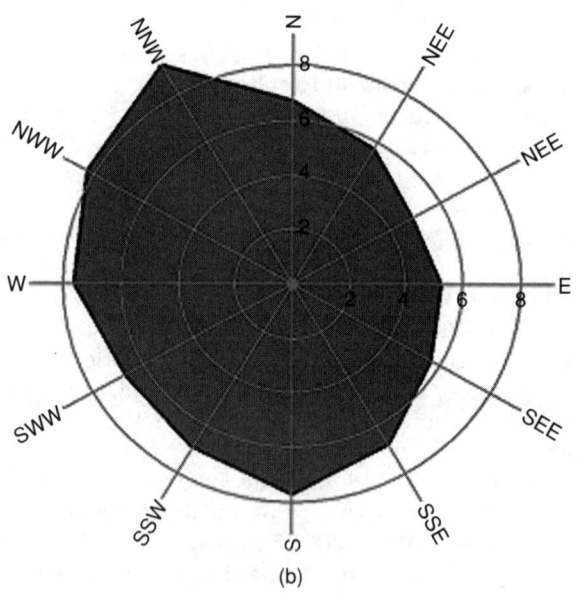

FIGURE 6-15 Wind rose for Valentine, Nebraska. (a) Frequency rose with relative frequency of different wind speed ranges in different sectors (see Fig. 6-12); (b) average wind speed in each sector; (c) energy density in each sector.

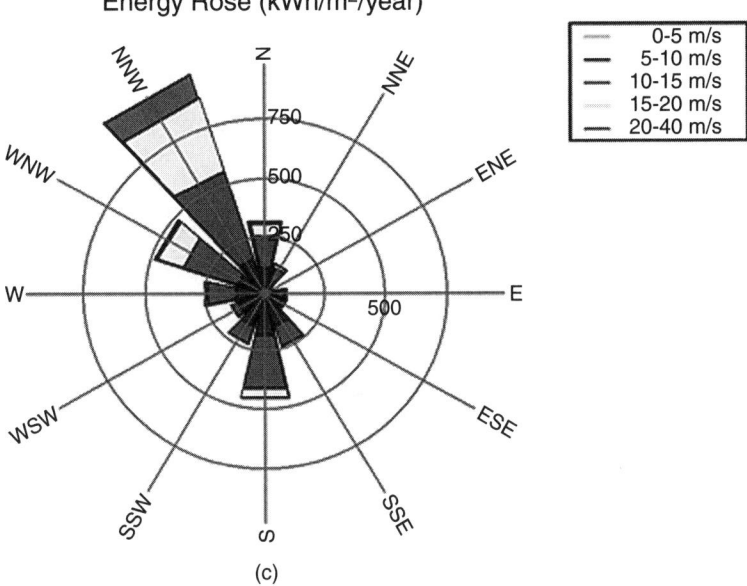

Energy Rose (kWh/m²/year)

FIGURE 6-15 (Continued)

turbulence intensity data, namely, mean (TI_{avg}), standard deviation (TI_{sd}), and $TI_{avg} + 1.28\ TI_{sd}$ are the jagged curves in Fig. 6-16. The location of the curve $TI_{avg} + 1.28\ TI_{sd}$ with respect to the NTM reference curves defines the turbine category. In this illustration, the hub height is 40 m. Turbine category along with turbine class (which is based on average wind speed at hub height) defines the design load for components of the turbine, so it directly affects the size of rotor and size of other components in the turbine. This is described further in Chapter 9.

Wind Shear

As described in Chapter 3, shear is used to extrapolate the horizontal component of wind velocity to different heights. Rearranging terms in Eq. (3-10), shear can be expressed as:

$$\gamma_{1,2} = \frac{\log\left(v_1/v_2\right)}{\log\left(h_1/h_2\right)} \tag{6-8}$$

where $\gamma_{1,2}$ is shear and v_1, v_2 are horizontal wind speeds at heights h_1, h_2. Since wind speeds are measured at three heights with redundant anemometers, there can be several values of shear. Normally, two values are computed $\gamma_{1,2}$ and $\gamma_{1,3}$, after the redundant time series have

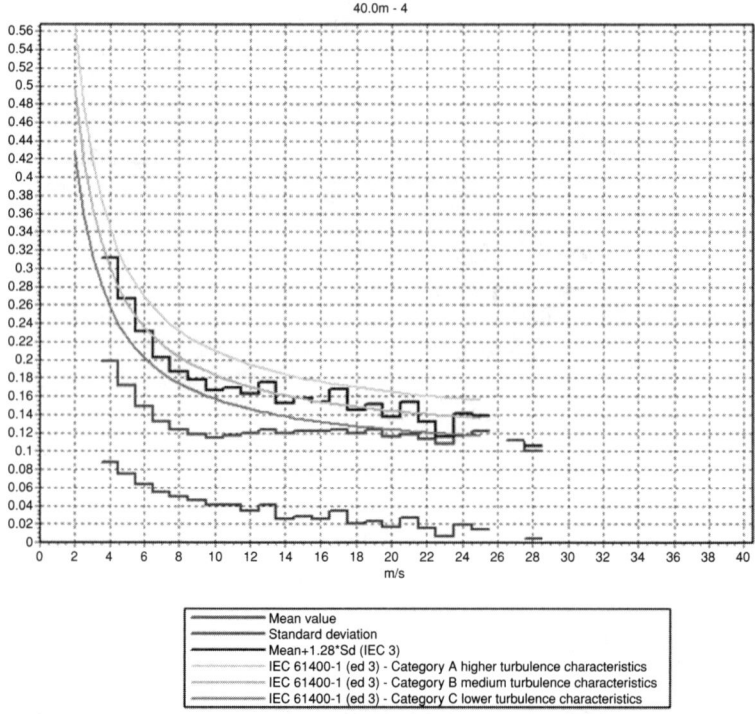

Figure 6-16 Plot of average and standard deviation of turbulence intensity as a function of wind speed at hub height. (Mean + 1.28*Standard_Deviation) curve is compared against IEC Category A, B, and C curves to determine the turbine category.

been consolidated into a single wind speed time series, where the index 1 corresponds to the highest anemometer and index 3 corresponds to the lowest anemometer.

For greater understanding of wind data, wind shear values are computed in each sector. In most locations, the value of shear has a very strong diurnal and seasonal variation. The diurnal and seasonal variations are because of convection from heating of earth during the day. Table 6-9 contains wind shear data for Valentine, Nebraska. At this location, highest energy density sectors are NNW, WNW, and S. Notice the difference in wind shear values in these sectors by day/night and summer/winter.

Air Density

The density of air is computed with Eqs. (3-19) and (3-20). Temperature, relative humidity, and barometric pressure are the measured quantities used in the equations.

Sector	All Data	Day	Night	Summer	Winter
N	0.124	0.09	0.171	0.135	0.081
NNE	0.126	0.076	0.177	0.158	0.057
ENE	0.153	0.096	0.211	0.176	0.122
E	0.183	0.119	0.244	0.243	0.128
ESE	0.216	0.163	0.261	0.263	0.189
SSE	0.176	0.139	0.216	0.192	0.195
S	0.17	0.135	0.214	0.171	0.181
SSW	0.128	0.1	0.163	0.136	0.124
WSW	0.185	0.158	0.226	0.166	0.183
W	0.22	0.179	0.27	0.202	0.221
WNW	0.166	0.13	0.212	0.152	0.215

TABLE 6-9   Average Wind Shear Values by Sector at Valentine, Nebraska

Power Density

The power density is computed with Eq. (3-9). A discrete version of the equation is:

$$PD = \left(\sum_{i=1}^{N} \rho_i v_i^3 \right) / 2N_y \qquad (6\text{-}9)$$

where N is the number of measurements, N_y is the number of years of measurement data, ρ_i and v_i are density, and wind speed at time i. A plot of energy density by sector is found in Fig. 6-1c.

Remote Sensing to Measure Wind Speed

As the hub heights and blade lengths of turbines have increased, met-tower based measurements at 30, 40, and 60 m, or sometimes 50, 60, and 80 m heights are inadequate to provide an accurate estimate for wind speed at the hub height, let alone over the entire turbine rotor. With both hub heights and rotor radius above 85 and 45 m, met-towers of height 130 to 150 m or more would be required to measure the wind speed over the entire turbine rotor. This would be cost prohibitive. Remote sensing provides a method to measure wind speed in this range of heights.

Ground-based remote sensing for wind measurements has been in regular use since 1990s; however, serious commercialization started in 2000s. There are two primary technologies, SODAR and LIDAR. Sonic Detection and Ranging (SODAR) is based on measuring Doppler shift in the frequency of the sound waves that are backscattered by temperature fluctuations in the atmosphere. Figure 6-17 is a picture of a

Figure 6-17 Picture of the Nomad™ SODAR unit. Size of the unit is 2 m × 2 m × 2 m and weighs between 350 and 450 kg. (Photo (copyright) Justin Arcangeli. Courtesy Second Wind Inc.)

SODAR unit. Light Detection and Ranging (LIDAR) is based on shift in frequencies because of reflection of light rays from small aerosol particles in the air. Figure 6-18 is a picture of a LIDAR unit.

SODAR[9,10] works with three or five beams that are generated from an array of speakers. The three beams measure the x, y, and z components of wind speed in a volume of air. The x and y components are measured by beams along the x and y axis at an angle of 10 to 20° from the vertical axis. The Doppler frequency shift is proportional to the wind speed aligned to the path of the sound wave, which yields the x, y, and z component of wind speed.

LIDAR uses light instead of sound and relies on the same Doppler shift principle. However, in this case, the radiation is scattered by aerosol particles like dust, moisture, and others.

Pros and Cons of Remote Sensing for Wind Measurements

The advantages of remote sensing compared to met-towers are numerous:[11]

- Ability to measure wind speed at heights higher than "economically" feasible by met-towers
- Ability to measure over large range of heights from 40 to 200 m
- Ability to measure wind shear directly over a large range of heights
- Ability to measure the vertical component of wind speed

FIGURE 6-18 Picture of the ZephIR® LIDAR unit. (Copyright Natural Power.)

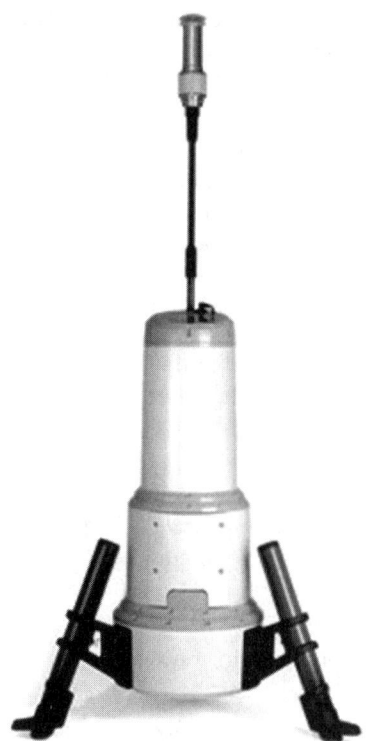

- No obstructions and shadow effects because of guy wires, booms, and tower
- Ease of deployment, specifically in areas where guy wires cannot be anchored. Offshore wind measurements commonly use remote sensing.
- Mobility of the remote sensing unit. This leads to flexibility in terms of rapid deployment and adjustments.
- Easy to repair.
- Unobtrusive and may be easily concealed as opposed to a tall met-tower that may be visible for miles. Most units are about 6 ft × 6 ft × 9 ft.

The limitations of remote sensing include:

- Data is not yet bankable because there are no standards for certifying SODAR and LIDAR data. Efforts are underway to establish standards. Data must be used in conjunction with met-tower data to improve the accuracy of computing wind speed over a large area. As experience and independent validation of remote sensing equipment increases, it is expected that, in the near future, data from these equipment will be accepted by the wind project financing community.

- SODAR dataset is usually not as complete as cup anemometer data. The reason is that data depends on backscatter of small fluctuations of the thermal refractive index in atmosphere. SODAR performs poorly when the atmosphere is thermally well mixed (afternoon), in the presence of precipitation and in the presence of ambient sound. Therefore, signal-to-noise filters are crucial; setting the filter too high will cause useful data to be rejected; setting the filter too low will cause noisy data to be included in the dataset.[11] Operator experience can be crucial in setting filters for accepting data with smaller signal-to-noise ratio. Raw SODAR data also undergoes complex data processing algorithms that are proprietary to the manufacturer.

- Wind speeds are measured and averaged over a larger volume, whereas, cup anemometer measures almost at a point. SODAR measurements at 100 m typically use 15,000 to 20,000 times the volume of air compared to cup anemometer.

- In a complex terrain, the measurements may not correlate well with met-tower data.

- Onsite calibration of SODAR is required. LIDAR usually does not require onsite calibration.

- SODAR created audio signal is a chirping noise, therefore it must be placed with an appropriate setback from residences and other inhabited structures. The chirp audio signal is also susceptible to echoes because of interactions with structures and trees.

SODAR and LIDAR can yield very useful wind speed and direction data for prospecting, even with the above limitations.

References

1. International Electrotechnical Commission (IEC). Wind Turbines—Part 12-1: Power Performance Measurements of Electricity Producing Wind Turbines, 2005. IEC 61400-12-1.
2. Dahlberg, J. A., Pedersen, T. F., and Busche, P. ACCUWIND—Methods for Classification of Cup Anemometers, Riso National Laboratory, Roskilde, 2006. Riso-R-1555(EN).
3. Young, M., Met Towers & Sensors—Science & Equipment Considerations, DNV Global Energy Concepts, AWEA Resource and Project Energy Assessment Workshop, Minneapolis, MN, 2009.

4. NRGSystems. NRG #40C Anemometer, Calibrated. *NRG Systems* [Online] 2010. http://www.nrgsystems.com/sitecore/content/Products/1900.aspx?pf=StandardSensors.

5. King, B., *The How and Why of Wind Sensor Mounting,* Conference 2009 Poster Presentation, Chicago, 2009.

6. WindGuard, Deutsche. Summary of Cup Anemometer Classification [Online] 12 2008. http://www.skypowerinternational.com/pdf/News%20and%20Media/new/First_Class_Advanced_Testing_Results_.pdf.

7. Taylor, M., Mackiewicz, P., Brower, M. C., and Markus, M., *An Analysis of Wind Resource Uncertainty in Energy Production Estimates,* EWEC, London, 2004.

8. National Renewable Energy Lab. Wind Resource Assessment Handbook [Online] 1997. www.nrel.gov/wind/pdfs/22223.pdf.

9. Crescenti, G. H., "A look back on two decades of Doppler Sodar comparison studies."*Bulletin of American Meteorological Society,* 78: 651-673, 1997.

10. Moore, K., and Bailey, B. *Recommended Practices for the Use of Sodar in Wind Energy Resource Assessment,* IEA Draft, Version 4, 2009.

11. Kelley, N., et al. *Comparing Pulsed Doppler LIDAR and SODAR and Direct Measurements for Wind Assessment,* National Renewable Energy Laboratory, Golden, CO, 2007. NREL/CP-500-41792.

12. Burton, T., Sharpe, D., Jenkins, N., and Bossanyi, E. *Wind Energy Handbook,* Wiley, Hoboken, NJ, 2001.

13. Hansen, M. O. L. *Aerodynamics of Wind Turbines,* EarthScan, Sterling, VA, 2008.

14. Wilson, R. E., Peter, B. S., and Lissaman, P. B. S. *Applied Aerodynamics of Wind Power Machines,* Oregon State University, Corvallis, 1974. http://ir.library.oregonstate.edu/jspui/bitstream/1957/8140/1/Applied_Aerodynamics.pdf.

15. Walls, L., and Sass, W. Viability of Sodar for Long-Term Resource Assessment, *Windtech International,* Sept. 2008.

CHAPTER 7

Wind Resource Assessment

A few observations and much reasoning lead to error; many observations and a little reasoning to truth.

—Alexis Carrel

Introduction

Wind resource assessment (WRA) is the discipline of estimating the strength of wind resources at a planned wind project site. The output of wind resource assessment, at a generic level, is wind conditions and annual energy production at a project site. A financial model uses this data to compute the financial performance of the wind project. WRA is, therefore, the core activity that determines viability of a wind project.

The chapter starts with an overview of the WRA process. This chapter will be divided into two parts. The first will deal with prospecting with no onsite measurement data. Before delving into resource estimation, exposition of sources of wind data is provided, followed by the most commonly used models. The second part of the chapter will deal with computations related to spatial extrapolation, long-term hindcast, and annual energy production.

Chapter 8 will cover advanced topics in WRA, loss estimation, and uncertainty associated with resource assessment.

Overview of Wind Resource Assessment

Figure 7-1 contains an overview of the WRA process. WRA starts with a preliminary assessment or prospecting. In this step, alternate sites are evaluated for adequate wind speed based on publicly available wind resource maps and wind data. The prospecting step

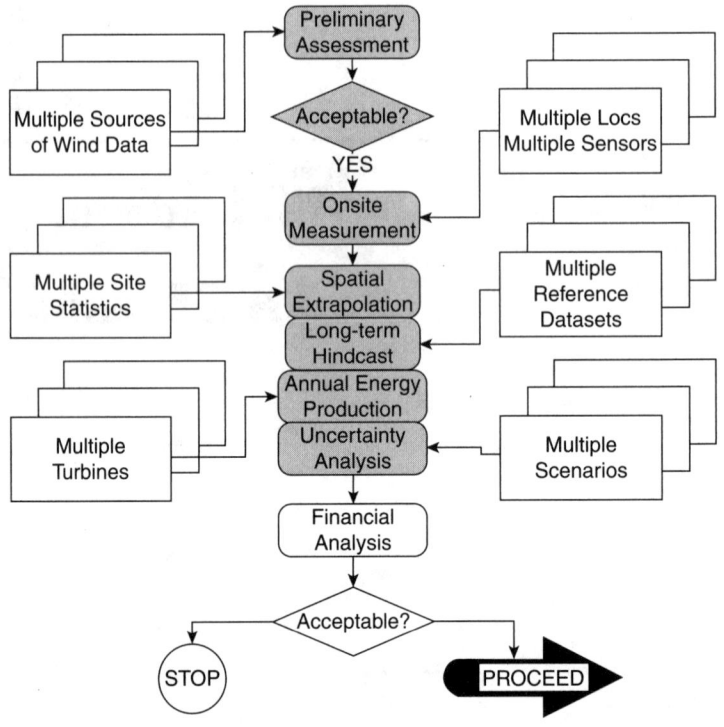

FIGURE 7-1 Overall process of Wind Resource Assessment (WRA) and financial analysis. The gray boxes are steps in WRA. Inputs to each step are indicated.

also involves a preliminary financial assessment to determine if a wind project is viable. If the site is acceptable, then an onsite wind measurement campaign is conducted. After wind data has been collected for sufficient period, typically one year or more, then a process of detailed WRA begins. It begins with spatial extrapolation, in which measured data at multiple locations within the project area are used to estimate wind speeds over the entire project area. This is extrapolation along the spatial dimension.

The next step in the detailed WRA is to extrapolate along the temporal dimension. A process called measure-correlate-predict (MCP) is used with multiple reference datasets as input. Reference datasets are long-term wind data from a variety of sources like reanalysis data from National Center for Atmospheric Research (NCAR), airports, and others. MCP extrapolates onsite measured data and generates a long-term dataset that covers the time period covered by the reference dataset. Next, annual energy production (AEP) is computed with several power production curves from different turbines. The last step is to compute uncertainty of AEP, which consolidates the uncertainty in each factor that influences AEP.

The output of the WRA is input to the financial analysis step, in which the financial viability of the project is assessed.

Source of Wind Data

There are three primary sources of wind data: Onsite measurement, network of weather stations, and numerical weather models. Onsite measurement is described in detail in Chapter 6. It forms the core input to the resource estimation and subsequent micrositing of wind projects.

The second source is the network of weather stations all around the world that provide wind speed and direction data. Various organizations collect and provide this data. Among them is the USA's National Climatic Data Center (NCDC) of National Oceanic and Atmospheric Administration (NOAA). It provides standardized quality-controlled wind speed data at a height of 10 m for almost all airports worldwide. The data is available from http://www.ncdc.noaa.gov/oa/mpp/, item DS3505. The dataset covers about 10,000 locations. For most US and European airports, hourly wind speed data is available for a fee. An annual subscription service is also available from NCDC that provides monthly downloadable wind speed data for about 1,000 US airports. Although this is valuable data, there are several significant issues with this dataset:

- Wind speed data is available at only one height, 10 m above the ground. Dataset includes, hourly average wind speed and wind direction; it does not contain standard deviation.

- The age, quality, height, precise location, and surrounding environment of the wind measurement instruments at each airport location are not known with certainty. Without this metadata pertaining to the instruments, the quality of raw data and the roughness surrounding the instruments cannot be determined. Common problems include height of anemometer is low; anemometer mast is surrounded by obstacles like buildings, trees, radar, and others; and roughness and contour are different at the airport compared to site of interest. Therefore, the uncertainty of calculations based on this dataset is high.

The third source of wind data is the long-term reanalysis data. According to Kalnay, et al.,[1] "The National Centers for Environmental Prediction (NCEP) and National Center for Atmospheric Research (NCAR) have cooperated in a project (denoted "reanalysis") to produce a retroactive record of more than 50 years of global analyses of atmospheric fields in support of the needs of the research and climate-monitoring communities. This effort involved the recovery of land surface, ship, rawinsonde, pibal, aircraft, satellite, and other data. These data were then quality-controlled and assimilated with a data assimilation system kept unchanged over the reanalysis period." NCAR data is available on 2.5° latitude by 2.5° longitude spatial grid and 6-h temporal grid. NCAR dataset is available in an easy-to-access form from a service provided by WindPRO, a wind assessment tool. It provides wind speed at 10- and 42-m heights, and wind direction. In addition to this coarse dataset, finer datasets are available for specific regions, for example, NARR (North American Regional Reanalysis)[2] covers most of North America and ECMWF (European Centre for Medium-Range Weather Forecasting)[3] covers most of Europe. The NARR dataset is available on 1° latitude by 1° longitude spatial grid and 3-h temporal grid. This dataset is not available in an easy-to-access method and format.

The large-scale nature of reanalysis data does not account for the local effects like elevation changes and roughness. Since it is a compilation of multiple data sources, the local effects are taken out from each data source to create a reanalysis dataset.

Resource Estimation Models

Given the three types of data sources, a variety of models are used to estimate wind speed. The most commonly used models are mesoscale, computational fluid dynamics (CFD), and microscale models. The purpose of these models is to use the wind data described in the previous section and create wind data for desired regions. The created wind data contains wind speed at multiple heights (or wind speed at single height and wind shear), and wind direction.

Mesoscale Models

Mesoscale scale models describe weather phenomena that have a spatial resolution of 20 to 2,000 km and a temporal resolution of hours to days. These models take as input reanalysis data, elevation, and roughness data. The reanalysis data provides the external forcing of the model through boundary conditions. The most common mesoscale model are KAMM (http://www.mesoscale.dk/), MM5 (www.mmm.uear.edu/mm5/mm5-home.html), and MC2 (www.cmc.ec.gc.ca/rpn/modcom/en/index en.html). The resolution of the models is a few kilometers and the coverage is a few hundred kilometers. The basic equations related to conservation of mass, momentum, and energy are solved in a finite element grid with a temporal resolution. Simulations with the mesoscale model yield a statistical picture of wind speed and wind direction.

Mesoscale models are combined with microscale models to provide useful results for prospecting.[4] The most common combination of meso- and microscale models are KAMM and WAsP. WAsP is a microscale model that is described later in this section. Several commercial companies use proprietary models to combine meso- and microscale models. The microscale model resolves the small-scale contour and roughness features.

CFD Models

Computational fluid dynamics (CFD) models are used to model: (a) Airflow in complex terrains, for example, airflow around and over mountains and (b) thermal effects. These models can provide a better understanding of turbulence when fine spatial resolution is used. CFD models are based on solving Reynolds-Averaged Navier-Stokes (RANS) equations along with turbulence models. Two popular 3-D CFD software programs are: WindSim and 3DWind. These programs are used to model both meso- and microscale wind projects. Depending on the detail in the model, the 3-D CFD models can be very computationally intensive. The output of the models is a steady-state time-independent solution of wind speed and direction.

Inputs to CFD models are: Digital terrain model, roughness map, and wind data at multiple locations for a year or more. The overall process is: The spatial domain is gridded into cells; the cells may be further nested to model features of interest; simulation is started, and the equations are solved iteratively until steady-state is reached. The output of the models is wind speed time series at each grid point.

WAsP, a Microscale Model

Wind Atlas Analysis and Application Program (WAsP)[5] is a microscale model developed by Risø in the late 1980s (http://www.wasp.dk/). The tool and the associated methodology have become the most popular method of performing wind assessments at the microscale. Several of the most prominent wind assessment tools like WindPRO and WindFarmer use the WAsP engine to perform wind assessments.

Definitions

- *Microscale Model.* The scale of applicability of microscale models is of the order of 100 km. Although the name may suggest otherwise, microscale is used to perform wind assessments for large wind farms that cover hundreds of square kilometers.
- *Regional wind climate (RWC).* The regional wind climate is a generalized wind statistic that describes the wind conditions over a much larger area (order of magnitude 100 km). A generalized wind statistic is generated from observed wind data

(site-specific) through a process that takes out site-specific in-fluences which include:

- Obstacles that are in 1 km radius from the observed wind site
- Roughness that is within 20 km radius from the site
- Contour or elevation of area inside a radius of 5 km
- Thermally driven flow

The distance ranges above are guidelines for normal conditions. RWC is generated through the process shown in Fig. 7-2 that follows the up arrow in the left side of the diagram. The output of this process is the RWC.

- *Wind atlas.* A collection of RWC for a large area that is of the order of thousands of kilometers.

- *Wind resource.* Site-specific wind speed or energy density at a specified height is called a wind resource. This is generated through the process in Fig. 7-2 that follows the down arrow in the right side of the diagram. The output of the process is the wind resource, shown at the bottom-right corner of Fig. 7-2.

- *Wind resource map (WRM).* WRM is a description of wind resources at a specific height for a specific granularity. Creation of WRM can be a very computationally intensive process for a finer granularity grid. Therefore, depending on purpose, appropriate granularity WRM is created. For instance, 200 m × 200 m granularity WRM may be created for prospecting, where as 50 m × 50 m WRM may be created for micrositing of turbines. WRM is created as a color plot of annual average wind speed or average energy density at a specific height; however, seasonal averages are becoming more popular. The WRM color map with the intensity of wind energy density is used to explore the variation in wind resources over an area.

- *Wind resource assessment (WRA).* WRA is a compilation of aggregate data that includes, annual or seasonal:
  - WRM
  - Wind rose, which is a plot of average annual wind energy density by direction
  - For each sector, average turbulence intensity
  - For each sector, average wind shear
  - For each sector, Weibull parameters for the distribution of wind speed
  - Average annual energy production for chosen turbines of a wind farm

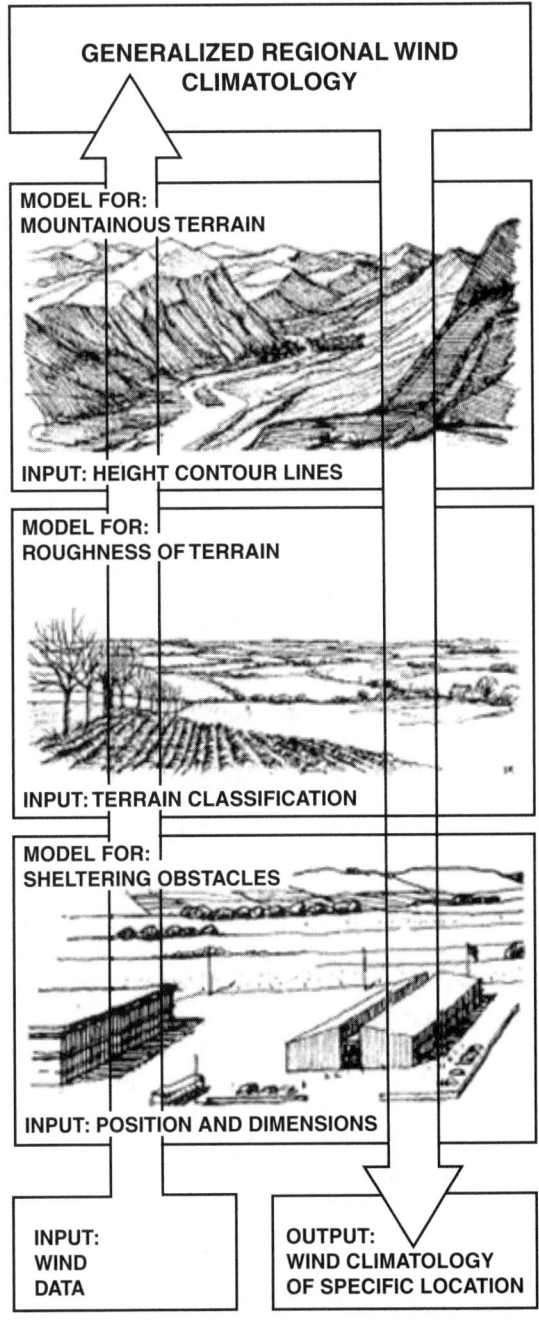

GENERALIZED REGIONAL WIND CLIMATOLOGY

MODEL FOR:
MOUNTAINOUS TERRAIN

INPUT: HEIGHT CONTOUR LINES

MODEL FOR:
ROUGHNESS OF TERRAIN

INPUT: TERRAIN CLASSIFICATION

MODEL FOR:
SHELTERING OBSTACLES

INPUT: POSITION AND DIMENSIONS

INPUT:
WIND
DATA

OUTPUT:
WIND CLIMATOLOGY
OF SPECIFIC LOCATION

FIGURE 7-2 WAsP model for wind resource assessment. (From Mortensen, N. G., et al. *Getting Started with WAsP 9,* Riso National Laboratory, Roskilde, Denmark, 2007. RISO-I-2571.)

Description of WAsP Model

The main objective of the WAsP model or any wind assessment model is to use measured wind data (or reanalysis data) and create wind resource estimates for desired areas and for desired heights. In essence, the models extrapolate (or interpolate) wind data from sites where it is available to sites with no wind data. In addition to the extrapolation along the horizontal plane, wind speed data is extrapolated along the vertical direction at multiple heights. This is the process of spatial extrapolation, along all three spatial axes. In sequel, sites with wind data are referred to as site A and sites where wind data is desired are referred to as site B. Site A may be a collection of measurement locations or a single measurement location with wind measurement instruments. Site B may be an area that is being prospected or a collection of proposed wind turbine locations. If site B is an area, then a grid has to be defined on which wind resources are estimated. For example, site B may be a 20 km × 20 km area with a 200 m × 200 m grid and wind resources are computed at all grid points. When multiple sites of type A are involved, then a weighted average of these sites is normally used to estimate wind speed at site B. One option is to use weights that are based on inverse distance between site B and multiple site As.

In Fig. 7-2, all the computational activities on the left-hand side of the diagram occur at site A, while all the activities on the right-hand side happen at site B. WAsP's approach is to sequentially remove the affects of obstacles, roughness, and elevation in order to derive a regional wind climate (RWC). The up arrow in Fig. 7-2 indicates this. The RWC is then translated to site B. The RWC is made location specific by applying elevation, roughness, and obstacles in sequence, as indicated by the down arrow in Fig. 7-2.

WAsP model takes as input:

> *Measured wind data at site A.* Typical sources of datasets are on-site measurement data from met-towers with instruments at multiple heights, airport data at 10 m, or any other source of onsite wind data.
>
> *Position and dimensions of obstacles near site A and near the assessment site, site B.* Obstacles are manmade structures like buildings, towers, row of trees planted as windbreakers, or naturally occurring structures like isolated trees and tall rock formations. The obstacles of relevance are structures taller than $1/4$ the hub height. In most cases, an area of radius 1 km is sufficient for modeling obstacles. Obstacles outside this area will not have a significant impact on the wind conditions. Most wind resource applications provide methods to specify dimension and orientation of obstacles.
>
> *Terrain roughness at sites A and B.* Roughness is a method to parameterize different types of terrains. An example of roughness classifications is found in Table 2-5. Most applications

provide drawing tools to mark areas of different roughness. Within a 5-km radius, detailed roughness classes should be manually specified and coarser roughness may be specified between 5 and 20 km radius. Very coarse-level roughness data is available online from satellite-based land cover and land use data. National Aeronautics and Space Administration (NASA) Modis satellite and global land cover characteristics (GLCC) database are two sources. The Modis roughness data is an indication of vegetation and water bodies only and does not account for roughness because of residential communities, towns, and cities.

Elevation contours at sites A and B. Contours with 10-m change in elevation are sufficient for preliminary assessment, while 5-m or finer contours may be required for detailed assessment of smaller areas. Online elevation data is available from SRTM (Shuttle Radar Topography Mission) project of NASA with a resolution of 1-arc second (approximately 30 m) for the continental United States and 3-arc second (approximately 90 m) for other areas.

Assumptions of the WAsP model are:[5]

- Sites A and B are subject to the same wind weather regime. In normal terrain, site B may be up to 100 km away from site A. For mountainous terrain, valid data may be estimated for site B that is only a few hundred meters away.
- The prevailing weather conditions are close to being neutrally stable.
- The measured wind data is reliable.
- The surrounding terrain of sites A and B is sufficiently gentle and smooth to ensure attached flows.
- The topographical models are adequate and reliable.

Guidelines for the applicability of WAsP type microscale models in different regions of the world are described in Table 7-1.

Output of WAsP Model

The WAsP model describes the regional wind climate (RWC) as a collection of Weibull parameters A and k for each of the following combinations:

- Roughness classes: 0, 1, 2, 3
- Heights: 10, 25, 50, 100, 200
- Directions: N, NNE, ENE, E, ESE, SSE, S, SSW, WSW, W, WNW, NNW

RWC consists of this set of 240 pairs of values.

Type of Region	Nature of Wind	Modeling Considerations	Other Factors
Arctic	Easterlies	Extrapolation beyond a few kilometers is difficult	Little data available.
Temperate plains and westerlies. Covers northern parts of North America, Europe, and Asia	Well suited for wind energy generation. Large-scale low-pressure systems dominate.	For noncomplex terrain, microscale models give good results when used for extrapolating wind speeds.	Significant wind energy production is in this area and most models are for these regions
Desert and semiarid areas	Seasonal winds	Roughness is low, so measured data can be translated easily.	Traditional models do not work well because of convection. Land-use intensity is low and construction is simple
Tropics	Seasonal wind systems like monsoon and trade winds	Models are quite reliable. High roughness due to forests or population density	High unmet demand for electricity
Open sea	Very high wind energy potential. Sparse wind data	Reliable and detailed wind maps exist in very few areas—areas that have offshore wind farms. Roughness is low.	Cost of gathering traditional met-tower based datasets is very high
Coastal areas	High wind energy potential. Sparse wind data	Impact of land on wind flow is pronounced in coastal areas.	Siting of wind projects is a challenge due to visual impact. Cost of gathering wind data is high.
Mountains	High wind energy potential. Sparse data. Wind regime is dominated by local effects.	Very difficult to model wind flow because of complex topography and local thermal effects.	Cost of collecting wind data is moderately high.

(From Redlinger, R. Y., Andersen, P. D., Morthorst, P. E. *Wind Energy in the 21st Century*, Palgrave, New York, 2002.)

TABLE 7-1 Characteristics of Regions with Respect to Wind Resources and Assessment

Height, m	Roughness Length (m)			
	0	**0.03**	**0.1**	**0.4**
10	9.6	6.6	5.7	4.5
25	10.5	7.9	7	5.8
50	11.2	9	8.2	7
100	12.1	10.4	9.6	8.3
200	13.1	12.6	11.5	10

*The four columns are for roughness classes 0, 1, 2, 4. The roughness lengths are in the title row. All the data in the table is mean wind speed in meters per second.

TABLE 7-2   Output of Wind Statistics Module of WAsP*

Table 7-2 shows the RWC, an output of the WAsP model for a single direction sector. The table was created in WindPRO. If the RWC is translated to site B with roughness length 0.03 (roughness class = 1) and the desired hub height is 100 m above the ground level, then the estimated average annual wind speed is 10.4 m/s.

A wind rose associated with the RWC is shown in Fig. 7-3. This is computed for "standard" conditions, roughness class = 1 and height = 50 m. Both wind speed rose and wind energy rose are displayed in Fig. 7-3.

The RWC described above is then translated to desired locations by applying affects because of local roughness, elevation, and obstacles. An example of this is in Fig. 7-4, which contains the RWC curve and the estimated wind speed curve at site B.

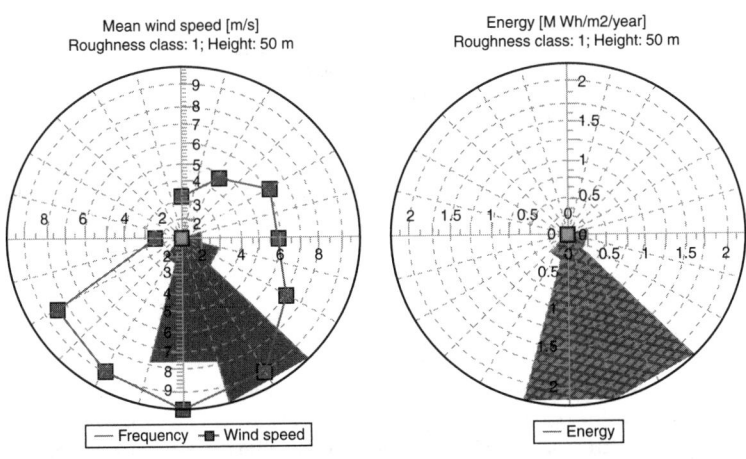

Mean wind speed [m/s]
Roughness class: 1; Height: 50 m

Energy [M Wh/m2/year]
Roughness class: 1; Height: 50 m

— Frequency -■- Wind speed

— Energy

FIGURE 7-3   Wind rose associated with the regional wind climate for height 50 m and roughness class of 1.

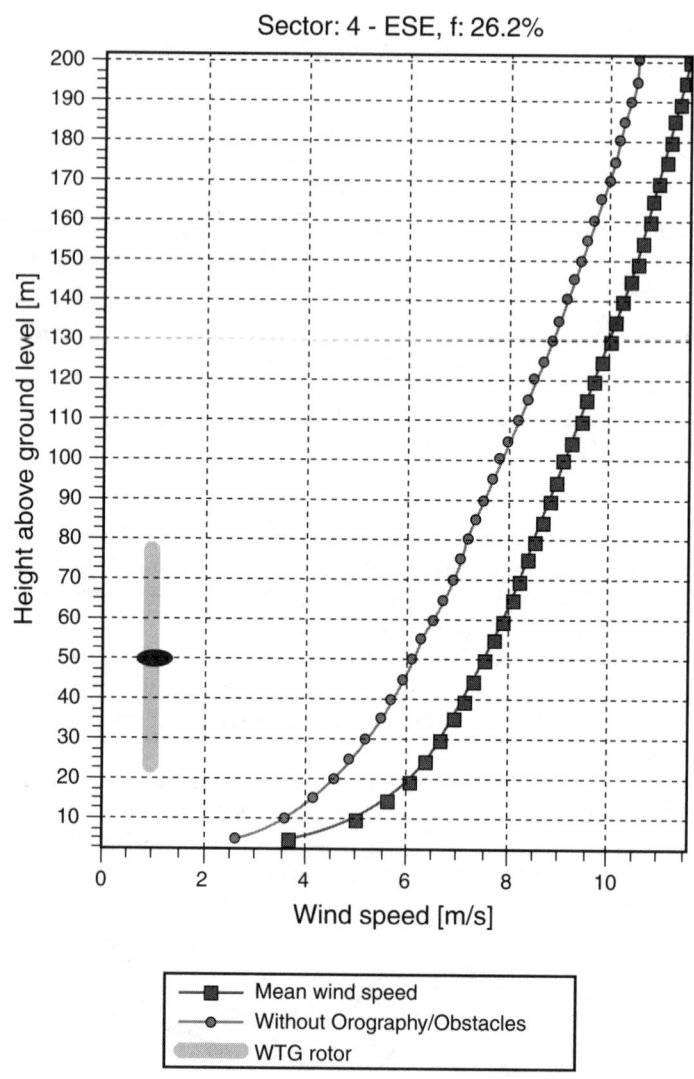

Figure 7-4 Graph for a specific direction sector with a plot of wind speed as a function of height with and without orography and obstacles.

Phases of Resource Assessment

In the rest of this chapter, the phases of resource assessment are described. Figure 7-5 describes the process. Preliminary assessment step comprises three steps. The duration of this step is 1–3 months. Step 2 is the onsite wind measurement task, which was covered in Chapter 6.

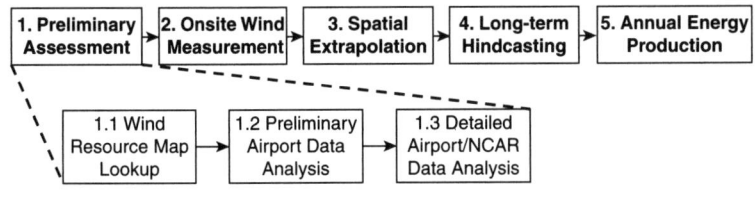

FIGURE 7-5   Phases of resource assessment.

It is the longest duration task that may take 1–3 years. Steps 3–5 are performed in sequence and the total duration of three steps is about 1–3 months.

Preliminary Wind Resource Assessment

Most projects start with no onsite measurement of wind data. A preliminary wind resource assessment is performed during the prospecting phase. It is a multistage process with the following steps:

1. Wind resource map lookup
2. Preliminary analysis of data from neighboring airports and other met-towers
3. Detailed analysis of wind data from neighboring airports and other met-towers

Wind Resource Map Lookup

In the prospecting phase of a wind project, a developer starts with a high-level color wind resource map like the one shown in Fig. 7-6. Most areas of the world have a wind resource map; a list of sources for maps is given below:

1. Coarse world wind resource maps may be found at http://na.unep.net/swera_ims/map/
2. US wind resource maps and a few international maps may be found at http://nrel.gov/wind/resource_assessment.html
3. European maps and international wind atlas may be found at: http://www.windatlas.dk.
4. A variety of online USA-centric GIS-based applications are also available that provide wind speed estimates. Two such applications are:
   a. 3Tier's http://FirstLook.3Tier.com
   b. AWS TrueWind's http://Navigator.awstruewind.com

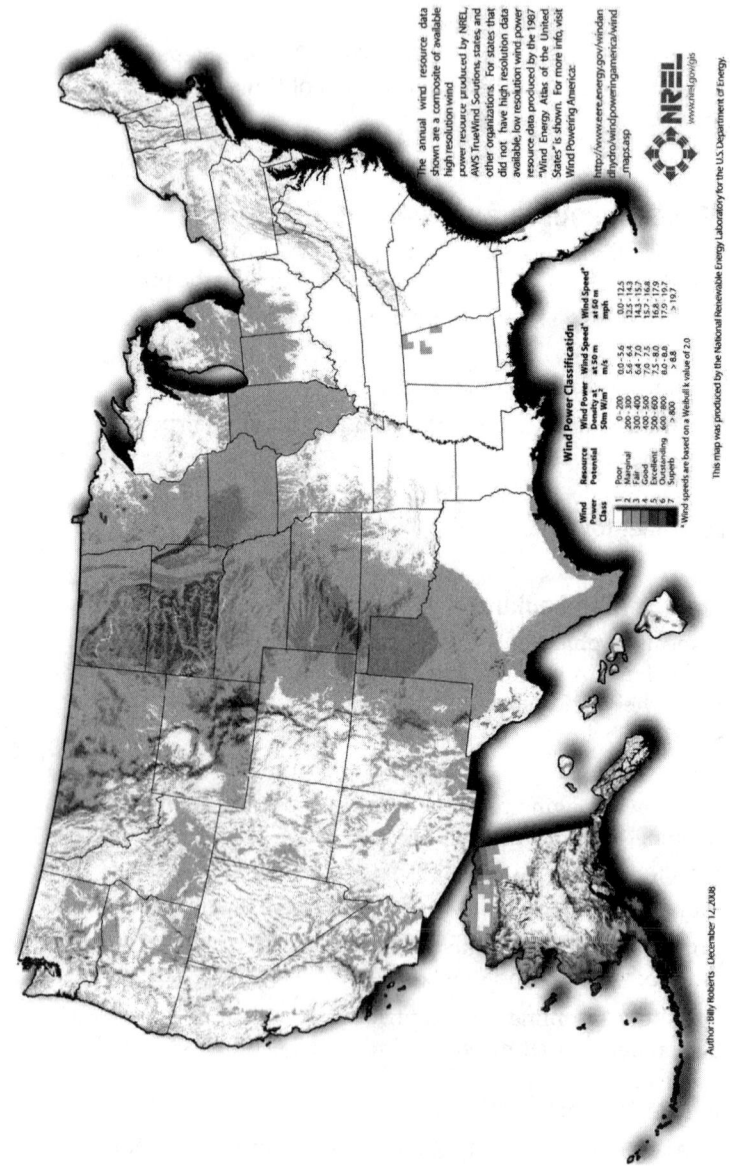

The annual wind resource data shown are a composite of available high resolution wind power resource produced by NREL, AWS TrueWind Soultions, states, and other organizations. For states that did not have high resolution data available, low resolution wind power resource data produced by the 1987 "Wind Energy Atlas of the United States" is shown. For more info, visit Wind Powering America:

http://www.eere.energy.gov/windandhydro/windpoweringamerica/wind_maps.asp

NREL
www.nrel.gov/gis

Wind Power Classification

Wind Power Class	Resource Potential	Wind Power Density at 50m W/m²	Wind Speed* at 50 m m/s	Wind Speed* at 50 m mph
1	Poor	0-200	0.0-5.6	0.0-12.5
2	Marginal	200-300	5.6-6.4	12.5-14.3
3	Fair	300-400	6.4-7.0	14.3-15.7
4	Good	400-500	7.0-7.5	15.7-16.8
5	Excellent	500-600	7.5-8.0	16.8-17.9
6	Outstanding	600-800	8.0-8.8	17.9-19.7
7	Superb	>800	>8.8	>19.7

* Wind speeds are based on a Weibull k value of 2.0

Author: Billy Roberts   December 12, 2008

This map was produced by the National Renewable Energy Laboratory for the U.S. Department of Energy.

Figure 7-6 Wind Resource Map of the United States. Billy Roberts, December 12, 2008. (This map was produced by the National Renewable Energy Laboratory for the US Department of Energy, and http://www.eere.energy.gov/windandhydro/windpoweringamerica/wind_maps.asp)

These sites are expanding rapidly to provide wind data for a large number of countries. The wind data is available for multiple grid sizes and heights.

These maps provide annual average wind speed categories or wind speed ranges over large areas. Most online maps are based on 2.5 km × 2.5 km grid while others are based on a coarser grid. These maps normally display wind speed at 50-m height. As expected, these maps display large-scale patterns with no temporal variation. The maps do not provide information about wind direction and shear. Therefore, the maps should be used as a tool to identify regions with good wind resources, and not for serious viability of a wind project.

Preliminary Analysis of Data from Neighboring Airports and Other Met-Towers

If the wind resource maps show promise, then the next task is to find neighboring airports within 100-km radius that have topography that is similar to the site under consideration. The simplest form of preliminary airport data analysis is available in RetScreen (www.RetScreen.net), a free MS-Excel based application. It contains a database of monthly average wind speed data from major airports at an elevation of 10 m above the ground level. Information about wind direction and shear is not available from this data source. The user has to input wind shear and input wind speed at the proposed site based on the airport data. That is, the user estimates wind speed at the proposed site based on a subjective assessment of how different the wind speed is at the proposed site compared to the airport. The user can select a turbine or create a turbine and its power curve. With this information, RetScreen computes average annual energy production. It also has a financial model that take as input capital costs, recurring costs, recurring revenue, debt/equity ratio, interest rates, and other parameters to compute payback period and other financial metrics. It produces pro forma income statement and cash flow statement. Although RetScreen has significant limitations, it is a good starting point for prospecting for developers that do not have access to more sophisticated tools.

Detailed Analysis of Wind Data from Neighboring Airports and Other Met-Towers

The WAsP methodology provides a more sophisticated level of prospecting analysis with a model for translating wind speed data from one location to another. It also takes as input elevation and roughness to estimate shear. Applications like WindPRO, WindFarmer, or

WAsP may be used to perform this analysis. The output of the analysis is wind speed time series data at the desired site and at the desired height, wind rose (relationship between wind speed and wind direction derived from wind data), approximation of shear, (based on surface roughness), and estimate of average annual energy production. This preliminary assessment involves:

1. Analysis of hourly wind speed data available at 10 m height from local airport
2. Creation of a GIS-based model of the airport site (site A) with contour elevation and terrain roughness
3. Computation of a generalized RWC from the airport data
4. GIS-based model of site B with modeling of elevation contour, roughness, and obstacles
5. Translation of the RWC to the desired wind project site (site B)
6. Customization of RWC to site B.

If no airport data is available, then reanalysis data from NCAR may be used in place of the airport data in the preliminary assessment steps described above. Even if airport data is available, NCAR data may be used to compare wind energy estimates from the airport data source.

Publicly available met-tower data, within 100 km of site must not be overlooked. Although it is rare, examples include met-tower data available from publicly funded wind data collection initiatives in Iowa, Illinois, Nebraska, and others. Any data source from lower-height met-towers (with anemometer less than 10 m in height) like those on highways, bridges, and buoys should be avoided because lower elevation wind data in greatly influenced by local surface conditions and therefore uncertainty is very high when used to predict wind speed at hub height.

Onsite Wind Measurement

This was covered in Chapter 6.

Spatial Extrapolation of Wind Resources from Measured Locations to Planned Wind Turbine Locations

For a single turbine installation, the wind measurement location (site A) is usually the same as the turbine location (site B). If the two locations are not the same, then this step is used. For a wind farm, there may be 10 or more turbine locations for each met-tower location. This step is, therefore, used to estimate wind speed at locations where measurements were not performed.

The models described in "Resource Estimation Models" are used to extrapolate wind resources from measured locations to planned wind turbine locations. WAsP is the most popular model that is used for this purpose. Micrositing CFD models may be used when assumptions pertaining to WAsP do not apply.

Hindcasting/MCP of Measured Data

Wind measurement for few years provides a window into the wind conditions at a site. However, this is only a short snapshot of the wind conditions. It is not the wind condition over the life of a wind project (typically, 20 years). Wind speed like any other weather parameter has cycles. Few years of measurement must, therefore, be placed in context of longer-term wind pattern. For instance, measurement may be in a low-wind speed period and, if this measurement were used to compute wind energy production over a period of 20 years, then the prediction would underestimate the energy production.

Hindcasting is a process of generating wind conditions that predict 20 or more years of wind data. It uses measurement data and long-term reference wind data. The process of hindcasting involves a technique called measure-correlate-predict (MCP), which has the following steps:

1. *Inputs:* Onsite wind speed measurement for a year or more, and long-term reference data for 20 or more years. The long-term reference data from neighboring airport or NCAR reanalysis data. Use as many long-term datasets as available.

2. Correlate measurement data with long-term reference datasets for concurrent time period. If correlation is acceptable, then choose these reference datasets for the next step.

3. Estimate wind speed for the historical period, which covers the duration of reference time series. This is the hindcasting step.

4. Convert the hindcast into a forecast, if necessary, or use the hindcast for energy computations.

In most practical applications, the correlate and predict steps are performed iteratively with several sources of long-term reference data. Experience has shown that long-term reference data is a key determinant to reduced uncertainty in prediction. Siddabathini and Sorensen[7] show that uncertainty can be reduced by proper choice of consistent data within the long-term reference data; in the example presented

Location	Coordinates	Heights, m	Time Resolution, h	Duration	Type
Valentine, NE	42.954N, 100.504W	10, 25, 40	1	3/21/1995 to 4/1/1999	Onsite met-tower
NCAR	42.500N, 100.00W	10, 42	6	4/1975 to 4/1999	Long-term reference

TABLE 7-3 Characteristics of Measured Data and Long-Term Reference Data for the MCP Example

in this paper, the authors show a significant reduction in uncertainty when 14 years of reference data is used versus 44 years of reference data. This is counterintuitive. However, the authors illustrate that there was a change in reference data about 14 years ago, probably because of change in instruments. This underscores the need for not only picking the right long-term reference data but also picking a time period that has consistent data.

The following example is used to illustrate the hindcasting/MCP process. Four years of measurement data is available from Valentine, Nebraska. This measurement wind data is from a tower that was managed by Global Energy Concepts for the Nebraska Power Association.[8] Raw data is available for download at http://hprcc1. unl.edu/cgi-bin/winddr.cgi. For long-term reference, NCAR reanalysis data for a duration of 25 years is used. Details of the datasets are in Table 7-3. In this example, the prediction period for hindcasting will be 3/21/1975 to 3/31/1995. The two time series are plotted in Fig. 7-7. Hindcasting will generate a time series that will extend the measured time series to the past periods.

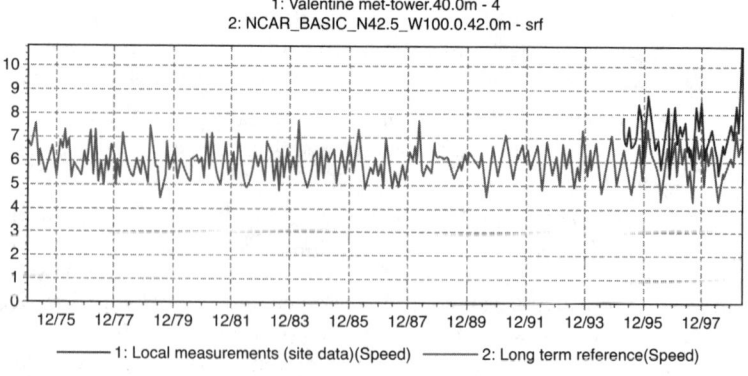

1: Valentine met-tower.40.0m - 4
2: NCAR_BASIC_N42.5_W100.0.42.0m - srf

——— 1: Local measurements (site data)(Speed)   ——— 2: Long term reference(Speed)

FIGURE 7-7 Plot of local measurement and long-term reference monthly average wind speed data. Concurrent period is from 4/1995 to 4/1999. Prediction period for hindcasting is from 3/1975 to 3/1995.

Correlate

The purpose of correlation is to understand if the measurement data and the long-term reference data are similar over the measurement (concurrent) period. If the two data series are similar, then the wind regimes are similar and, therefore, a valid hindcast can be generated. The meaning of correlation and its use is described next.

Consider two data series: M_i is the measured data series and L_i is the long-term reference data series. If

$$M_i = aL_i + b \text{ then correlation is } + 1$$

$$M_i = -aL_i + b \text{ then correlation is } - 1$$

$$M_i, L_i \text{ are unrelated random series, then correlation is } 0$$

where i is an index that goes from 1 to N, and N is the number of points in the data series that correspond to concurrent time periods. Correlation is 0 when two data series are independent.

In general, correlation index is defined as:

$$\rho = cov(M, L)/(\sigma_M.\sigma_L) = E[(M - \mu_M).(L - \mu_L)]/(\sigma_M.\sigma_L)$$

$$= \sum_{i=1}^{N} (M_i - \mu_M)(L_i - \mu_L)/(N.\sigma_M.\sigma_L) \tag{7-1}$$

where ρ is the correlation between M and L time series, $cov()$ is the covariance function, $E[]$ is the expected value function, σ_M, σ_L are standard deviation, and μ_M, μ_L are mean of M and L.

Often the time series M_i, L_j do not have the same measurement interval. For example, a typical interval for measurement data is 10 min whereas interval for airport reference data is 1 h and interval for reanalysis NCAR data is 6 h. If a 10-min interval and 1-h interval data are correlated, then there are two options for synchronizing the time series:

Compute hourly average of the 10-min interval data and align with L_j, or pick data points in M_i that have the shortest time difference with data points in L_j.

The choice of method depends on how the long-term data is collected and recorded. If L_j contains average wind speed data, then the first method is appropriate. However, in most cases, information about method of long-term data collection and recording is not available. In such cases, both methods may be tried and the method that yields the higher correlation should be chosen.

Correlation values of 0.9 or above are considered excellent correlation. To provide context to value of correlation, consider correlation between Valentine met-tower data from anemometers at 40 and 25 m, and 45- and 10-m heights. Correlation values are in Table 7-4.

Valentine Data Series	Aggregation	Correlation
40 m and 25 m	None	0.978
	Daily	0.984
	Weekly	0.984
	Monthly	0.983
40 m and 10 m	None	0.941
	Daily	0.966
	Weekly	0.963
	Monthly	0.973

TABLE 7-4 Correlation of Wind Speed Data Measured at Different Heights and with Different Methods of Aggregation

Aggregation of data (daily, weekly, and monthly) removes noise and the correlation is higher for more aggregate data. Also, as expected, the correlation between 40 and 10-m measurements are lower because the 10-m measurement is influenced by ground level roughness.

Correlations between measured and reference time series are computed in the following manner.

- The two time series are synchronized to the time step of the series with the larger time interval
- The two time series are averaged if daily or weekly correlations are desired; otherwise, there is no averaging for raw correlations
- The two time series are split into 16 (or 12) direction sectors
- Data pairs are filtered out if there are errors or significant disagreements between wind directions. One criterion for filtering data may be: Disable data pairs if the absolute value of difference in direction is more than 99°.
- Compute the correlations for each of the 16 (or 12) sectors.
- Compute the weighted mean of sector-wise correlations. The weights are proportional to the number of points in each sector.

This process is repeated for all available long-term wind-speed time series. Long-term time series with acceptable correlations are chosen for the subsequent step of prediction. The guidelines in Table 7-5 may be used to determine acceptable correlations. Although there are no hard and fast rules, this table provides guidelines for correlation factors. If the correlation does not meet the criteria mentioned, then the prediction may not be meaningful. As mentioned earlier, care must be exercised in choosing the appropriate level of aggregation. For instance, correlation between reanalysis data and 10-min measurement data may not be meaningful because of the large difference between

Time Series 1	Time Series 2	Correlation of Raw Data	Correlation of Daily Average Data
10-min met-tower data (site A)	10-min met-tower data (site B)	>0.65	>0.75
10-min met-tower data	Hourly airport data	>0.6	>0.75
10-min met-tower data	6-h reanalysis data	>0.55	>0.75

TABLE 7-5 Guideline for Determining if Two Wind Speed Time Series Share the Same Wind Climate

measurement intervals, 10 min versus 6 h. Instead, correlation of daily averages of the two time series may be an appropriate measure of predictability.

The correlation results for the Valentine example using WindPRO are in Table 7-6. Correlations are computed for 12 sectors. Second column contains the number of data points in each sector. Third and fourth column contain statistics of measured wind speed. Fifth and sixth column contain statistics of wind speed from NCAR data. Seventh and eighth column contain statistics of wind speed ratio (onsite/long-term reference). Ninth and tenth column contain statistics of difference between wind direction between measured and reference data. Final column contains correlations of the raw data. Since onsite measurement data is hourly and NCAR is every 6 h, raw data points with the shortest time difference are chosen. That is, every sixth point of Valentine data is matched with NCAR data; for instance, on 3/21/1995 1:00 AM Valentine data is paired with NCAR data for the same hour, Valentine data from 2:00 AM to 6:00 AM is ignored, 7:00 AM Valentine data is paired with NCAR data for the same hour, and so on. Correlations for different aggregates are presented in Table 7-7. As expected, the aggregate correlations are much higher than the raw correlation. Monthly average wind speed data is plotted in Fig. 7-8.

Other than correlation, other simple tests can determine if the long-term reference data is suitable for prediction:

- *Comparison of diurnal and monthly pattern.* If the measured data and reference data do not have similar diurnal and monthly patterns, then the reference dataset may not be a good choice
- *Comparison of wind rose.* Sectors with the highest wind speed and highest frequency may not be the same between measured and reference data, but should be within +/− 1 sector. If there are larger differences, and the differences cannot be explained by examining the contour and roughness, then the reference data may not be a good choice. (See Fig. 7-9a to c.)

132

| Sector | Count | Site Wind Speed | | Ref Wind Speed | | Ratio (site/ref) | | Wind Veer (site-ref) | | Correlation |
		Mean	StdDev	Mean	StdDev	Mean	StdDev	Mean	StdDev	
All	5888	7.06	3.5	5.90	3.2	1.209	0.353	12.6	32.2	0.6829
N	514	6.55	3.3	6.29	3.1	1.093	0.347	3.7	32.7	0.5375
NNE	286	5.33	3.0	5.52	2.9	1.020	0.337	8.3	36.1	0.5574
ENE	277	4.78	2.4	4.73	2.5	1.015	0.319	12.0	39.1	0.4558
E	322	5.12	2.5	4.73	2.5	1.062	0.299	16.0	40.3	0.5480
ESE	520	6.24	2.7	5.44	2.5	1.168	0.341	17.9	33.6	0.5546
SSE	773	7.34	3.1	6.00	2.7	1.235	0.284	17.6	28.8	0.6978
S	684	7.29	3.0	5.91	2.8	1.217	0.301	15.6	31.3	0.7276
SSW	363	6.31	2.5	4.63	2.2	1.250	0.361	15.7	32.7	0.4354
WSW	333	6.66	2.8	4.54	2.0	1.414	0.356	15.8	32.8	0.6382
W	418	7.49	3.5	5.31	2.9	1.358	0.386	12.1	32.7	0.7158
WNW	655	8.80	4.2	7.02	3.9	1.279	0.399	9.5	27.7	0.7615
NNW	743	8.58	4.4	7.53	4.0	1.195	0.346	7.8	27.4	0.7342

*Computed in WindPRO.

TABLE 7-6 Summary of Wind Speed Statistics, Wind Direction Statistics, and Correlation for the Valentine Example*

Wind Speed Aggregate	Correlation
Daily average	0.782
Weekly average	0.795
Monthly average	0.884

TABLE **7-7** Correlation between Valentine and NCAR Wind Speed Data for Average Over Different Time Periods

Thus far, only two long-term reference data sources have been considered: Airport and reanalysis. For analysis of large wind farm in complex terrain, virtual mast data may be used as long-term reference data. Virtual mast data is generated from mesoscale models, CFD simulation, or other similar models.

Predict

The objective of this step is to use the short-term onsite wind measurement data to predict long-term wind data at proposed WTG sites. As mentioned before, predict here means estimating the past. This prediction is important to obtain a lower uncertainty estimate for the average annual energy production over the 20-plus year life of a wind turbine. Three most commonly used methods for prediction will be described: Regression, Weibull Parameter Scaling, and Matrix Method. In addition to the prediction methods, a statistical method is used to create the long-term time series.

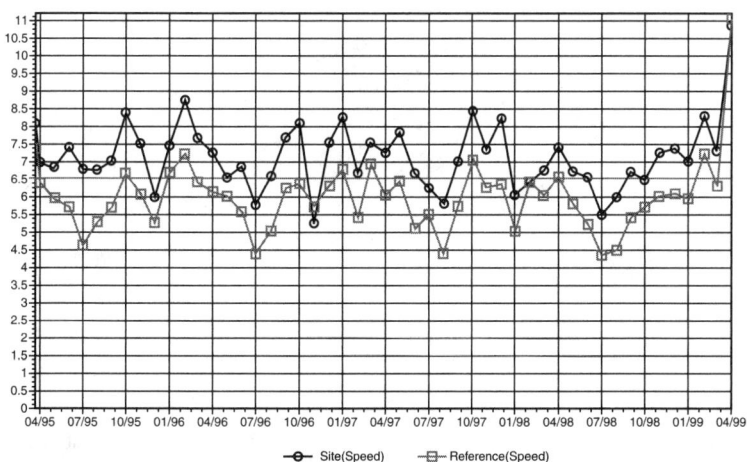

FIGURE **7-8** Plot of concurrent measured and reference monthly average wind speed data. Computed in WindPRO.

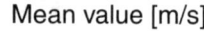

Mean value [m/s]

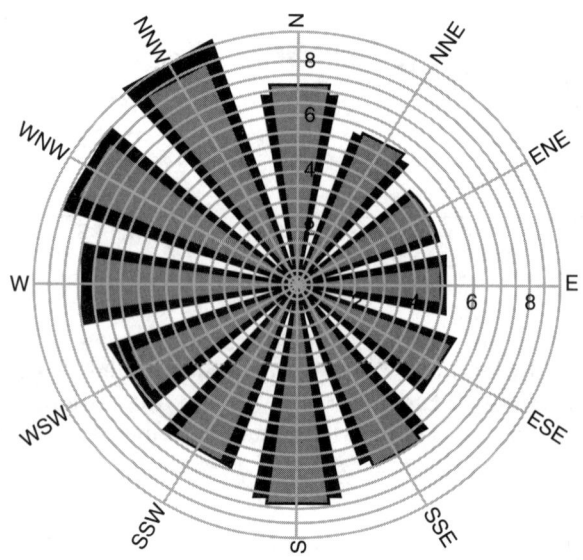

—— Concurrent site data   —— MCP result (long-term site data)

(a)

Energy [W/m2]

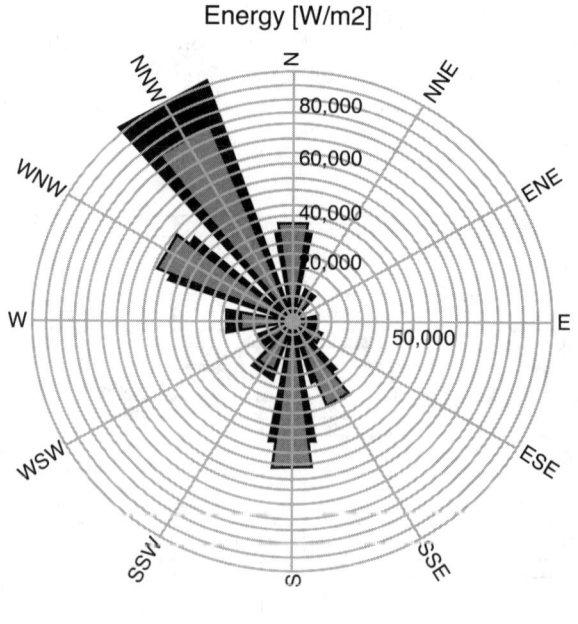

—— Concurrent site data   —— MCP result (long-term site data)

(b)

Figure 7-9 a–c Comparison of Valentine and NCAR data in each sector. Computed in WindPRO.

Frequency [%]

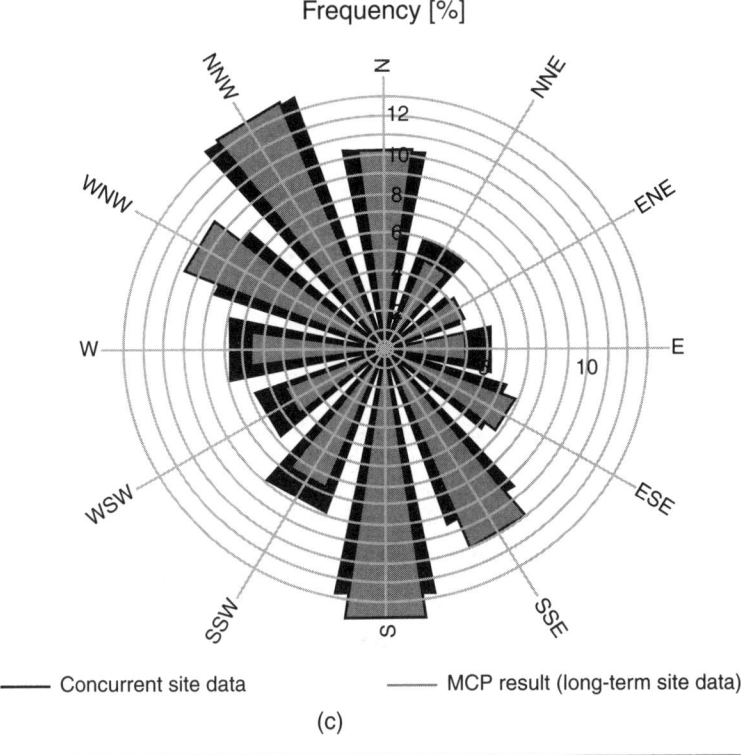

—— Concurrent site data   —— MCP result (long-term site data)

(c)

FIGURE 7-9 *(Continued)*

Generic Prediction Method

The purpose of prediction is to compute a transfer function, $f(x)$:

$$Y = f(X) + \varepsilon \qquad (7\text{-}2)$$

where Y is the predicted time series for onsite wind data, X is the reference time series (long-term wind data), and ε is the residual. Prediction is a two-step process:

- *Computation of transfer function, $f()$.* The function $f()$ is determined based on a forecasting method that is applied to 1 (or more) year of onsite measurement data and concurrent long-term reference data.

- *Creation of the predicted time series using $f()$ and ε.* The function $f()$ is applied to long-term reference data and the residual term is added to the result. The residual term ε is important because ignoring it can cause the energy production to be underestimated by up to 10%.[9] The reason will be explained later in the section.

In the Valentine example, X is the 25 years of NCAR wind data at 42 m elevation and Y is the 25 years of wind speed prediction (hindcasting) at Valentine at 40 m elevation. Note, in this example, the resolution of prediction is 6 h.

If hourly airport is used as the long-term reference data instead of NCAR data, and 10-min measurement data is available, then a process similar to the example may be deployed, except the resolution of prediction will be 1 h.

The statistical distribution of the residual term is computed using measurement data by:

$$\varepsilon_i = YM_i - f(X_i) \qquad (7\text{-}3)$$

where YM_i is the onsite measurement data and $f(X_i)$ is the predicted data for the measurement time period i. ε_i should be a random time series, that is, the distribution should be Gaussian with zero mean. The quality of the prediction method is reflected in the distribution of ε. When prediction methods are applied to each of the 12 or 16 direction sectors (of the reference time series), then each sector's ε should be independent of wind speed. For example, it should not be the case that at lower wind speeds ε has a positive mean and at higher wind speeds ε has a negative mean.

The final prediction time series is created using Eq. (7-2). Note not only is the transfer function f applied to X, but the residual is added. Residual is computed using a randomization method, like Monte Carlo simulation. This is called the long-term corrected wind distribution; it is an artificial time series that has been randomized by adding the residual.[9] This addition is not in vain, because energy is a cubic function of wind speed; a positive residual contributes much more to increasing energy compared to the contribution of negative residual in lowering energy.

In the following sections, four prediction methods are described.

Regression

The most common regression models used for wind speed estimation are linear and quadratic. Within the two models there are two variations: Regression through $(0, 0)$ and regression with no constraints (see Table 7-8).

The regression parameters a_1, a_2, a_3 are computed using least squares algorithm. Although regression through $(0, 0)$ is intuitive, it is not a good representation at higher wind speed, which is where energy production happens. An unconstrained regression method is, therefore, used. As stated above, in addition to the regression parameters, the distribution of the residual is computed and added to the prediction.

Regression Method	Equation
Linear regression	$Y = a_1 X + a_2$
Linear regression through (0,0)	$Y = a_1 X$
Quadratic regression	$Y = a_1 X^2 + a_2 X + a_3$
Quadratic regression through (0,0)	$Y = a_1 X^2 + a_2 X$

Source: WindPRO.

TABLE 7-8   Regression Methods Commonly Used

Wind direction is normally predicted using linear regression of the form:

$$Y = X + a_2 \qquad (7\text{-}4)$$

The regression methods are applied sector-by-sector. Figure 7-10a and b graphically illustrate the use of linear regression in two sectors. The straight line is the prediction. Error is plotted in Fig. 7-11a and b.

Weibull Parameter Scaling

In this method, the Weibull parameters A and k are computed for wind speed in each of the 12 or 16 wind direction sectors for both the onsite measured data and reference data. In this simple method, the following is used to scale the parameters:[9]

$$\lambda_{site}^{long} = \lambda_{site}^{short} \cdot \left(\lambda_{ref}^{long} / \lambda_{ref}^{short} \right) \qquad (7\text{-}5)$$

where λ_{site}^{long} is predicted parameters A or k for onsite wind data. The quantity in parenthesis is the correction factor (see Table 7-9).

The frequency of each sector also needs to be computed by normalizing the sum to 100%.[9]

$$f_{sitei}^{long} = \left[\frac{f_{sitei}^{short}}{f_{refi}^{short}} \right] f_{refi}^{long} \Big/ \sum_{i=1}^{N} \left[\frac{f_{sitei}^{short}}{f_{refi}^{short}} \right] f_{refi}^{long} \qquad (7\text{-}6)$$

where f_{sitei}^{long} is the prediction of onsite frequency in sector i, N is the number of sectors.

This method is appropriate only when:

- Both time series are primarily Weibull distributions, and the scaling of the two parameters is not very large.

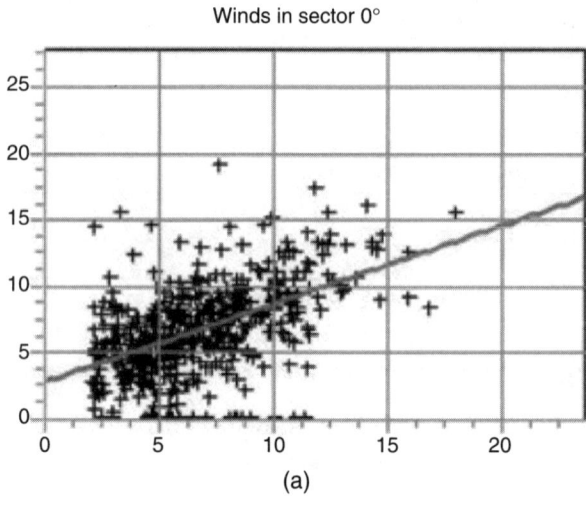

(a)

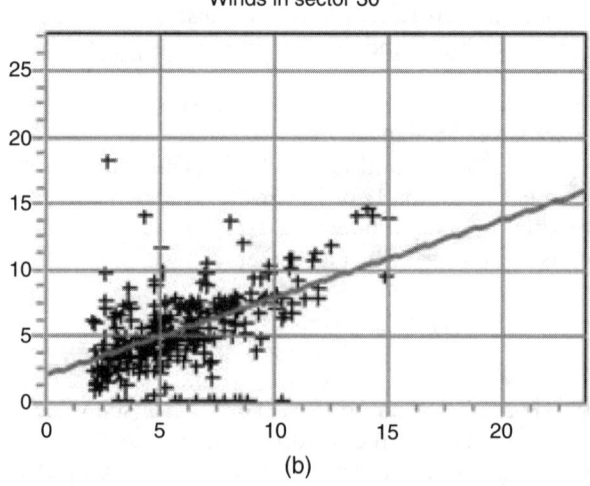

(b)

FIGURE 7-10 a and b Linear regression in two sectors to predict wind speed. x-axis is the measured and y-axis is the reference. Plus represents pairs of measured and reference.

- The sector-wise frequency of the measured is very similar to the long-term reference.

Table 7-9 shows the calculations of this method.

Matrix Method

As the name suggests, this method is applied to a matrix of wind speed ranges and direction sectors from the reference time series. An

Wind Residuals (Sector 0°)

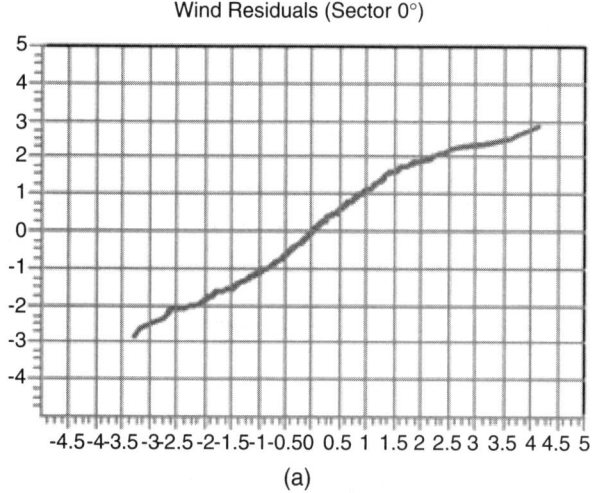

(a)

Wind Residuals (Sector 30°)

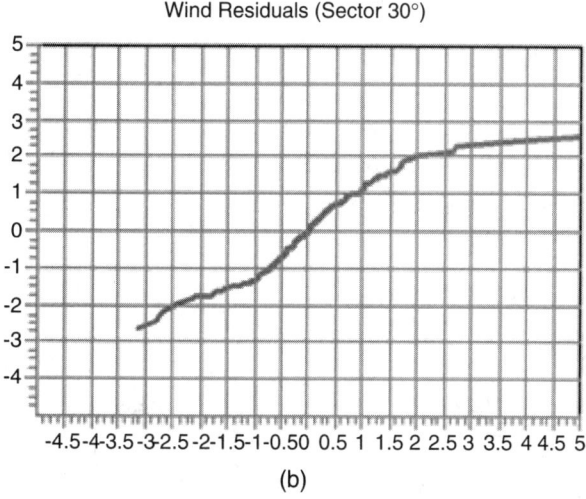

(b)

Figure 7-11a and b Error in wind speed prediction in two sectors. A straight line indicates that the residuals are normally distributed.

example of the matrix is in Table 7-10. In each cell, the following values from the two concurrent time series are computed: Mean and standard deviation of wind speed-up, which is $v_{site} - v_{ref} = \Delta v$ and wind veer, which is $\theta_{site} - \theta_{ref} = \Delta\theta$. In the table, mean and standard deviation are two quantities in the parenthesis. See Figs. 7-12a and b, and 7-13a and b for a graphical illustration of the binning and computation of the delta quantities—wind speed-up and wind veer.

Reference full: NCAR_BASIC_N42.5_W100.0
Height: 42.00

sector	0-N	1-NNE	2-ENE	3-E	4-ESE	5-SSE	6-S	7-SSW	8-WSW	9-W	10-WNW	11-NNW	Mean
A-parameter [m/s]	7.23	6.15	5.32	5.46	6.30	7.33	7.01	5.29	4.98	5.86	7.50	8.13	6.72
Mean wind speed [m/s]	6.40	5.45	4.72	4.84	5.58	6.51	6.22	4.69	4.41	5.20	6.65	7.20	5.96
k-parameter	2.036	2.003	1.934	1.998	2.111	2.543	2.467	2.369	2.131	1.902	1.954	2.071	2.025
frequency [%]	8.80	5.42	4.68	5.05	7.60	14.06	11.86	6.19	5.47	6.81	11.35	12.71	100.00

Reference Concurrent: NCAR_BASIC_N42.5_W100.0
Height: 42.00

sector	0-N	1-NNE	2-ENE	3-E	4-ESE	5-SSE	6-S	7-SSW	8-WSW	9-W	10-WNW	11-NNW	Mean
A-parameter [m/s]	7.15	6.21	5.41	5.41	6.08	6.87	6.80	5.25	5.06	5.99	7.81	8.55	6.69
Mean wind speed [m/s]	6.33	5.51	4.79	4.79	5.39	6.09	6.03	4.65	4.48	5.32	6.94	7.58	5.93
k-parameter	2.115	1.976	2.006	2.005	2.149	2.520	2.411	2.176	2.225	1.866	1.829	1.973	1.949
frequency [%]	8.73	4.86	4.70	5.47	8.83	13.13	11.62	6.17	5.66	7.10	11.12	12.62	100.00

Local Site Concurrent: Valentine met tower
Height: 40.00

sector	0-N	1-NNE	2-ENE	3-E	4-ESE	5-SSE	6-S	7-SSW	8-WSW	9-W	10-WNW	11-NNW	Mean
A-parameter [m/s]	7.69	6.45	5.58	5.75	6.48	7.65	8.73	7.87	7.55	8.40	9.57	10.33	8.11
Mean wind speed [m/s]	6.82	5.72	4.95	5.13	5.75	6.79	7.78	7.02	6.78	7.52	8.49	9.15	7.19
k-parameter	1.941	1.897	2.393	1.671	2.567	2.649	2.852	2.992	3.411	3.136	2.424	2.229	2.231
frequency [%]	10.34	6.03	3.77	5.23	6.27	9.51	12.77	8.83	6.84	7.80	9.02	13.59	100.00

Correction Factors

sector	0-N	1-NNE	2-ENE	3-E	4-ESE	5-SSE	6-S	7-SSW	8-WSW	9-W	10-WNW	11-NNW
A-parameter	1.011	0.990	0.985	1.009	1.037	1.06E	1.031	1.007	0.985	0.979	0.960	0.951
k-parameter	0.962	1.014	0.964	0.997	0.982	1.009	1.023	1.089	0.958	1.020	1.069	1.050
frequency	1.008	1.117	0.994	0.923	0.861	1.071	1.021	1.004	0.967	0.960	1.020	1.007

Local MCP-Corrected Data: Valentine met tower
Height: 40.00

sector	0-N	1-NNE	2-ENE	3-E	4-ESE	5-SSE	6-S	7-SSW	8-WSW	9-W	10-WNW	11-NNW	Mean
A-parameter [m/s]	7.78	6.39	5.50	5.80	6.71	8.17	9.00	7.93	7.44	8.23	9.19	9.83	8.12
Mean wind speed [m/s]	6.91	5.66	4.87	5.18	5.96	7.26	8.03	7.10	6.67	7.37	8.16	8.71	7.19
k-parameter	1.868	1.923	2.307	1.666	2.522	2.673	2.918	3.258	3.267	3.197	2.591	2.340	2.313
frequency [%]	10.40	6.72	3.74	4.82	5.38	10.16	13.01	8.85	6.61	7.47	9.18	13.66	100.00

*Correction factor is ratio of reference full and reference concurrent. Correction factor is then multiplied to local site concurrent to obtain prediction.

TABLE 7-9 Computations for the Weibull Parameter Scaling Method*

Sectors	Wind Speed Bins (reference)					
	0–1 m/s	1–2 m/s	...	7–8 m/s	...	24–25 m/s
N	$\Delta v = (0.5,0.4)$ $\Delta\theta = (14,4)$	(0.55,0.45) (16,5)		(0.92,0.5) (22,5)		No data
NNE	$\Delta v = (0.3,0.1)$ $\Delta\theta = (15,3)$	(0.45,0.35) (16,5)		(0.92,0.5) (22,5)		No data
...						
WNW	$\Delta v = (0.3,0.1)$ $\Delta\theta = (13,3)$	(0.35,0.25) (19,5)		No data		No data

*Wind speed bins and wind direction sector bins form the matrix. Each cell of the matrix contains mean and standard deviation of wind speed-up and wind veer.

TABLE 7-10   Illustration of the Matrix Method for Prediction*

In order to have a full set of transfer functions such that any wind speed and direction combination in the long-term reference dataset can be converted into onsite predicted dataset, regression functions are defined. This will provide values for cells in the matrix with "No data" or cells with a small dataset from which meaningful statistics cannot be computed. It also smoothes the changes between bins. A pictorial representation of this is in Fig. 7-14a and b.

Two regressions are performed for each direction sector. The first regression function computes the wind speed-up as a function of wind speed.

$$\Delta v_j^{long} = f_j^v \left(v_{ref,j}^{long} \right) + \varepsilon_j^v \tag{7-7}$$

where $j = 1$ to 12 or 16 is the direction sector index, and ε_j^v is the residual term.

The second regression function computed the wind veer as a function of wind speed.

$$\Delta \theta_j^{long} = f_j^\theta \left(v_{ref,j}^{long} \right) + \varepsilon_j^\theta \tag{7-8}$$

These regression functions become transfer functions that use long-term reference data to compute onsite long-term prediction of wind speed and wind direction. It is assumed that the relationships hold for all periods. The next step is to add the residual term to the predicted wind speed time series. A simple method would be to assume independence between the distribution of wind speed-up and wind veer. Under this assumption, the standard deviation for wind speed-up and wind veer in each cell is used to create the corrected time series for wind speed and wind direction. A more sophisticated method is to

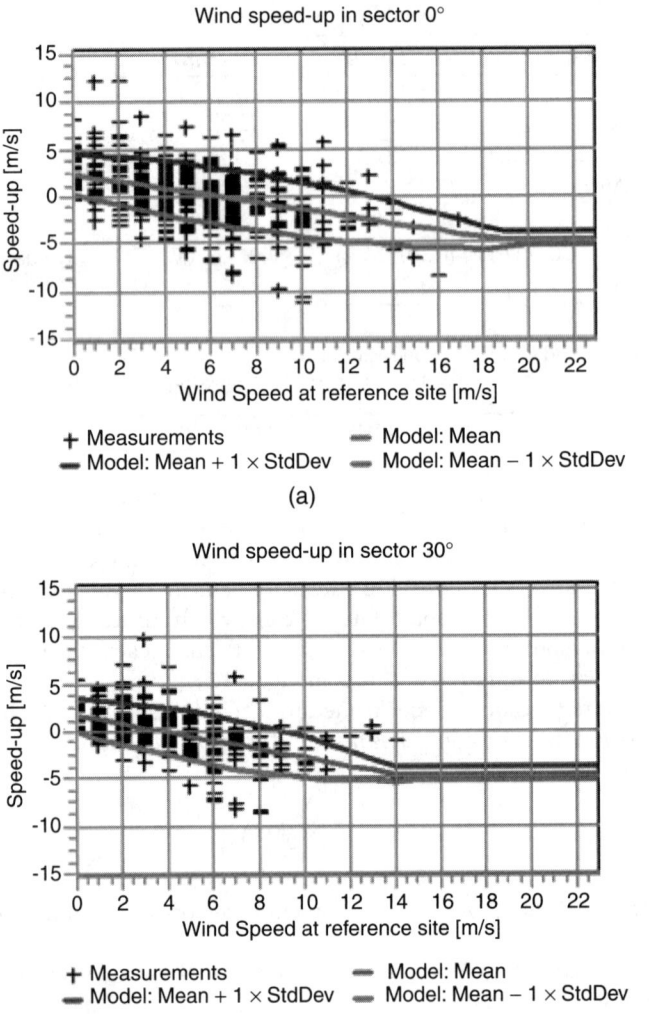

Figure 7-12 a and b Matrix method computes mean and standard deviation of wind speed-up in two sectors.

assume that the two distributions are dependent, and to use a joint distribution. For details, see WindPRO[9].

Choosing a Method for Prediction
Several methods for prediction are described above. In order to find the best predictor method, the following is estimated by examining the distribution of the residual term for the concurrent measured and

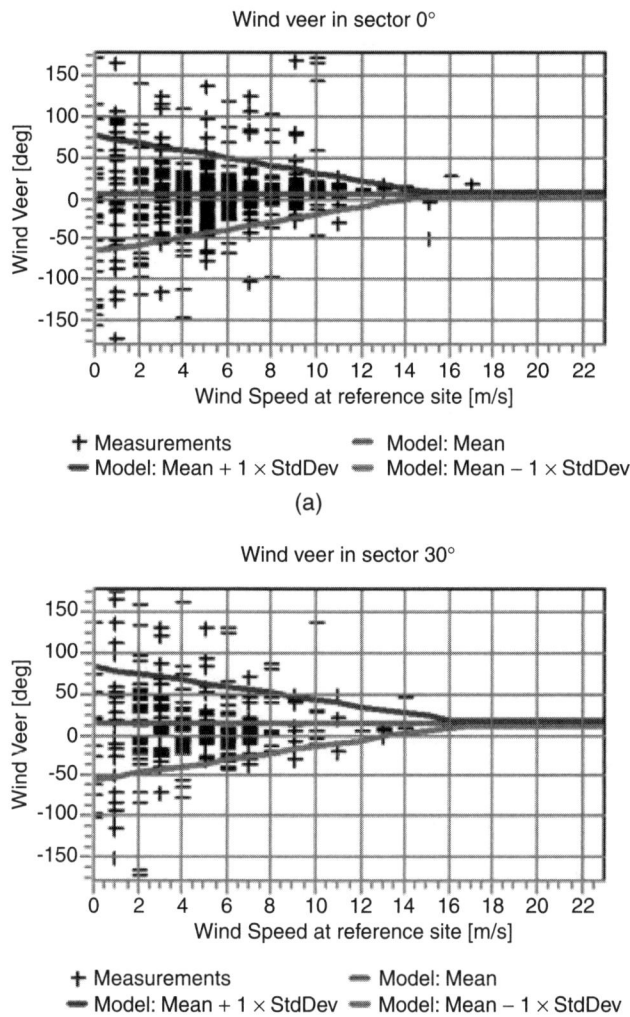

Figure 7-13a and b Matrix method computes mean and standard deviation of wind veer.

corresponding predicted time series.

$$\varepsilon_i = YM_i - f(X_i) \qquad (7\text{-}9)$$

where i is the index of the measured time intervals, ε_i is the wind speed residual, YM_i is the measured time series, X_i is the concurrent long-term time series, and $f(X_i)$ is the predicted time series. The distribution of ε should be a Gaussian distribution with mean zero

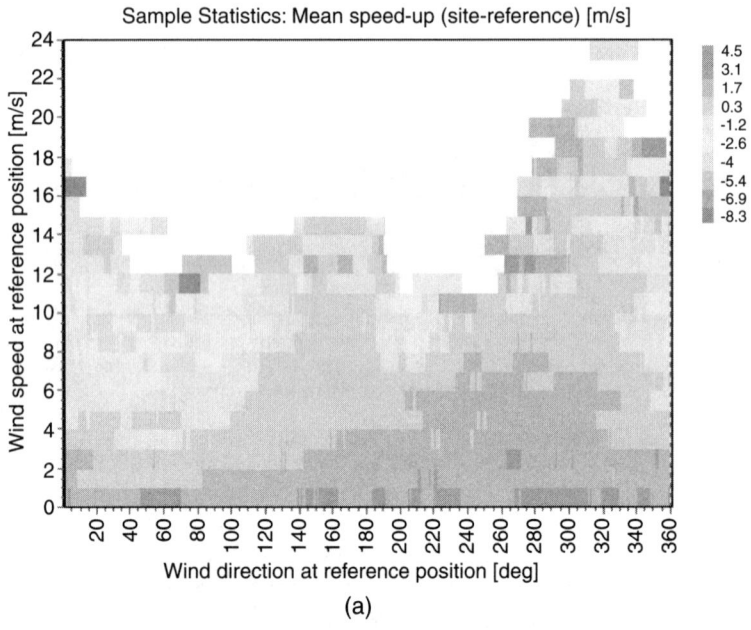

(a)

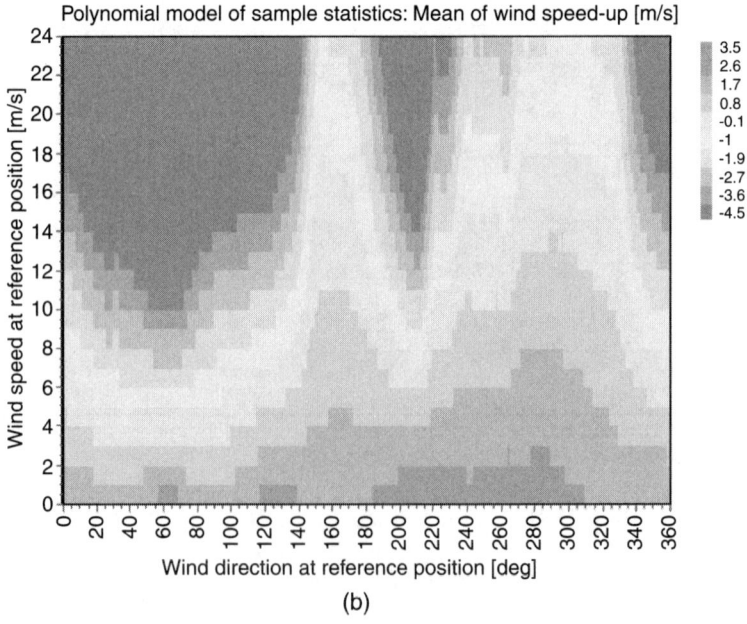

(b)

FIGURE 7-14   The shades in the plots are wind speed-up in a cell of the matrix. A cell is 1° direction by 1 m/s wind speed. Plot of concurrent data series is in (a) with some empty cells. Linear regression is used to fill up the empty cells and smooth the variation, as shown in (b).

and standard deviation of σ. The standard deviation also called the standard error is an estimate of the uncertainty of the prediction. Another estimate of uncertainty is the correlation between YM and $f(X)$ time series. Both estimates of uncertainty are provided by tools at the end of prediction using any of the methods listed above. A prediction method with lowest σ and the highest correlation is a candidate for prediction. Care must be exercised to ensure that a candidate method is not chosen just because it has the lowest standard deviation. For example, if the correlation post prediction (between $f(X)$ and X) is lower than correlation pre-prediction (YM and X) for the measurement period, then the prediction method may not be valid.

To improve prediction, the following changes to the prediction method must be evaluated for better predictability:

- Filtering out of wind speed data below 3 m/s. This data is not important and it can skew the linear regression.

- If the long-term reference data is hourly and measured data is 10-min, then using hourly average of measured data versus point estimates may yield a higher correlation.

After the transfer functions have been determined, predicted time series is computed by adding the residual term in Eq. (6.3) using the Monte Carlo method. The predicted time series is then used to compute annual energy production, as described below.

Annual Energy Computations

There are two common metrics used to compute energy production. The first is power density

$$\text{Power density} = \sum_{i=1}^{N} \frac{1}{2} \rho_i v_i^3 \qquad (7\text{-}10)$$

where ρ_i, v_i are air density and predicted wind speed in time period i; each time period is of length Δt hours, and N is the number of data points in the predicted long-term wind speed time series. Unit of measure of power density is: kilowatt per square meter.

$$\text{Annual energy production (AEP) of specific turbine} = \frac{\sum_{i=1}^{N} PC(v_i)}{N/8760}$$
$$(7\text{-}11)$$

where $PC(v_t)$ is the power curve of the chosen turbine, (for example see Fig. 2-8).

References

1. Kalnay, E., et al. "The NCEP/NCAR 40-year reanalysis project," *Bulletin of the American Meteorological Society*, 77: 437–471, 1996.
2. Reanalysis, North American Regional. North American Regional Reanalysis Homepage. *North American Regional Reanalysis* [Online] 2007. http://wwwt.emc.ncep.noaa.gov/mmb/rreanl/.
3. Gibson, R., Kållberg, P., and Uppala, S. "The ECMWF re-analysis (ERA) project," *ECSN Newsletter*, 5: 11–21, 1997.
4. Landberg, L., Myllerup, L., Rathmann, O., Petersen, E. L., Jørgensen, B. H., Badger, J., and Mortensen, N. G. "Wind resource estimation—an overview," *Wind Energy*, 6: 261–271, 2003.
5. Mortensen, N. G., et al. *Getting Started with WAsP 9*. Riso National Laboratory, Roskilde, Denmark, 2007. RISO-I-2571.
6. Redlinger, R. Y., Andersen, P. D., and Morthorst, P. E. *Wind Energy in the 21st Century*, Palgrave, New York, 2002.
7. Siddabathini, P., and Sorensen, T. *Uncertainty of Long Term Correction of Wind Speed: The Importance of Consistency of Data*, EWEC, Marseille, 2009.
8. Global Energy Concepts, Inc. Nebraska Wind Energy Site Data Study, Final Report. [Online] May 1999. www.nlc.state.ne.us/epubs/E5700/B047-1999.pdf.
9. Thogersen, M. L., Motta, M., Sorensen, T., and Nielsen, P. *Measure-Correlate-Predict Methods: Case Studies and Software Implementation*, European Wind Energy Conference, Milan 2007. http://emd.dk/WindPRO/WindPRO%20Documentation,%20Energy%20Calculations.

CHAPTER 8

Advanced Wind Resource Assessment

*As far as the laws of Mathematics refer to reality, they are not certain,
and as far as they are certain, they do not refer to reality.*
—Albert Einstein

Introduction

This chapter will cover advanced topics in resource assessment. The first topic is extreme wind speed (EWS). EWS is an estimate of the wind speed that is exceeded once every 50 years. As a part of resource assessment, EWS is used to determine the Class of wind turbine and used by the turbine manufacturer to determine if the turbine has the capability to withstand loads associated with EWS. The second topic is RIX, an enhancement to the WAsP model for rugged terrain. The third topic is wake of turbines and the associated losses in energy production. Alternate methods for modeling wake are presented. The next section presents the definition of International Electrotechnical Commission (IEC) Classes of wind turbines.

Investors in a wind project are keen to understand if all the losses have been accounted for and the uncertainty of a project has been modeled in a resource assessment. The final sections deal with losses, uncertainty, and the role of uncertainty in a bankable wind resource assessment.

Extreme Wind Speed (EWS)

Chapter 7 focused on wind assessments from the point of view of energy production. Other aspects of a wind assessment that are of interest to turbine manufacturers are the dynamic loads on the tower, blades, and other components of the turbine caused by extreme wind conditions. In addition to distribution of wind speed and turbulence, EWS is an important parameter. The estimated EWS is the maximum wind speed that is likely to occur in 50 years. There are several intervals of interest: 3-s, 10-s, and 10-min. For instance, v_{50y}^{3s} is the 3-s average estimated EWS that is exceeded once every 50 years; v_{50y}^{10m} is the 10-min average estimated EWS that is exceeded once every 50 years. These quantities are estimated statistically because, in most cases, 50-year time series of 3-s wind speeds are not available and, furthermore, any single 50-year time series may not be representative. Gumbel distribution is commonly used to model extreme values of time series data. It is a two-parameter (a, b) distribution of the form:[1]

$$F(v) = e^{-e^{-(v-b)/a}}$$

(8-1)

This is the annual cumulative probability that wind speed v is exceeded. Mean and standard deviation of the distribution are:

$$\mu = b + \gamma a$$

(8-2)

$$\sigma = \frac{\pi a}{\sqrt{6}}$$

(8-3)

where γ is the Euler's constant and is equal to 0.5772. The process of finding the parameters of the distribution are:

1. Choose a wind speed time series that spans 10 years or more. This is recommended to obtain a higher degree of confidence in the estimate of EWS. Smaller time series increase uncertainty in the estimate of EWS. The granularity of the time series will also determine the granularity of the EWS. For example, if the time series is hourly average wind speed, then EWS will be the extreme hourly average wind speed in, say, 50 years.

2. Specify a threshold for sampling extreme events. For example, a threshold of 15 m/s may be used for identifying extreme wind events.

3. Identify a collection of extreme points in a time series with wind speeds above the threshold value. The collection must contain at least 20 points for higher degree of confidence in

Date	EWS	Prob (EWS)	ln(-ln(Prob))
1/15/1997	20.2	0.05	1.11
1/4/1997	20.3	0.10	0.86
12/30/1998	20.4	0.14	0.67
3/26/1999	20.5	0.19	0.51
7/1/1997	20.6	0.24	0.36
11/22/1998	20.7	0.29	0.23
10/19/1995	20.8	0.33	0.09
12/17/1996	20.8	0.38	−0.04
4/18/1995	20.9	0.43	−0.17
10/11/1997	21.4	0.48	−0.30
4/6/1997	21.7	0.52	−0.44
1/18/1996	22	0.57	−0.58
12/30/1997	22	0.62	−0.73
12/8/1995	22.2	0.67	−0.90
3/24/1996	23.2	0.71	−1.09
10/27/1995	23.5	0.76	−1.30
11/10/1998	24.1	0.81	−1.55
4/25/1996	24.7	0.86	−1.87
10/29/1996	25	0.90	−2.30
2/10/1996	28	0.95	−3.02

TABLE 8-1   Listing of the Extreme Events for Example with 4-Year Time Series

estimate of EWS. Smaller number of points increases uncertainty in the estimate of EWS.

Use Eqs. (8-2) and (8-3) to solve for values of a and b. An alternative is to assign probability to each event and create a Gumbel plot, as described in example below and illustrated in Table 8-1 and Fig. 8-1. A linear regression method may be used in the Gumbel plot to estimate values of a and b by fitting a straight line to the extreme points.

Compute 50-year and other n-year extreme values by assuming that Eq. (8-1) is the annual extreme value distribution, and a n-year event occurs with an annual probability of $1/n$. Therefore, the probability of not exceeding the EWS in n years is:

$$\text{prob (EWS, } n) = \left(1 - \frac{1}{n}\right) \tag{8-4}$$

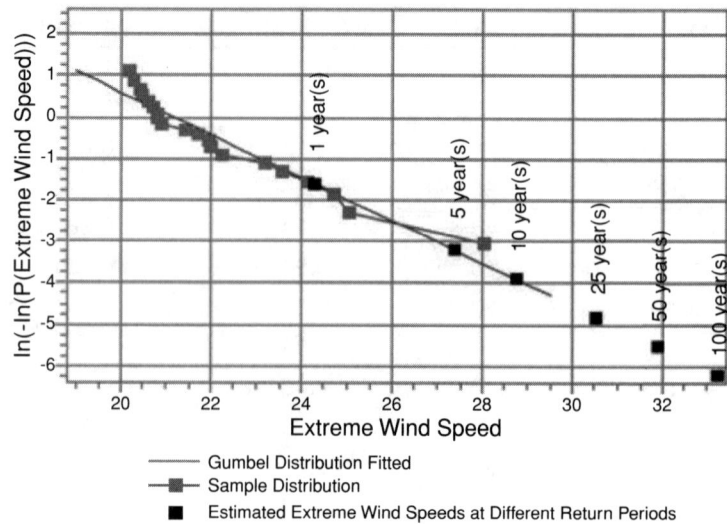

Figure 8-1 Plot of the extreme wind speed with a linear regression fit to compute parameters *a*, *b* of the Gumbel distribution. (Created in WindPRO.)

If the number of extreme points in step 3 suggest multiple extreme events per year (*epy*), then

$$\text{prob (EWS, } n) = \left(1 - \frac{1}{n.epy}\right) \tag{8-5}$$

Inverting Eq. (8-1) yields:

$$\text{EWS} = b - a\ln(-\ln(\text{prob (EWS, } n))) \tag{8-6}$$

As an example, consider the 4-year time series of hourly wind speed data from Valentine, Nebraska. Twenty events with wind speed of 20 m/s or higher are identified in Table 8-1; these are the extreme wind speed events. The wind speeds are sorted and probabilities of occurrence not exceeding the EWS from the minimum to maximum are computed as multiples of $1/(20 + 1)$. The values of $\ln(-\ln(\text{probability(EWS)}))$ is computed and plotted (see Fig. 8-1). In this example, the extreme wind speed sample has the following properties:

- Threshold wind speed is 20 m/s

$$\text{Sample size} = 20 \text{ over a 4-year period, and } epy$$
$$= 5 \text{ events per year}$$

- Mean of sample = 22.26 m/s, standard deviation = 2.4949 m/s

$$a = 1.945, \quad b = 21.137 \, \text{m/s}$$

Return Period Year(s)	Prob (EWS, n)	ln(-ln (prob))	EWS m/s	StdDev/ Mean %	EWS + 1 × StdDev m/s	EWS + 2 × StdDev m/s
1	0.8	−1.50	24.3	4.2	25.3	26.3
5	0.96	−3.20	27.4	6.7	29.2	31.1
10	0.98	−3.90	28.7	7.7	30.9	33.2
25	0.992	−4.82	30.5	8.9	33.2	35.9
50	0.996	−5.52	31.9	9.7	34.9	38.0
100	0.998	−6.21	33.2	10.4	36.7	40.1

Source: Created in WindPRO.

TABLE 8-2 Extreme Wind Speed Values for Various Spans

The 1-, 5-, 10-, 25-, 50-, and 100-year EWS are marked in Fig. 8-1 and computed in Table 8-2. As an example, 50-year EWS is computed using Eqs. (8-5) and (8-6):

$$\text{prob (EWS, } n) = \left(1 - \frac{1}{50*5}\right) = 0.996$$

$$\ln(-\ln(\text{prob (EWS, } n))) = \ln(-\ln(0.996)) = -5.52$$

$$v_{50y}^{60m} = \text{EWS} = b - a\ln(-\ln(\text{prob (EWS, } n))) = 21.137 + 1.945*5.52$$
$$= 31.9 \, \text{m/s}$$

Therefore, 31.9 m/s is the 60-min mean extreme wind speed that will not be exceeded in 50 years, with a probability of 99.6%. The standard deviation of EWS, EWS + Std_dev, and EWS + 2Std_dev are computed in Table 8-2.

WAsP Model in Rugged Terrain

A linear model like WAsP encounters poor prediction in rugged terrain. Examples of poor prediction are:

- Met-tower is in a relatively flat area, but the planned turbine locations are in a rugged area, and vice-versa
- The ruggedness at the met-tower location in the 12 directional sectors is different compared to the ruggedness of turbine locations.

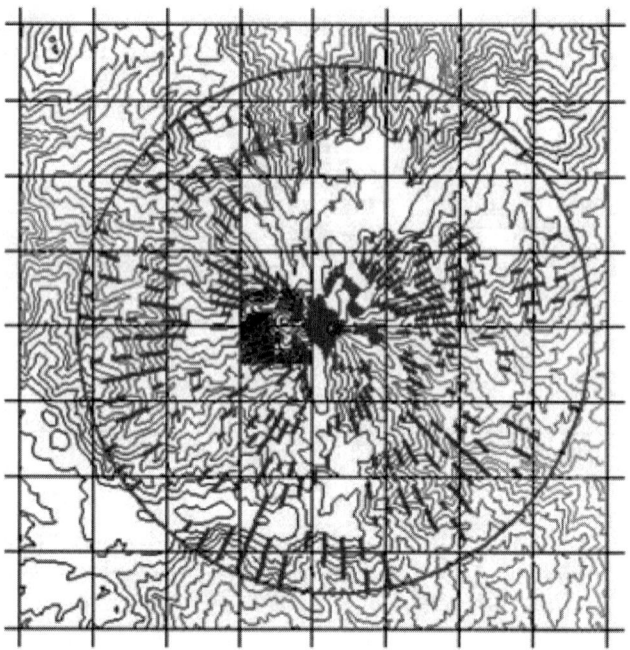

Figure 8-2 Illustration of a site with elevation contours, circle of radius r, 72 radii emanating from the center and dark radial segments indicating segments with slope greater than a threshold. (From Mortensen, N. G., Bowen, A. J., and Antoniou, I. *Improving WAsP Predictions in (too) Complex Terrain*, EWEC, 2006.)

Ruggedness is a measure of the change in elevation or slope of terrain. Quantitatively, ruggedness index (RIX) is defined as: Fraction of terrain surface that is steeper than a critical slope θ_c. There are three parameters to computing RIX:

- Calculation radius around a site center, normally a radius of 3.5 km is used
- Critical slope, normally a value of $\theta_c = 0.3$ is used
- Number of radii, normally 72 radii are used

Figure 8-2 illustrates the process of computing RIX; 72 radii are drawn from the center and for each radius, segments are identified (darker segments in the figure) with slopes above the critical slope. RIX is calculated as the ratio of the sum of segment lengths to sum of the radii. If 72 radii are used, then:

$$\text{RIX} = \frac{\sum_{i=1}^{72} s_i}{72r} \tag{8-7}$$

where s_i is the length of each segment with slope above the critical slope.

The difference between RIX at the turbine location and RIX at the met-tower location is called the delta RIX:

$$\Delta RIX = RIX_{WTG} - RIX_{met}$$

In terms of RIX, spatial extrapolation using WAsP is recommended only when the RIX between measured and predicted sites is close to zero or, at most, 5%.[2] For higher values of RIX, Mortensen et al.[2] report a log-linear relationships between ratio of wind speed and delta RIX based on empirical data for a site in Portugal with a high coefficient of determination, R^2. Furthermore, it is reported that the relationship is insensitive to radius, critical slope (0.3 – 0.45), and sector-specific ΔRIX.

$$\ln\left(\frac{v_p}{v_m}\right) = 1.508\ \Delta RIX \qquad (8\text{-}8)$$

$$\ln\left(\frac{AEP_p}{AEP_m}\right) = 5.175\ \Delta RIX \qquad (8\text{-}9)$$

where v_p, v_m are predicted and measured wind speed, and AEP_p, AEP_m are predicted and measured average annual energy production. The above equations provide a method for spatial extrapolation for cases with difference in ruggedness.

The universality of the equations and the constants is uncertain and site-specific validation is recommended. Dependence of the constants on the directional sectors must also be validated.

Wake of Turbines

Consider a simple wind farm with a single wind direction, and turbines arranged in a grid as shown in Fig. 8-3. The second row of

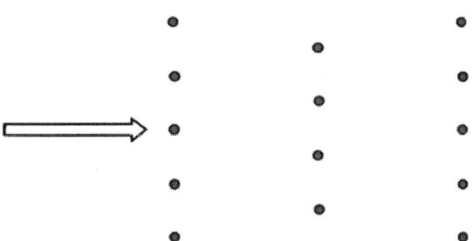

FIGURE 8-3 Illustration of a wind farm with three rows of turbines. The arrow indicates the direction of wind. The distance of $9D$ and $3D$ along the direction of wind and perpendicular to the direction of wind is commonly used.

turbines is in the wake of the first row of turbines. The third row of turbines is in the wake of the first and second row of turbines. Wake impacts turbines in two primary ways: Lower wind speed and increase in turbulence. Both the affects result in reduced energy production, while an increase in turbulence causes greater structural loading of the turbines. The reduction in energy in a wind farm because of wake can be in the range of 2 to 20% depending on the distances and ambient turbulence. The impact of wind speed deficit and turbulence are essentially eliminated at a distance of $20D$ in the wake of the rotor, where D is the diameter of the rotor. However, for efficient use of property, guidelines that are more practical have been developed. In wind farm design, a guideline of $9D$ distance along the primary direction of wind and $3D$ distance perpendicular to the primary direction of wind is used in the industry to locate turbines.

There are two prominent models for computing the deficit in wind speed: Linear model by N.O. Jensen, and eddy viscosity model by Ainslie.

N.O. Jensen Model for Wake

According to the actuator disk theory in Chapter 2, the thrust on a turbine is given by Eq. (2-31). Restating the equation in terms of the thrust coefficient C_T yields:

$$F = 2\rho A_r v_0^2 a(1-a) = \frac{1}{2}\rho A_r v_0^2 C_T \qquad (8\text{-}10)$$

where

$$C_T = 4a(1-a) \qquad (8\text{-}11)$$

From Eq. (2-30)

$$v_2 = (1-2a)v_0 = \sqrt{1-C_T}\,v_0 \qquad (8\text{-}12)$$

where v_0 is the free stream horizontal wind speed, v_2 is the downstream wake wind speed, and a is the axial induction factor.

$$\text{Deficit in speed} = 1 - \frac{v_2}{v_0} = 1 - \sqrt{1-C_T} \qquad (8\text{-}13)$$

Assuming a linearly expanding wake with slope of k, the deficit as a function of x becomes:

$$1 - v_x/v_0 = (1 - \sqrt{1-C_T}) . /(d + 2kx)^2 \qquad (8\text{-}14)$$

where d is the rotor diameter, k is the slope or wake decay constant, and x is the distance from the rotor. Onshore value of $k = 0.075$ and for offshore value of $k = 0.04$ are commonly used.[3]

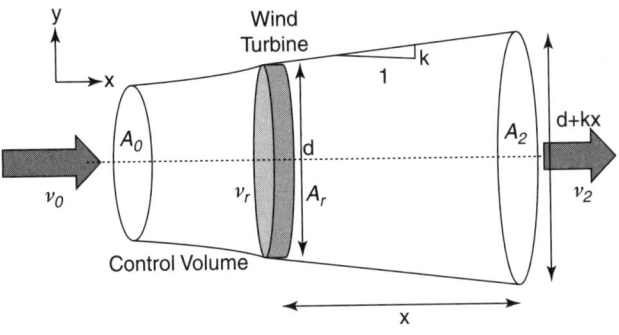

FIGURE 8-4 Illustration of the assumptions in N.O. Jensen's wake model.

Figure 8-4 contains a pictorial representation of the Jensen wake model's wind speed in the wake of a turbine.

Ainslie's Eddy Viscosity Model

The Jensen model assumes that there is a clear demarcation between the wake and the normal wind speed throughout the wake. The Ainslie eddy viscosity model is a more sophisticated model. Turbulence in the wake has two components: Shear-generated turbulence and tip vortices shed by the blades. The tip vortices are high frequency and decay quickly. The shear-generated turbulence is created by the substantial difference in wind speed at the outer edge of the rotor and the free-stream fluid flow just outside—simplistically, the axial wind speed reduces significantly in the volume behind the rotor, but just outside this volume, the wind speed is normal. The energy in the turbulence is dissipated as heat.

Ainslie created an axis-symmetric formulation of the time-averaged Navier-Stokes equation in cylindrical coordinates with eddy viscosity closure to model the wake. It is, in part, a theoretical and, in part, an empirical model. The theoretical details may be found in the WindPRO Users manual.[1]

Combining Wind Speed Deficits from Multiple Turbines

In most wind farms, there are multiple rows of turbines. The wind speed deficit at turbine is impacted by the upstream wakes. The most popular model is to compute the combined effect through square root of the sum of squares of the deficit.

$$\delta v_n = \sqrt{\sum_{k=1}^{n-1} \delta v_{kn}^2} \qquad (8\text{-}15)$$

where $\delta v_n = 1 - v_n/v_0$ is the wind speed deficit at the nth turbine that has $n - 1$ upstream turbines, and δv_{kn} is the wind speed deficit because of turbine k on turbine n.

Turbulence Modeling

In addition to wind speed deficit, there is an increase in turbulence because of wake. Modeling of turbulence is primarily done through empirical models as theoretical models are less developed. Turbulence intensity in wind resource assessments is defined in terms of 10-min wind speed data as:

$$I_0 = \sigma(h)/v_{10}(h) \tag{8-16}$$

where I_0 is the ambient turbulence intensity at height h, $\sigma(h)$ is the standard deviation, and $v_{10}(h)$ is the average of 10-min wind speed at height h. If average and standard deviation are not measured at hub height, then the following may be used to extrapolate turbulence to hub height:[1]

$$I_0(h_2) = v_{10}(h_1)I_0(h_1)\left(\frac{h_2}{h_1}\right)^{-\gamma} \tag{8-17}$$

where $I_0(h_2)$ is the ambient turbulence intensity at hub height; h_1, h_2 are measurement and hub heights and γ is the shear.

The total turbulence intensity is computed using:

$$I_T^2 = I_0^2 + I_+^2 \tag{8-18}$$

where I_T and I_+ are total turbulence intensities and additional turbulence intensity because of wake.

There are several empirical turbulence models to compute I_+. The model suggested by IEC 61400-1, edition 3 is:[4,5]

$$I_+^2 = 0.9/(1.5 + 0.3 d \sqrt{v_{hub}})^2 \tag{8-19}$$

where d is the distance between two turbines normalized by rotor diameter. Additional details are available in IEC 61400-1.[5]

Optimal Layout of Turbines in Wind Farm

The objective is to obtain a layout that produces the highest annual energy while taking into account wake losses, and a variety of layout related constraints. The prerequisites for an optimal layout of turbines in a wind farm are:

- Wind resource map at a detailed level of granularity, a 50 m x 50 m granularity is recommended for wind farm layout. This resource map specifies the mean wind speed in each 50 m x 50 m cell.

- Turbine details like rated power, tower height, rotor diameter, and others

- Wind farm details like desired size of wind farm in terms of megawatt and type of layout (regular grid versus random).

- Constraints of the wind farm. There will be several site-specific constraints, a few examples are presented below:

  - Boundary of the wind farm

  - Outline of roads, transmission lines, microwave paths, water bodies, residential/office/industrial/recreational areas, and other features that require a setback

  - Setback distances from each marked feature. In some areas, local wind ordinances specify the setback distances from different features. For instance, setback from property boundary may be $1.25H$; setback from roads, transmission lines, and water bodies may be $1.5H$; inhabited areas may be 500 m, where H is the total height = tower height + blade radius. The setback conditions are discussed in more detail in Chapter 11.

  - An elliptical exclusion zone around a turbine. The ellipse is defined using distances between turbines perpendicular to and parallel to the predominant direction of wind. To speed up the layout optimization, these constraints are specified so that the algorithm treats the elliptical region around the center of a turbine as an exclusion zone for placing other turbines. As mentioned in the wake section, $9D$ and $3D$ are guidelines for the major and minor radius of the ellipse.

  - Exclusion areas because of environmental constraints like noise, shadow flicker, visual impact, wetlands, wildlife, and others.

  - Exclusion areas because of flight safety and interference with long-range radars

Wind resource assessment applications, like WindPRO and WindFarmer, have implemented tools to specify the constraints and compute an optimal layout using discrete optimization methods. A general optimization problem of this type will have multiple local maxima and the challenge is to find the global maximum or an acceptable local maxima. An optimization of large 100+ MW wind farm may take a few hours to run. The results include a layout of turbines, and average annual energy production and wake losses for each turbine. Wake loss of 10 to 12% is acceptable; turbines with higher wake losses are progressively removed from the layout and the layout is then reoptimized. For instance, consider a first pass wind farm optimization of a 120-MW wind farm with 60 turbines (each turbine is 2 MW). Suppose this layout has 10 turbines with wake losses of more than 10%. In order to reduce the wake losses, the above layout will be re-optimized by removing the five lowest energy-producing turbines and setting a constraint of 110 MW. If required, additional reoptimizations may be performed in order to achieve acceptable wake loss.

As with most real-life modeling problems, at the start of the optimal layout process, not all constraints are available. Even for available constraints, not all information is available with specificity. The optimal layout is, therefore, executed dozens of times. Optimality of energy production is important because even small increases (about 1%) in energy output may lead to additional annual revenue per turbine of $50,000 to $100,000.[6]

Wind Turbine Class Selection

An integral part of wind resources assessment is to determine the wind turbine class (Table 8-3) that is suitable for the site. This step involves verifying that the actual site conditions are less severe than the design conditions for the class of turbine. IEC 61400-1[5] provides the design conditions. For convenience, the design conditions are split into two broad categories:

- Wind conditions: Extreme wind speed, turbulence, and wake effects
- Other conditions like terrain, soil conditions, and seismic

Wind turbine class must be chosen such that the following conditions are satisfied:[4,5]

WTG Class		I	II	III
	V_{ref} (m/s)	50	42.5	37.5
A	I_{ref}		0.16	
B	I_{ref}		0.14	
C	I_{ref}		0.12	

Source: From *Wind Turbine, Part 1 Design Requirements.* International Electrotechnical Commission, Geneva, 2005. IEC 61400-1 Edition 3.

TABLE 8-3 Wind Turbine Class Definition

1. Onsite 50-year extreme wind speed computed from 10-min averages is less than the WTG's V_{ref}.

2. The probability density function of onsite 10-min average wind speed at hub height is less than the probability density of Rayleigh distribution with $V_{ave} = 0.2V_{ref}$ for all wind speed in the range of 0.2 to 0.4 times WTG's V_{ref}. For most Weibull distributions with the value of k close to 2, this may be simplified to: The 10-min average wind speed at hub height is less than $0.2V_{ref}$.

3. Onsite effective turbulence intensity (including wake effect) is less than WTG's I_{ref}. The onsite turbulence intensity that is of interest is for wind speed in the range of 0.6 times rated wind speed to cut-out wind speed of the turbine.

4. Onsite average wind shear at hub height must be positive and less than 0.2.

5. Onsite flow angle with respect to the horizontal plane must be $\pm 8°$

In complex terrain, the longitudinal TI must be multiplied by a correction factor[5] C_{ct}:

$$C_{ct} = \begin{cases} \dfrac{1}{1.375}\sqrt{1 + \dfrac{\sigma_2^2 + \sigma_3^2}{\sigma_1^2}} \\ 1.15 \text{ if } \sigma_i \text{ is not available} \end{cases} \qquad (8\text{-}20)$$

where $\sigma_1, \sigma_2, \sigma_3$ are longitudinal, vertical, and transversal standard deviation of wind speed. Complex terrain is defined as one that does not meet the conditions defined in Table 8-4.

Distance Range from Wind Turbine	Max Slope of Fitted Plane	Maximum Terrain Variation from a Disc with Radius 1.3 z_{hub} Fitted to the Terrain
<5 z_{hub}	<10°	<0.3 z_{hub}
<10 z_{hub}		<0.6 z_{hub}
<20 z_{hub}		<1.2 z_{hub}

*z_{hub} is the hub height. A terrain that does not meet these conditions is a complex terrain

Source: From *Wind Turbine, Part 1 Design Requirements*. International Electrotechnical Commission, Geneva, 2005. IEC 61400-1 Edition 3.

TABLE 8-4 Definition of Simple Terrain with Respect to a Fitted Plane that Passes Through the Turbine Base*

The choice of turbine classes can get complex. In cases that involve parameters that straddle two classes or cases that involve insufficient data to compute the parameters, turbine manufacturers should be consulted to determine the WTG Class.

Estimation of Losses

There are two key concepts that have not been discussed up to this point: Losses and uncertainty. These two mutually exclusive concepts, however, are often mentioned together, which causes confusion. A wind resource assessment is not bankable without a rigorous analysis of losses and uncertainty.

Losses are estimates of decrease in energy output that is known. As an example, consider energy loss because of transmission of electrical energy from generator to grid. Suppose an electrical engineer estimates the electrical losses to be 2.2%. This is the expected loss in energy. Uncertainty, on the other hand, is a statistical concept that describes the unknowns associated with estimates. In the same example, several unknowns may cause the losses to be 2.1 or 2.3%. The uncertainty in the loss estimate is ±0.1%. In addition to uncertainty in loss estimates, there are uncertainties associated with estimates of the annual average energy production. This section will focus on losses and the next section will focus on uncertainty.

Following are categories of losses:

- *Wake losses.* This is loss in energy production because of reduced energy and increased turbulence in the wake of turbines. Wake was covered in the second section of this chapter. Wake losses may be due to internal turbines within the wind farm, or because of external turbines in adjacent wind farms. Wind resource assessment software, like WindPRO and WindFarmer, contain a variety of wake models to compute wake losses for individual turbine and for the entire wind farm.

- *Plant availability.* This is loss in energy production because of unavailability of the plant even though the wind resource is available. Reasons for unavailability of plant are:

  1. Scheduled and unscheduled maintenance of turbine

  2. Scheduled and unscheduled maintenance of balance of plant

  3. Grid unavailability, which may be caused by fault on the grid because of external reasons

- *Electrical losses.* The difference in energy produced at the generator and the energy delivered to the grid is the electrical loss. Subcategories of electrical losses are:

1. Transformer losses

2. Transmission losses from generator to grid

3. Turbine and wind farm internal power consumption

- *Turbine performance.* This is a decrease in energy production because of aerodynamic, mechanical, or electrical performance of the turbine and high wind hysteresis. The first three are usually lumped into power curve loss. Subcategories are:

    1. Power curve loss due to soiling of blades, deterioration in performance of gearbox, other mechanical components and the generator

    2. *High-wind hysteresis.* When the 10-min average wind speed exceeds the cutoff speed, the turbine will shut down. Subsequently, the turbine controller monitors the wind speed and does not restart as soon as the wind speed is below the cutoff wind speed. Instead, the controller waits for the wind speed to drop by a certain amount below the cutoff before restarting. This loss in energy is high-wind hysteresis.

    3. *Wind modeling.* This accounts for sources of losses that were not modeled in wind resource assessment like inflow angle, rough terrain, and others.

- *Environmental.* Several environmental factors can cause lower energy production. Some of the environmental effects may be accounted for by the preprocessing of wind data and wind resource assessment software. If these are accounted for, then the losses are already included in the annual energy production estimate, and the losses should not be double counted. Examples of environmental factors include:

    1. Shut down and performance degradation because of icing

    2. Extreme weather conditions that are outside the range allowed by manufacturer for operating a turbine. For example, turbine manufacturers specify operational range of temperatures outside which the turbine is shutdown. Other weather related events that cause shutdown include cyclones, hurricanes, and tornadoes.

    3. Seasonal activity of migratory birds, bats, and other species. In order to reduce the impact on wildlife, turbine operators deploy methods (including radar) to detect presence of seasonal migratory birds and bats. When the conditions are detected, the turbines are slowed or shut down. Sometimes these are voluntary and other times it is a condition for obtaining an environmental permit.

4. *Lightning strikes.* Direct lightning strikes can take a turbine offline until it is checked for damage.

5. *Change in roughness because of growth.* Over time, growth of forests and tall trees will increase the roughness in the area leading to lower energy production.

- *Curtailment of energy production.* Energy production may be curtailed because of either grid constraints or high degree of turbulence from adjacent turbines. The subcategories are:

  1. Grid operator may order curtailment of wind energy production for a variety of reasons, including, low demand, high supply from other sources, grid failure, etc.

  2. *Wind sector management.* Layout of turbines may be such that on rare occasions the direction of wind causes turbines in the wake to experience a high degree of turbulence. Under such conditions, the turbines subject to high turbulence are shut down. As an example, consider a wind farm in which the shortest distance between turbines is three times the rotor diameter along the direction that is perpendicular to the primary wind direction. In this example, wind sector management would be deployed when the direction of wind is perpendicular to the primary direction of wind and the wind speed is large enough to cause high degree of turbulence. Wind sector management improves the longevity of the turbine by eliminating large fatigue loads.

- *Others.* A variety of locale-specific losses may exist including:
  1. *Earthquakes.* An earthquake will cause a wind farm to shut down for safety reasons. It may not resume until an inspection is completed to check the integrity of foundation and other structural elements. Although there is uncertainty in the timing of earthquakes, over a lifetime of the wind farm, the average loss may be predicted.

Estimation of the losses may require extensive research into locale-specific performance of equipment, weather conditions, grid conditions, and others. Objective estimates of losses may be substituted for subjective estimates of local experts, but a temptation to broadly use industry averages should be avoided when preparing a bankable resource estimate. For preliminary wind resource assessment, a loss estimate of 10% is commonly used as a placeholder. Table 8-5 lists generic estimates for losses. These losses are applied to the annual energy production estimate that are computed from wind speed estimate and power performance curve.

Loss Category	Loss Estimate	Comments
Wake losses	5–15%	WindPRO and WindFarmer have tools to compute wake losses
Plant availability		
Turbine related	TS	Unavailability due to schedule and unscheduled maintenance of turbine is difficult to estimate as new technologies evolve. Also, consult O&M operator.
BPO related	TS	Consult O&M operators.
Grid unavailability	LS	Use data from local utility regarding frequency, duration, and types of faults on grid to compute losses
Electrical losses	2–4%	
Transformer losses	TS	Equipment supplier of transformer and switch gear will provide loss estimates.
Transmission losses	TS	Modeling of conductor and its length; soil resistivity and soil conductance; and average temperatures during the year will yield amount of losses
Internal power consumption	TS	Consult turbine manufacturer
Turbine performance	1.5–5%	
Power curve loss	TS	Consult O&M operators
High wind hysteresis	TS, LS	Wind data analysis along with turbine controller parameters may be used to compute the losses
Wind modeling	LS	Inflow angle modeling and RIX modeling may be used to model losses
Environmental	1–3%	
Icing	LS	Analysis of weather data may be used to estimate losses
Outside operating range	TS, LS	Analysis of weather data along with turbine specifications may be used to estimate losses
Wildlife	LS	Analysis of migratory patterns and other movement patterns may be used to estimate losses
Lightning	LS	Analysis of lightning data may used along with time to inspect in order to estimate losses
Roughness change	LS	Roughness change may be used to compute losses
Curtailment	1–3%	
Grid	LS	PPA terms and analysis of demand and supply may be used to model losses
Wind sector	LS	Wind farm layout and wind speed data by sector may be used to estimate losses
Other		
Earthquake	LS	Seismic database may be used estimate frequency

*LS = Locale-specific; TS = technology-specific.

TABLE 8-5 Loss Categories, Generic Estimates, and Comments*

Uncertainty Analysis

Wind resource assessment is not complete without uncertainty analysis. The reason is wind resource assessment provides information about wind speeds, which is transformed into average annual energy production (AEP), which is transformed into average revenue of the project. Uncertainty in wind resource is translated into uncertainty in AEP, which is translated into uncertainty in revenue. Any financial analysis of a wind project must consider not just the average revenue, but also the uncertainty in revenue. The most common method of incorporating uncertainty is to compute P50, P84, P95, and P99 estimates of revenue. Consider an example that has mean annual revenue of $2000 and standard deviation of $300. Table 8-6 contains the values for PN. Most financial decisions are made based on P95 or higher estimates of revenue. Therefore, a financier is keenly interested in the estimate of σ and the methodology used to compute it. A bankable wind assessment should contain revenue estimates for various levels of PN along with a methodology for estimation of uncertainty.

Typical sources of uncertainty in energy production estimates include:

1. *Sensors.* This was discussed in detail in Chapter 6. The uncertainty includes sensor's inaccuracy, calibration inaccuracy, and inaccuracy because of mounting and setup, and other factors (like, overspeeding for cup anemometers, large volume-based measurement for Sonic Detection and Ranging (SODAR))

2. *Shear model.* This was also discussed in Chapter 6. This is the uncertainty because of extrapolation from sensor heights to hub height.

3. *The spatial distribution model.* This pertains to the uncertainty introduced when wind data is translated from measured locations to the individual wind turbine locations.

PN Formula	PN Revenue Estimate	Meaning of PN
$P50 = \mu$	$2000	Revenue of at least $2000 will be realized with 50% certainty
$P84 = \mu - \sigma$	$1700	Revenue of at least $1700 will be realized with 84% certainty
$P95 = \mu - 1.645\,\sigma$	$1506	Revenue of at least $1506 will be realized with 95% certainty
$P99 = \mu - 2.326\,\sigma$	$1302	Revenue of at least $1302 will be realized with 99% certainty

TABLE 8-6   Uncertainty Expressed as *PN*

Terrain Type	RMS Error (%)
Simple terrain, two met-towers	5
Simple terrain, three met-towers	2
Rolling hills, three met-towers	7
Complex terrain, two met-towers	15
Complex terrain, three met-towers	8

TABLE 8-7 Root Mean Square Error in Wind Speed Estimation Using a Variety of Models for Spatial Extrapolation

4. *Long-term climate adjustment.* This pertains to the prediction model used to produce a long-term wind estimate from short-term measurements. In addition, it includes ability of hind-casted data to predict the wind conditions in the future.

5. *Plant loss uncertainty.* Note, plant loss uncertainty is not the same as plant losses. This item deals with uncertainty associated with estimates of losses.

The focus of this section will be on items (3) and (4).

In the spatial distribution model, uncertainty is pertinent if the measurement location and planned turbine location is not the same. Studies have shown the error in wind speed estimates varies by type of terrain and number of met-towers. Results from one such study by DNV Global Energy Concepts[7] is shown in Table 8-7.

Site-specific uncertainty because of long-term climate adjustment is difficult to estimate. Uncertainty of 3 to 5% is to be expected. A procedure for estimation is described by Siddabathini,[8] but it requires several years of measurement data.

Estimating Uncertainty of Annual Energy Production: Framework for Combining Uncertainty

There are two aspects to understanding uncertainty: Amount of uncertainty and its impact on the quantity of interest (in this case, it is AEP or revenue). The impact is also called the sensitivity of a variable to the quantity of interest. Quantity of interest may be annual energy production (AEP) and financial performance such as payback period or net-present value (NPV). In the rest of this discourse, AEP will be the quantity of interest. The square of the combined uncertainty on AEP can be written in a generalized form as:[9]

$$U_{AEP}^2 = \sum_{i=1}^{N} \sum_{j=1}^{N} \rho_{i,j} c_i c_j u_i u_j \qquad (8\text{-}21)$$

where u_i, u_j are uncertainty of the components, $\rho_{i,j}$ is the correlation coefficient between components i and j, and c_i, c_j are the sensitivity coefficients of components i and j to AEP.

When components i, j are uncorrelated, that is, $\rho_{i,j} = 0$ when $i \neq j$, then Eq. 8-21 becomes:

$$U_{AEP}^2 = \sum_{i=1}^{N} c_i^2 u_i^2 \tag{8-22}$$

When i, j are perfectly correlated, that is, $\rho_{i,j} = 1$, then:

$$U_{AEP} = \sum_{i=1}^{N} c_i u_i \tag{8-23}$$

Table 8-8 presents a few sources of uncertainty, sensitivity factors, and net uncertainty.[9] Note the sensitivity of AEP to wind speed is:

$$c_1 = \left. \frac{\partial U_{AEP}}{\partial v} \right|_{v=v_{ref}} / \left(v_{ref} / U_{AEP} \right) \tag{8-24}$$

Theoretically, c_1 should be 3; in reality, the power curve determines the sensitivity, which is in the range of 1.25 and 2. For most of the other uncertainty factors, the uncertainty is expressed directly in terms of its

Component of Uncertainty	Sensitivity Factor	Amount of Uncertainty (%)	Net Uncertainty of AEP Because of Component (%)
Wind speed measurement	1.5	5	7.5
Wind speed spatial extrapolation	1.5	3	4.5
Wind speed long-term correction	1.5	3	4.5
Wind shear, height extrapolation	1.5	2	3
Air density	1	0.3	0.3
Power curve	1	0.6	0.6
Wake losses in wind farm	1	1.7	1.7
Unaccounted for Loss	1	1	1
Total uncertainty of AEP assuming components are uncorrelated is square root of sum of squares			10.5%

TABLE 8-8   Parameters Used in Uncertainty Computation

impact on AEP; therefore, sensitivity is one. Table 8-8 illustrates the computations with generic values; actual values are project-specific and must be computed as part of a comprehensive assessment.

Nonbankable versus Bankable Resource Estimates

One of the purposes of resource assessment is to quantify the economics of a project in a manner that would satisfy an investor. It does not matter if the project is funded with internal resources or funded by an external investor. In either case, the assessment must be done with a rigor that ensures that:

a. High-quality calibrated instruments are used

b. Data has been collected properly

c. Data has been analyzed properly

d. An audit trail exists such that items (b) and (c) can be verified independently

e. Losses and uncertainties associated with the wind resource over the long-term are identified and quantified

For purposes of discussion of bankable and nonbankable, assume that the wind resources are sufficient. As an example, consider an area in Class 4 wind category with wind speed in the range of 7 to 7.5 m/s at 50-m elevation above the ground. In this area, the difference between bankable or nonbankable resource estimate is the amount of due diligence that has been performed while performing steps (a)–(e) above in order to reduce the uncertainty associated with wind resources. The difference between bankable or nonbankable is not whether the resource estimate results in good versus poor wind conditions.

Why is uncertainty so important in determining bankability of a wind project? The reasons are wind energy depends on the cube of wind speed (in theory, and quadratic, in practice), therefore, small changes in wind speed estimates can cause the return on investment of a wind project to fluctuate significantly. A bankable resource estimate is, therefore, one in which enough verifiable data is available to quantify the uncertainty in wind resource at the planned wind project location. With uncertainty quantified, an investor can compute returns on a project for various risk scenarios.

A bankable resource estimate is one that is:

- Based on actual onsite measurement data with quality instruments that are calibrated, and not based on anecdotal data

- Based on at least 1 year of wind measurement data and not based on short-term measurement data

- Based on measurements at multiple heights and in which the highest anemometer is close to planned hub height, and not based on lower height measurements

Nonbankable resource estimate is a preliminary assessment of wind conditions normally used during prospecting for wind sites. It is normally the first stage of resource estimation when not all the uncertainties in wind conditions are completely understood or completely quantified and, therefore, the return on investment has a high degree of uncertainty. Nonbankable resource estimates may be sufficient for self-financed small and micro wind projects.

References

1. Nielsen, Per. *WindPRO 2.5 Users Guide*, EMD International, Aalborg, Denmark, 2006.
2. Mortensen, N. G., Bowen, A. J., and Antoniou, I. *Improving WAsP Predictions in (too) Complex Terrain*, EWEC, Athens, 2006.
3. WAsP. *Wake Effect Model*. WAsP [Online] 2007. http://www.wasp.dk/Products/WAsP/WakeEffectModel.html.
4. Nielsen, M., Jørgensen, H. E., and Frandsen, S. T. *Wind and Wake Models for IEC 61400-1 Site Assessment*, EWEC, Marceille, 2009. http://www.ewec2009proceedings.info/allfiles2/529_EWEC2009presentation.pdf.
5. International Electrotechnical Commission. *Wind Turbine Part 1 Design Requirements*, International Electrotechnical Commission, Geneva, 2005. IEC 61400-1 Edition 3.
6. Schlez, W., and Tindal, A. Wind farm siting and layout design [Online] Garrad Hassan and Partners, Ltd. http://www.wwindea.org/technology/ch02/en/2_4_2.html.
7. VanLuvanee, D., Rogers, T., Randall, G., Williamson, A., and Miller, T. *Comparison of WAsP, MS-Micro/3, CFD, NWP, and Analytical Methods for Estimating Site-Wide Wind Speeds*, AWEA Wind Resource Assessment Workshop, Minneapolis, MN, 2009.
8. Siddabathini, S., and Sorensen, T. *Uncertainty of Long Term Correction of Wind Speed: The Importance of Consistency of Data*, EWEC, Marseille, 2009.
9. Derrick, A., *Uncertainty, the Classical Approach.* AWEA Wind Resource Assessment Workshop, Minnesota, MN, 2009.
10. Burton, T. Sharpe, D., Jenkins, N., and E. Bossanyi, E. *Wind Energy Handbook*, Wiley, Hoboken, NJ, 2001.
11. DNV/Risoe. *Guidelines for Design of Wind Turbines*, Det Norske Veritas and Risø National Laboratory, Copenhagen, 2002.

CHAPTER 9

Wind Turbine Generator (WTG) Components

Introduction

In this chapter, the main components of a utility-scale horizontal axis wind turbine are described. Wind turbine generators (WTG) have evolved into complex machines, much like simpler car engines have evolved into complex machines. At a high level, there are three major systems:

1. *Rotor system.* This includes blades that capture energy and a rotor hub that connects the blades to the shaft, along with pitch mechanism that assists in efficient capture of energy.

2. *Nacelle.* This contains all the components that sit on top of the tower, except the rotor system. It includes main shaft, gearbox, generator, brake, bearings, nacelle frame, yaw mechanism, auxiliary crane, hydraulic system, and cooling system. Depending on the design of the turbine, only some of the components may be present.

3. *Tower and foundation.* These structural elements carry all the forces and moments to the ground.

At the end of the chapter two additional topics are covered: Design loads and certification of turbines. The chapter concludes with technical specifications of four WTGs.

Rotor System

The rotor system captures wind energy and converts into rotational kinetic energy. This is accomplished through blades that connect to a rotor hub that is connected to the main shaft. In large utility-scale turbines, the rotor hub has mechanisms to pitch the blade, that is, rotate along the longitudinal axis of the blade.

Blades

Although turbine blades are, in principle, similar to airplane wings in terms of generating lift, there are significant design differences.

- Twist along the longitudinal axis of the blade. As discussed in Chapter 4, in order to achieve a constant angle of attack along the entire length of the blade, a twist is added to the blade.
- Turbine blades are thinner and longer because it yields enhanced performance in lower wind speed.
- Stall characteristics are different. Wind turbines continue to operate under stall conditions between rated and cut-out wind speed, whereas an airplane avoids stall conditions.
- Soiling of wind turbine blades with dust, dead insects, and others can cause significant loss of power. Coupled with the high expense of cleaning blades, this leads to a design challenge.

The cross section of a turbine blade is in Fig. 9-1. The components of a blade are:

- The core of the blade is made of balsa wood or foam; the core gives the blade its shape. This is also called the spar, which is like a long tubular beam along the length of the blade.
- Upwind and downwind aerodynamic shell made of fiberglass and epoxy resins. These two are glued at the leading and at the trailing edge. The shells are glued to the spar with an adhesive.
- Root of the blade is a metallic cylinder with bolts to connect the blade to the rotor hub.

FIGURE 9-1 Cross section of a blade.

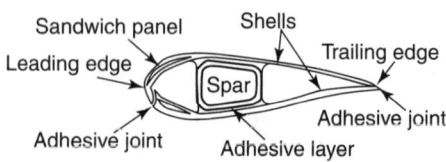

Sandwich panel   Shells
Leading edge   Trailing edge
Spar
Adhesive joint
Adhesive joint   Adhesive layer

- Lightning protection system, with lightning receptors all along the length of the blade connected to conductors that ground the lightning.
- Sensors in the blade to monitor stress, strain, acoustic emissions, and other signals.
- Overspeed control mechanism like pitchable tips. Modern large turbines do not have this feature.

There are two predominant methods for the manufacturing of large blades: Epoxy prepregmolding and vacuum-assisted resin transfer molding (VARTM).[1] In the epoxy prepreg molding, fiberglass impregnated with epoxy is laid out in layers and placed in a mold. The layers are pressed and then cured at elevated temperature. In the VARTM, fiberglass is laid in a pre-form and placed in a closed mold. In this mold, epoxy resin is sucked in using vacuum and then cured to form a blade. VARTM has resulted in a simpler process, although it is still time intensive. Application of epoxy resin on such a large structure without imperfections, like air pockets and without resin-rich pockets, is challenging. These imperfections cause stress concentrations leading to fatigue failure. The percentage of material by weight in a blade is in Table 9-1.

Blades of large utility-scale turbines are 35-plus meters in length. The total weight of a 40-m blade can be 5,700 kg. (See Tables 9-5 to 9-8 for weight of blades of five commercial utility-scale turbines.) One of the largest mass-produced blades is a 61.5-m blade for a 125-m rotor diameter offshore turbine with a weight of 18 tons. For this LM Glasfiber blade, about 30% reduction in weight was achieved with combination of carbon and glass fibers.[3] Carbon fiber-reinforced plastics are lighter weight, possess about three times the stiffness of glass fiber-reinforced plastics, and possess significantly better fatigue properties.

Component of Blade	Weight, %
Fiberglass	51
Epoxy resin	33
Bonding adhesive	7.5
Sandwich core	4
Miscellaneous—bolts and lightning protection	4.5

Source: From Ashwill, T. *Blades: Trends and Research Update.* [Online] 5 12, 2008. http://www.sandia.gov/wind/2008BladeWorkshop/PDFs/Mon-02-Ashwill. pdf.

TABLE 9-1 Weight of Components in a Blade

Figure 9-2 Forces
and moments
acting on a blade.

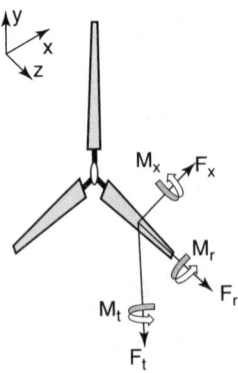

Forces and Moments

Since the blade is the exposed element, it takes the most complex set of forces and moments, along each of the six axes. For convenience, all the forces and moment are indicated along axial, radial, and tangential directions, as shown in Fig. 9-2.

- *Axial force, F_x.* The lift and drag forces have components in the tangential and axial directions. The change in the net axial force is a function of the wind speed. This force is the primary contributor to the thrust force on the rotor.

- *Radial force, F_r.* There are two primary sources of radial force: Gravity and inertia. The weight of the blade is a constant force acting vertically downward. From the point of view of the blade, the force is cyclical—changing from radial to tangential with a frequency equal to the frequency of rotation. The inertial force has two components, centrifugal force because of rotation and forces during braking and acceleration. The blade experiences radial acceleration (because of change in direction of velocity) and, therefore, radial force. For a constant-speed rotor, this force is constant when the turbine is rotating and zero when the rotor is stationary.

- *Tangential force, F_t.* There are four primary sources of tangential force: Lift, drag, and gravity, and inertial forces. The inertial forces are during startup and braking of the rotor. This is the force that produces torque on the rotor.

- *Bending moment along the axis of rotation, also called edgewise bending moment, M_x.* This bending moment arises because of lift forces that increase radially when moving from hub to tip of blade. This force reaches a peak close to the tip. Another contributor to the bending moment is gravity. Measured at the root of the blade, this bending moment oscillates between positive and negative value with a frequency equal to the frequency of rotation.

- *Bending moment in the plane of rotation along the tangential direction, also called the flapwise bending moment, M_t.* This bending moment arises because of drag forces that are the highest at the tip of the blade and lowest at the hub. The frequency of the moment depends on the frequency of wind speed. Measured at the root of the blade, this oscillates with much smaller amplitude and stays positive. This moment is dependent on the aerodynamic load and, therefore, it varies with the variation in wind speed.

- *Twisting moment along the longitudinal axis of blade.* This twisting moment is due to the lift force. The frequency of the moment depends on the frequency of wind speed. Measured at the root of the blade, this is a smaller component of moment.

No mention of loads is complete without the mention of fatigue loads. Wind turbines are subject to extreme levels of cyclic loads. For instance, the root of the blade is subject to load that cycle every rotation. At 30 rpm and 20-year life of turbine, the number of cycles is $30*60*8760*20 = 0.3 \times 10^9$ cycles.

Rotor Hub

The next major component of the rotor is the hub. Blades are radially bolted to the hub. On the axial end, the rotor hub is connected to the drive train, which may be one or more of the following: Main shaft, gearbox, and generator. The hub is made of high-quality cast iron. It transfers load from the blades to the nacelle frame and to the drive train. The manner of transferring loads from the hub to rest of the components in the nacelle depends on the turbine configuration—direct drive or with gearbox.

Alternative Configurations of Turbines

In the beginning of this section, two configurations are discussed: With and without gearbox. At the end of the section, a two-blade turbine with a rotor lifting and lowering mechanism is described.

A typical turbine with gearbox has a fixed-speed generator that produces power at grid frequency. Although newer turbines allow some variation in speed of generator, nevertheless the rotor hub rpm is stepped up from about 20 to 30 rpm to generator speed, which is 50 to 75 times. Generator speed is typically 1,000 rpm or more.

Figure 9-3 is a schematic of a traditional turbine with gearbox. Label 1 in the schematic illustrates two types of pitch drives: Motor and hydraulic. Rotor hub is labeled 2, which is connected to the main shaft (labeled 3). Main shaft is a forged component made from hardened

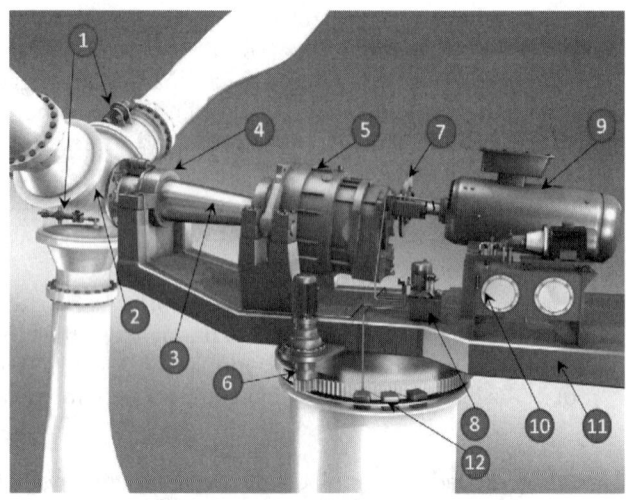

Figure 9-3   Schematic of the components of a wind turbine. 1. Pitch drives of two types—motor and hydraulics. 2. Rotor that connects blades to main shaft. 3. Main shaft. 4. Main bearing. 5. Gearbox. 6. Yaw drive. 7. Disk brakes. 8. Brake hydraulic system. 9. Generator. 10. Main hydraulic system. 11. Nacelle frame. 12. Yaw brakes. (With permission from Bosch Rexroth, maker of gearbox, pitch drive, yaw drive, and hydraulic system.)

and tempered steel and it rests on the main bearing labeled 4. The main bearing rests on the nacelle frame (labeled 11). The main shaft connects the hub to the gearbox (labeled 5). The rotational torque is transferred from the hub to the main shaft, which transfers the torque to the gearbox. The bearing transfers all three components of force on the rotor hub—parallel to axis of tower, parallel to axis of rotation, and transverse force—to the bearing, which transfers to the nacelle frame. The nacelle frame (labeled 11), rests on the tower and transfers all the forces to the tower. The yaw drive (labeled 6), rotates the nacelle along the vertical axis. The gearbox is connected to generator (labeled 9). The shaft connecting the two has disk brake (labeled 7). The brake hydraulic system (labeled 8), controls the brakes.

An alternative configuration for a turbine with gearbox is to directly connect the rotor hub and gearbox. This is seen in Fig. 9-4. Here, a double roller-tapered bearing is integrated with rotor hub and nacelle frame, thereby eliminating main shaft to carry power from rotor to gearbox. The outer ring of the bearing is connected to the nacelle frame and the inner ring is connected to a short rotor shaft. It eliminates the need for two bearings. This configuration makes the nacelle compact and reduces the overall weight of the turbine. However, maintenance of gearbox may be challenging.

In direct drive turbines, there is no gearbox and the hub is directly connected to a variable speed generator (see Fig. 9-5). In such a

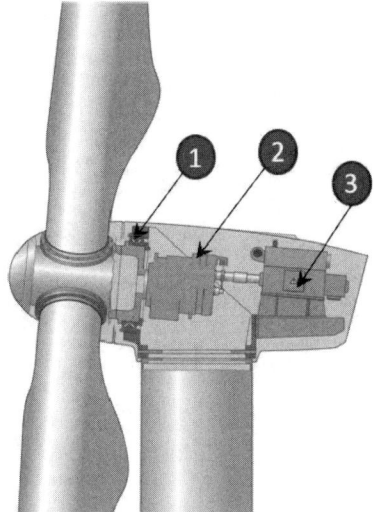

Figure 9-4 Schematic of a turbine with gearbox. Double roller tapered bearing is integrated with hub and nacelle frame, thereby eliminating the main shaft to carry power from rotor to gearbox. 1. Bearing mounted on hub. 2. Gearbox, 3. Generator. (With permission from SKF, maker of bearings.)

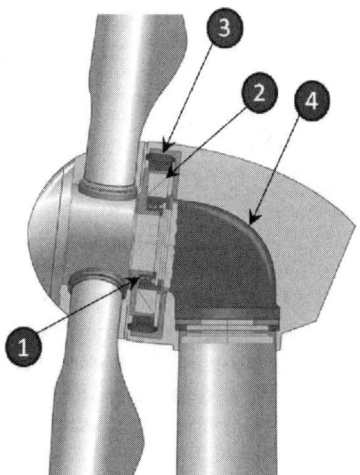

Figure 9-5 Schematic of a direct-drive turbine. Double roller tapered bearing is integrated with hub and nacelle frame, thereby eliminating main shaft. 1. Bearing mounted on hub. 2. Rotor of generator. 3. Stator of generator. 4. Nacelle frame. (With permission from SKF, maker of bearings.)

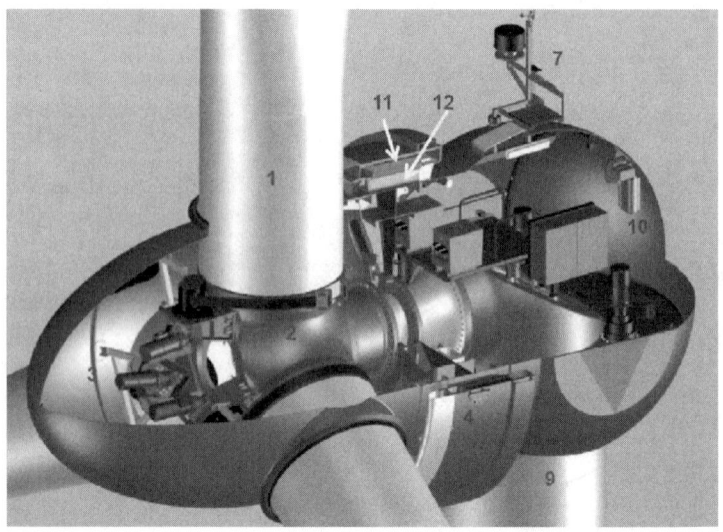

Figure 9-6 Schematic of a direct-drive turbine. 1. Rotor blade. 2. Hub. 3. Blade pitch system. 4. Generator rotor. 5. Generator stator. 6. Yaw drive. 7. Wind measurement instruments and obstruction light. 8. Main frame of nacelle. 9. tower; 10. auxiliary crane; 11. permanent magnets; 12. stator windings. Area between 11 and 12 is the air gap. (With permission from Vensys Energy AG.)

configuration, there is no main shaft. The rotor hub directly drives the rotor of the generator. Notice the bearing's inner ring is directly connected to the nacelle frame and outer ring is connected to rotor. Double roller taper bearing transfers all loads, other than torsional load, from the rotor and generator to the nacelle frame.

A more detailed nacelle schematic for direct drive turbine from Vensys Energy AG is in Fig. 9-6. Rotor blades (labeled 1) turn the rotor hub (labeled 2). This turbine has a pitch drive (labeled 3) with three motors that drive toothed belts. The rotor hub drives the rotor of the generator (labeled 4) with permanent magnets. The stator of the generator is labeled 5; note, it is inside the rotor of the generator; The permanent magnet on the rotor (labeled 11) and stator windings (labeled 12) can be seen; gap between labels 11 and 12 is the air gap of the generator. Yaw drive is labeled 6, the main frame of nacelle is labeled 8, and the tower is labeled 9.

A schematic of Vergnet HP 1 MW, a two-blade turbine is seen in Fig. 9-7. The advantage of a two-blade turbine is lower cost and lower weight. This comes with a small reduction in efficiency. Two-blade turbines require a teetering hub to manage the changing axial loading of blades. When the blades are in a vertical position, the axial load on the blade pointing up is higher than the load on the blade pointing

Figure 9-7 Schematic of a two-blade turbine. 1. Rotor blade. 2. Pitch drive. 3. Teetering hub. 4. Rotor brake system. 5. Yaw. 6. Drive train. 7. Generator. 8. Auxiliary lifting system. 9. Gearbox. 10. Lifting/lowering beam. (With permission from Vergnet, a 275-kW and 1-MW turbine maker.)

down (because of wind shear), which creates a bending moment that tries to turn the rotor over the nacelle. When the blades are horizontal, the axial loads are the same. A teetering hub (labeled 3), manages this moment. This turbine contains a gearbox (labeled 9), and an auxiliary crane. The auxiliary crane's lifting system and the lowering beam are labeled 8 and 10.

Pitch

The pitch mechanism controls the angle of the blades with respect to the plane of rotation. Smaller turbines do not have a pitch mechanism; instead, these turbines rely on stall to regulate the speed of rotation at high wind speed, as described in Chapter 4. Control of pitch allows a turbine to capture energy at low wind speed and to capture a constant amount of energy at wind speed higher than the rated wind speed. There are several methods of controlling the pitch angle; all involve a mechanism that controls the angles of the blades. Control algorithm continuously monitors wind speed and energy production and adjusts the pitch of blades. Blades are pitched significantly when the wind speed is higher than the rated speed to change the angle of attack and induce stalling. There are several pitching mechanisms, the most commonly used are:

- Hydraulic system in which a hydraulic cylinder that is controlled by an oil pump which acts on the blades

- Rack and pinion in which rack is circular and meshes with pinions of the two or three blades. The rack is driven by a motor through a worm gear
- Individual motor to control each blade. This is the most common method of controlling pitch.

The pitch mechanism is at the front of the rotor hub, as shown in schematics Figs. 9-3 and 9-6. A toothed gear is at the interface of the rotor hub and the root of blade. An external motor operates the pitch with a gear (see Fig. 9-3) or a toothed belt (see Fig. 9-6). The external motors and any other part of the pitch mechanism are attached to the hub and rotate with it.

Nacelle
The primary components of nacelle are the main shaft, bearings, gearbox, generator, brake, nacelle frame, hydraulic systems for brakes and lubrication, and cooling systems.

Gearbox
Gearbox is a standard component in nondirect drive turbines. The purpose is to increase the rotational speed. There are different types of gears in a gearbox: Planetary gears and spur/helical gears.

As an example, consider a turbine with speed of rotor hub = 24 revolutions per minute (rpm) and speed of a generator = 1,800 rpm. This speed conversion will require a gearbox of 1:75 ratio, which can be accomplished in a three-stage gearbox. Stage one and two are planetary gears in series, each with ratio of 1:5. This yields speed ratio of 1:25. An additional gear stage consisting of spur gear of ratio 1:3 will further increase the speed ratio to 1:75. Planetary gears are compact, typically about one-third the size of spur/helical gears. Two disadvantages of planetary gears are: (i) Overheating due to reduced surface area to release heat generated because of friction, and (ii) higher noise because the teeth are not beveled.

A schematic of gearbox is in Fig. 9-8. The life of a wind turbine gearbox is about 10 years and unexpected failure can lead to significant downtime. Therefore, gearbox condition monitoring systems are recommended. A comprehensive condition monitoring system will measure vibration, temperature, properties of lubricating oil, wear particles in oil, and others in order to provide early warning.

Yaw Drive
The yaw mechanism allows the turbine to face the wind, that is, align the plane of rotation to be perpendicular to the direction of wind.

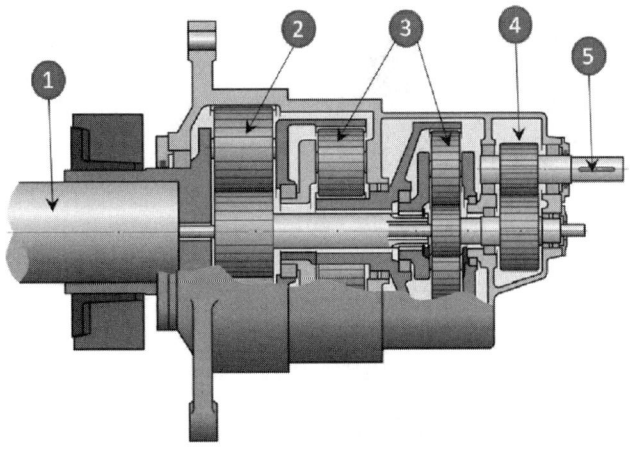

Figure 9-8 Schematic of a multistage gearbox. 1. Main/input shaft. 2. Differential gears. 3. Planetary gears. 4. Spur gears. 5. Output shaft to generator. (With permission from Bosch Rexroth, a gearbox manufacturer.)

Smaller turbines (and some older large turbines) use a passive yaw, which are of two types: Tail vane to orient the plane of rotation and downwind turbine where the wind flows over the nacelle before turning the blades.

Almost all large utility-scale turbines are upwind turbines with active yaw. Active yaw is more expensive because it controls the yaw using an electromechanical drive and a control system that monitors wind direction. The yaw motor is in the nacelle frame and its gear connects to a large gear that connects the nacelle to the tower (see Figs. 9-3 and 9-6). The yaw mechanism also has yaw brakes to lock the position of the yaw.

Improper control algorithm or improper working of the yaw drive leads to an average nonzero angle between the wind direction and axis of rotation. This leads to lower energy production and high nonsymmetrical loads.

Nacelle Housing and Frame

The nacelle housing covers all the components inside the nacelle from the elements of nature. It is usually made of glass-reinforced plastic. The nacelle frame transfers all the loads, other than the useful torque load, to the tower. The useful torque load produces energy in the generator. All other loads are transmitted to the nacelle through bearings and bolted components, like gearbox and generator. Bearings are a critical component to transfer rotational motion to gearbox and/or generator and transfer all other loads to the nacelle frame. For location of the main bearing, see Figs. 9-3, 9-4, and 9-5. SKF's wind

FIGURE **9-9** Double roller taper bearing. 1. Double rollers. 2. Taper. (With permission from SKF.)

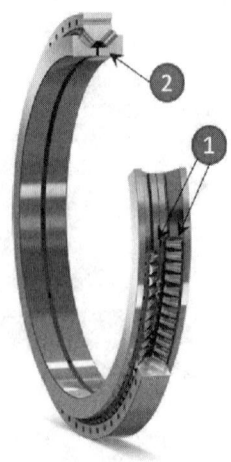

turbine bearings (see Fig. 9-9) are a double row of tapered roller bearings, with an outer diameter as large as 2.4 m.[3] The trend is toward replacing two bearings with a single large radius bearing and replacing the bearing housing with a design that integrates the bearing into nacelle frame.

Lifting/Lowering Mechanism

Vergnet's 1-MW turbines have a unique mechanism to lower and raise the upwind part of nacelle, including rotor blades and hub. The intent is to: (a) reduce the cost of heavy cranes during erection and maintenance, and (b) improve safety and reduce damage during a cyclone.

The Vergnet turbine has a split nacelle, a downwind part with the auxiliary crane and generator, and an upwind part with the blades, rotor hub, and gearbox. The tower components and downwind part of nacelle are raised using a self-erecting system called VeriLift. After the auxiliary crane is in place, it lifts the upwind part of the nacelle. The lifting mechanism is shown in Fig. 9-10. More details are found in Table 9-6.

Towers

Towers are tubular tapered steel structures. Steel plates of thickness $1\frac{1}{2}$ to $\frac{1}{2}$ in. thick are rolled into cylinders and seam is then longitudinally welded. At the ends of the tower sections, flanges are welded to the inside of the cylinders for onsite bolting during erection. Normally, three tower sections are constructed and shipped to the site. There is a logistical reason that limits the diameter of a tower. In the United States, the minimum height of an overpass is $16\frac{1}{2}$ to 17 ft. This limits the diameter of the bottom section of tower that can be transported by truck to be less than 4.3 m. Rail transit has even more stringent

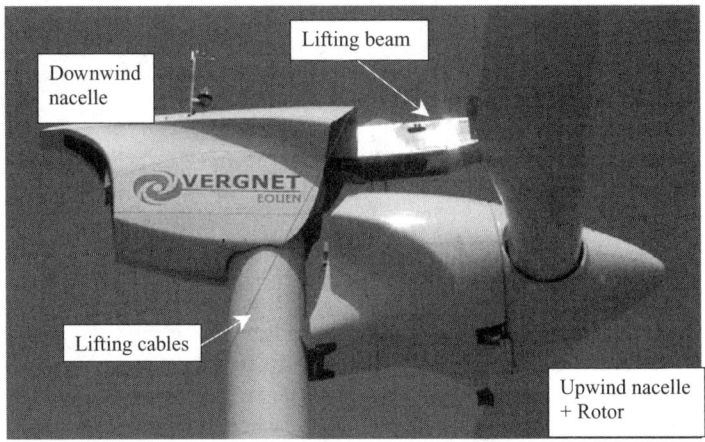

Figure 9-10 Schematic of Vergnet's BirdLike mechanism that allow the gearbox and rotor with blades to be lowered or raised. The downwind part of nacelle is lifted and installed first. It then lifts or lowers the upwind part. (With permission from Vergnet.)

height and width restrictions for transporting towers. This limitation of diameter of tower leads to a limitation on the weight of the nacelle, rotor, and tower. A practical guideline to emerge is that turbines of size 3 MW or less with tower height less than 100 m can be transported within the United States.

If towers of larger diameter are required, then the alternative is to fabricate the base of the tower onsite, since it is the largest diameter section. One option is to cast a concrete tower as the base and then install steel tower sections above the concrete base tower section. Another option is to transport sections of the tower that are welded on site.

Cranes assemble the three or more tower sections vertically. The joint between two tower sections is usually a bolted joint. The outside surface of the tower is smooth and conical; each tower section has inside flanges that are bolted together. Towers have an access door at the bottom and a "man-lift" system to transport construction and repair crew to the nacelle. It also has stairs with platforms for rest.

The weight of a wind turbine tower is considerable. (See Tables 9-5 to 9-8 for dimensions and weight of towers for five commercial turbines.)

Foundation

There are two primary drivers to foundations design: Soil conditions and loads on the turbine. The forces and moments acting on the foundation of a turbine are illustrated in Fig. 9-11. An example of the primary loads on a 2.5-MW turbine[5,6] foundation is seen in Table 9-2. Turbine manufacturers provide a bolt cage or an embed ring that is

FIGURE 9-11 Loads on a foundation.

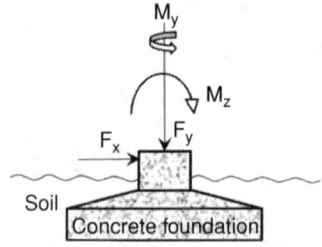

embedded into the foundation. It also defines how the tower will interface with the foundation. Most turbine manufacturers also provide a sample foundation design for "normal" soil conditions.

Weight of the structure is the largest force that must be overcome by foundation. The bending moment because of thrust force applied at the hub height is a large moment that must be overcome. The bending moment acts to overturn the entire turbine; the foundation provides the necessary resistance. This bending moment causes the upwind side of the foundation to be tension and the downwind side to be in compression.

At a high-level, there are two types of foundations spread-footing and deep foundations.

Spread-Footing Foundation

This is the most common type of foundation. As the name suggests, it has a large diameter and short depth. The weight of the foundation and soil resistance provide the strength. These foundations are suitable for soils with good strength characteristics. At the bottom of the spread footing, the soil is compacted with gravel and other materials. The soil density and strength are measured at the bottom before the foundation construction is begun. A spread foundation for 1.5-MW turbine is typically 17 m in diameter and placed at a depth of 1.5 to 3.5 m (see Fig. 9-12.)

Type of Load	Extreme Wind Loads
Vertical load, F_y	2,500 kN
Lateral load, F_x	800 kN
Unfactored extreme moment, M_z	70,000 kN –m
Factored tower bearing pressure	35,000 kN/m^2
Foundation volume	500 m^3

TABLE 9-2 Example of Loads on a 2.5-MW Class Turbine

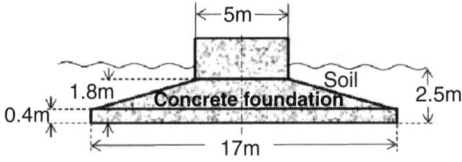

FIGURE 9-12 Approximate dimensions of a spread foundation for typical 1.5-MW size turbine.

Deep Foundation

There are different types of deep foundations for wind turbines: Piles, drilled shaft, caisson, piers, and others. Traditionally, a deep foundation is built on weaker soil by either driving down a solid concrete or steel beam into the ground or drilling a shaft and pouring concrete. However, a deep foundation, like the one described below, is used for normal soil conditions as well. As an illustration, the Patrick and Henderson (P&H) tensionless foundation is described in this section. A schematic of the foundation is in Fig. 9-13 and an image of a P&H foundation under construction is in Fig. 9-14. Approximate dimensions

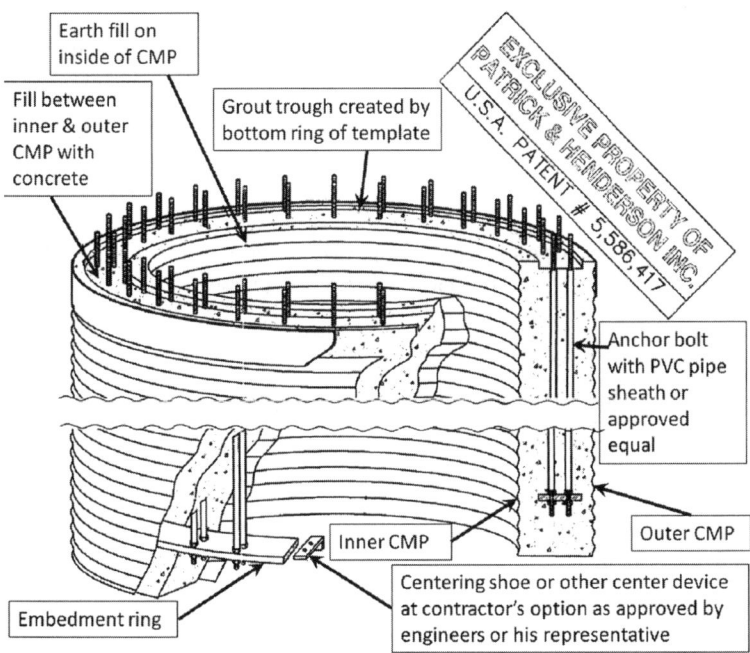

FIGURE 9-13 Isometric view of the patented P&H tensionless foundation. For illustration, the view is cut horizontally and longitudinally. (With permission from Patrick and Henderson, Inc.)

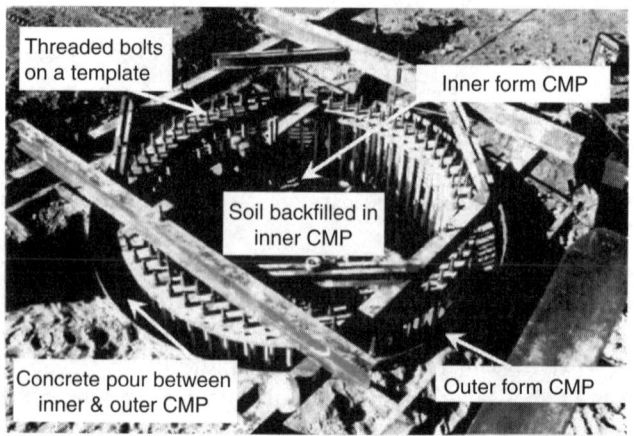

Threaded bolts on a template

Inner form CMP

Soil backfilled in inner CMP

Concrete pour between inner & outer CMP

Outer form CMP

Figure 9-14   P&H Pier foundation for a Vestas 660-kW in 1999 at a site in Nebraska. (With permission from Patrick and Henderson, Inc.)

for a Siemens 2.3-MW IEC IA turbine with 101-m rotor and 80-m hub height are seen in Table 9-3. The construction process of a P&H tensionless foundation consists of:

- Excavate foundation hole that is at least 24 in. larger than the outer corrugated metal pipe (CMP) diameter.
- A CMP is placed in the hole to provide a mold for casting concrete.
- Sand and cement slurry is put to backfill the area between the soil and outside surface of the CMP.
- An embedment ring is placed at the bottom of the CMP. A template ring is placed at the top of the CMP. Holes in the embedment ring and template match the bolt pattern in the

Parameters	Dimension
Outer CMP diameter	16 ft
Inner CMP diameter	12 ft
Height of foundation	34 ft
Number of anchor bolts	160 (80 in each ring)
Outside diameter of anchor bolt	1.5-inch ASTM A-615

Source: With permission from Patrick and Henderson, Inc.

Table 9-3   Approximate Dimensions for P&H Tensionless Foundation for Siemens 2.3-MW IEC IA

flange at the bottom of the tower. Threaded steel rods encased in PVC sleeves are then bolted to the embedment ring and the template.

- The bolt cage is lifted to attach a rebar hoop, one around the outside bolts and another around the inside bolts. The bolt cage is then lowered into the CMP while ensuring that the cage is leveled and the cage is concentric with the outer CMP.

- An inner CMP is placed concentric to the outer CMP. A plug of concrete is cast at the bottom of the inner CMP. Electrical conduits are installed. The inner CMP is then backfilled with soil.

- In the concentric area between the outer and inner CMP, foundation concrete is poured.

- Concrete floor slab and top collar are cast with the threaded rods extending above the concrete.

- After curing of concrete, the grout trough is grouted. Then, the tower base is placed on grout and leveled.

- The anchor bolts are then post-tensioned to keep the concrete in compression for all conditions of loading.

Design Loads of Wind Turbines

IEC 61400-1, Wind Turbines-Part 1: Design requirements,[7] provides the minimal design requirements for wind turbines. In this section, readers are exposed to some parts of the design requirements; however, readers are referred to the IEC 61400-1[7] for details. The IEC design requirements are specified using a case-based methodology. Overall, 12 cases are identified by IEC; each case is a combination of the following four conditions:[7]

1. *Design situations that represent the combinations of modes of operation of a turbine.* There are eight design situations specified: Power production; power production plus occurrence of fault; start up; normal shutdown; emergency shutdown; parked (standing still or idling); park and fault conditions; transport, assembly, maintenance, and repair. This is the minimal set, additional modes of operation may be added that have a reasonable probability of occurrence and significant impact.

2. *Wind conditions for analyzing the design situations.* Several types of wind conditions are used for the purposes of design: Normal wind profile model (NWP), normal turbulence model (NTM), extreme wind speed model (EWM), extreme turbulence model (ETM), extreme wind direction change (EDC),

extreme coherent gust with direction change model (ECD), extreme wind shear (EWS), and extreme operating gust (EOG). The above wind conditions are paired with the appropriate design situation. Two wind conditions are explained in more detail below.

3. *Two types of design analysis are performed, ultimate strength (U) and fatigue (F).* Ultimate strength analysis is static analysis against the strength of materials of extreme loads because of forces, moments, and deflections. Fatigue loads are computed by: (a) dividing into discrete stress cycles each with a mean, range, and frequency, (b) estimating duration of each stress cycle for the 20-year life of the turbine, and (c) estimating damage and summing up the damage using Palmgren-Miner rule.

4. *Partial safety factors.* IEC 61400-1 defines three classes of components depending on the consequences of failure. It also defines three types of loads conditions: Normal (N), abnormal (A), and transportation and erection (T). Partial safety factors for ultimate strength analysis and fatigue analysis are defined for three classes of components and the different load conditions.

Table 9-4 shows the various groupings of design situation, wind condition, type of design analysis, and partial safety factor. This is the minimal set of design cases that must be analyzed. Turbine manufacturer must document the design analysis for each component and the system as whole for at least the 12 design cases.

Design Wind Conditions

In order to illustrate the design process, two types of wind regimes are considered: Normal operations and extreme conditions. The extreme conditions fall into two categories with respect to the frequency of wind events: 1- and 50-year events. Depending on the classification of the turbine that is being designed (Class IA, ..., IIIC) appropriate wind conditions are used, as described in Table 8-3.

Normal Wind Profile Model (NWP)

For computing design loads under normal wind conditions the following parameters are used, per IEC 61400-1:[7]

$$V_{ave} = 0.2 \, V_{ref} \qquad (9\text{-}1)$$

Design Situation	Wind Condition	Other Conditions	Type of Analysis	Partial Safety Factors
Power production	NTM $V_{in} < V_{hub} < V_{out}$	For extrapolation of extreme events	U	N
	NTM $V_{in} < V_{hub} < V_{out}$		F	*
	ETM $V_{in} < V_{hub} < V_{out}$		U	N
	ECD $V_{hub} = V_r - 2$ m/s, V_r, $V_r + 2$ m/s		U	N
	EWS $V_{in} < V_{hub} < V_{out}$		U	N
Power production plus occurrence of fault	NTM $V_{in} < V_{hub} < V_{out}$	Control system fault or loss of electrical network	U	N
	NTM $V_{in} < V_{hub} < V_{out}$	Protection system or preceding internal electrical fault	U	A
	EOG $V_{hub} = V_r \pm 2$ m/s, V_{out}	External or internal electrical fault including loss of electrical network	U	A
	NTM $V_{in} < V_{hub} < V_{out}$	Control, protection, or electrical system faults including loss electrical network	F	*
Start up	NWP $V_{in} < V_{hub} < V_{out}$		F	*
	EOG $V_{hub} = V_r \pm 2$ m/s, V_{out}		U	N
	EDC $V_{hub} = V_r \pm 2$ m/s, V_{out}		U	N
Normal shutdown	NWP $V_{in} < V_{hub} < V_{out}$		F	*
	EOG $V_{hub} = V_r \pm 2$ m/s, V_{out}		U	A
Emergency shutdown	NTM $V_{hub} = V_r \pm 2$ m/s, V_{out}		U	N
Parked (standing still or idling)	EWM 50-year recurrence period		U	N
	EWM 50-year recurrence period	Loss of electrical network connection	U	A
	EWM 1-year recurrence period	Extreme yaw misalignment	U	N
	NTM $V_{hub} < 0.7 V_{ref}$		F	*
Parked and fault conditions	EWM 1-year recurrence period		U	A
Transport assembly, maintenance and repair	NTM V_{maint} to be stated by the manufacturer		U	T
	EWM 1-year recurrence period		U	A

Source: International Electrotechnical Commission. *Wind Turbine Part 1 Design Requirements* . Geneva, 2005. IEC 61400-1 Edition 3. With permission from IEC.

TABLE 9-4 Design Load Cases from IEC 61400-1

Probability distribution of wind speed at the hub is assumed to be a Rayleigh distribution:

$$p\,(V_{hub}) = 1 - e^{-\pi(V_{hub}/2\,V_{ave})^2} \tag{9-2}$$

Wind profile as a function of height is assumed to follow the power law with shear of $\gamma = 0.2$.

$$V\,(z) = V_{hub}(z/z_{hub})^{\gamma} \tag{9-3}$$

The turbulence standard deviation under normal conditions is assumed to be:

$$\sigma_1 = I_{ref}(0.75\,V_{hub} + 5.6) \tag{9-4}$$

Extreme Wind Speed Model (EWM)

For computing design loads under steady extreme wind conditions the following parameters are used or 50-year and one-year extreme wind speed:

$$V_{e50}\,(z) = 1.4\,V_{ref}(z/z_{hub})^{0.11} \tag{9-5}$$

$$V_{e1}\,(z) = 0.8\,V_{e50}(z) \tag{9-6}$$

For computation of steady extreme loads, a misalignment of $\pm 15°$ is assumed.

For turbulent extreme wind the following wind conditions are assumed:

$$V_{50}\,(z) = V_{ref}(z/z_{hub})^{0.11} \tag{9-7}$$

$$V_1\,(z) = 0.8\,V_{e50}(z) \tag{9-8}$$

$$\sigma_1 = 0.11\,V_{hub} \tag{9-9}$$

Similar to the normal wind profile model above, IEC 61400-1 defines wind conditions for all other cases EWM, NTM, ETM, EDC, ECD, EWS, and EOG. The wind conditions are used to compute the loads on various components. The safety factors for ultimate strength and fatigue are then used to size the components. Computations of design loads has evolved to use of software programs that model aeroelasticity of turbines subject to the wind regimes defined in the design cases. Software programs in this domain include FLEX 5 (developed by Stig Øye, Danish Technical University), HAWC2[8] (developed by Risoe, Denmark), Bladed[9] (developed by Garrad Hassan, UK) and others. In these programs, the blades, nacelle, and tower are discretized to create a finite element model (FEM). The structural dynamics of this system is then simulated in the time domain by applying: (a) Wind loads as a function of time, (b) using the equations of motion, and (c) properties of materials to resist forces, moments, and torsion.

Turbine Certification

Certification of a utility-scale wind turbines has become a necessity. It gives the buyers, financiers, and grid operators confidence that a certified WTG has undergone design review, field testing, and other processes to ensure that there is integrity to the claims of the manufacturer. Specifically, certification provides an independent verification of manufacturer's claims about Class of turbine (Class I, II, III; A, B, C), power production curve, noise emission curve, structural integrity, and safe operations.

There are four entities involved in certification:

- Turbine manufacturer that wishes to obtain a "type certification." At the end of the certification process for a specific type of turbine, a manufacturer will get a type certificate for the turbine.
- International Electrotechnical Commission (IEC), the international standards organization. IEC defines and publishes the standards for almost all aspects of a wind turbine.
- *Accredited Certification Agent.* A turbine manufacturer contracts with an accredited certification agent to certify a turbine. There are several companies that certify utility scale turbines, the leading companies of which are: DNV in cooperation with Risoe, Germanischer Lloyd, TUV Nord, and others.
- *Accredited Design Evaluation and Testing Agent.* A certification agent does not perform all the design evaluations and it may not perform testing. Specialized engineering companies independently evaluate designs and, if necessary, perform detailed finite element method (FEM) analysis of specific components. Specialized testing companies perform independent testing of turbines and the results are submitted to the certification agent.

Type certification process, according to IEC WT01, involves three mandatory modules and one optional module:[10,11]

- *Design Evaluation Conformity Statement.* The intent of this module is to evaluate the conformance of design and documentation with established standards (like IEC 61400-1).

The design assessment verifies that all the design cases, load assumptions, and safety factors used in design of all the major components, blades, drive train, and tower meet the established standard. It also verifies the controls, protection systems and electrical components, manufacturing plan, installation plan, operations and maintenance plan, and personnel safety plan. In addition, a review of documentation pertaining to the turbine operations, commissioning, and maintenance is performed.

- *Type Testing Conformity Statement.* In this module, the turbine is functionally tested and the design loads on key components are verified through measurement in a realistic test environment. NREL's National Wind Turbine Center (NWTC) is one of the centers in the United States that provides type testing services. In addition to safety, functional tests, and verification of performance curves the following are also performed: Dynamic and static blade tests and load measurements of other key components under a variety of loading scenarios— different wind speeds, different turbulence, startup, normal shutdown, emergency shutdown, variety of failure modes, and others.

- *Manufacturing Conformity Statement.* In this module, the manufactured turbine and the manufacturing processes are evaluated to ensure that (a) the manufacturing processes conform to the specifications in the design document about how components should be made, (b) all the other associated processes like procurement and quality assurance are adequate to ensure quality in manufacturing. ISO 9001 accreditation of the manufacturing processes, although not required, takes care of most of the requirements.

- *Type Characteristics Measurements (optional).* In this module, measurements are performed to validate the power production curve, power quality parameters, and noise emissions of the turbine.

In the Design Evaluation Conformity Statement module of the certification process, the assumptions, inputs, and results of the aeroelasticity analysis are submitted to certification agency for verification. The certifying agency or the design evaluation agency verifies the calculations in some cases repeats the calculations with a different software package.

After the conduct of the above evaluations, if the turbine complies with the standards, then a verification report and type certificate is issued by the certification entity.

Vensys Energy, AG, Neunkirchen, Germany

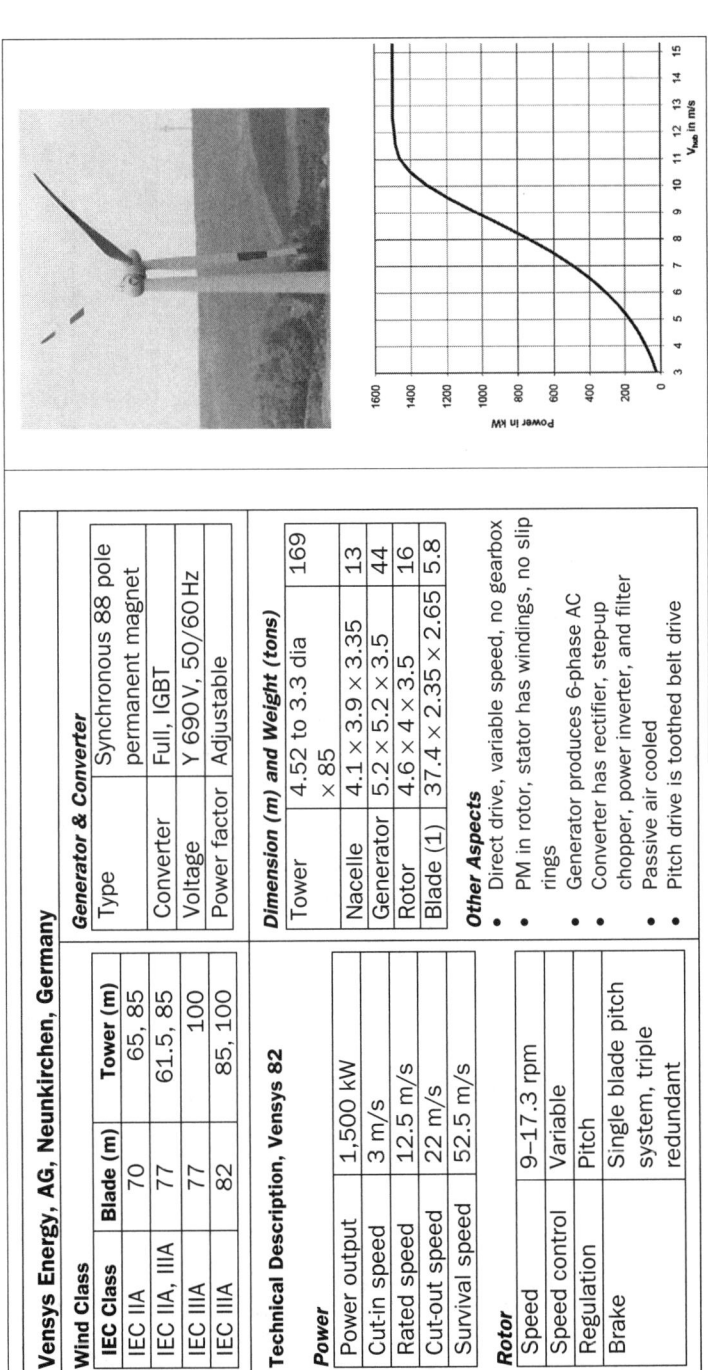

Wind Class

IEC Class	Blade (m)	Tower (m)
IEC IIA	70	65, 85
IEC IIA, IIIA	77	61.5, 85
IEC IIIA	77	100
IEC IIIA	82	85, 100

Technical Description, Vensys 82

Power

Power output	1,500 kW
Cut-in speed	3 m/s
Rated speed	12.5 m/s
Cut-out speed	22 m/s
Survival speed	52.5 m/s

Rotor

Speed	9–17.3 rpm
Speed control	Variable
Regulation	Pitch
Brake	Single blade pitch system, triple redundant

Generator & Converter

Type	Synchronous 88 pole permanent magnet
Converter	Full, IGBT
Voltage	Y 690V, 50/60 Hz
Power factor	Adjustable

Dimension (m) and Weight (tons)

Tower	4.52 to 3.3 dia × 85	169
Nacelle	4.1 × 3.9 × 3.35	13
Generator	5.2 × 5.2 × 3.5	44
Rotor	4.6 × 4 × 3.5	16
Blade (1)	37.4 × 2.35 × 2.65	5.8

Other Aspects

- Direct drive, variable speed, no gearbox
- PM in rotor, stator has windings, no slip rings
- Generator produces 6-phase AC
- Converter has rectifier, step-up chopper, power inverter, and filter
- Passive air cooled
- Pitch drive is toothed belt drive

Source: With permission from Vensys Energy AG.

TABLE 9-5 Detailed Specifications of Vensys 1.5-MW Turbine

Vergnet Eolien, Ormes, France

Wind Class

IEC Class	Rotor (m)	Tower (m)
IEC I	55	60
IEC II	58	70
IEC III	62	70

Generator & Converter

Type	Asynchronous squirrel cage
Converter	Full, IGBT
Voltage	Y 690 V, 50/60 Hz
Power factor	Adjustable

Technical Description GEV HP 1 MW

Power, Class III

Power output	1,000 kW
Cut-in speed	3 m/s
Rated speed	14.5 m/s
Cut-out speed	25 m/s
Survival speed	52.5 m/s operating 86 m/s lowered

Rotor

Speed	12–23 rpm
Speed control	Variable speed through PLC
Regulation	Pitch
Brake	Disc
Gearbox	3-stage planetary & spur gear

Dimension (m) and Weight (tons)

Tower	70	78
Nacelle	14.8 × 4.6 × 4.9 m³	65
Generator		4.7
Gearbox		6.5
Blade (1)	30	4.5

Other Aspects

- Two-blade rotor
- Delta-3 teetering hub with elastic damper to reduce variability in thrust load
- Self-erecting, requires 90-ton crane
- Bird-like lowering system to lower upwind part of nacelle with rotor blades during hurricane and for maintenance
- Except blade, all parts fit into sixteen 40 ft containers

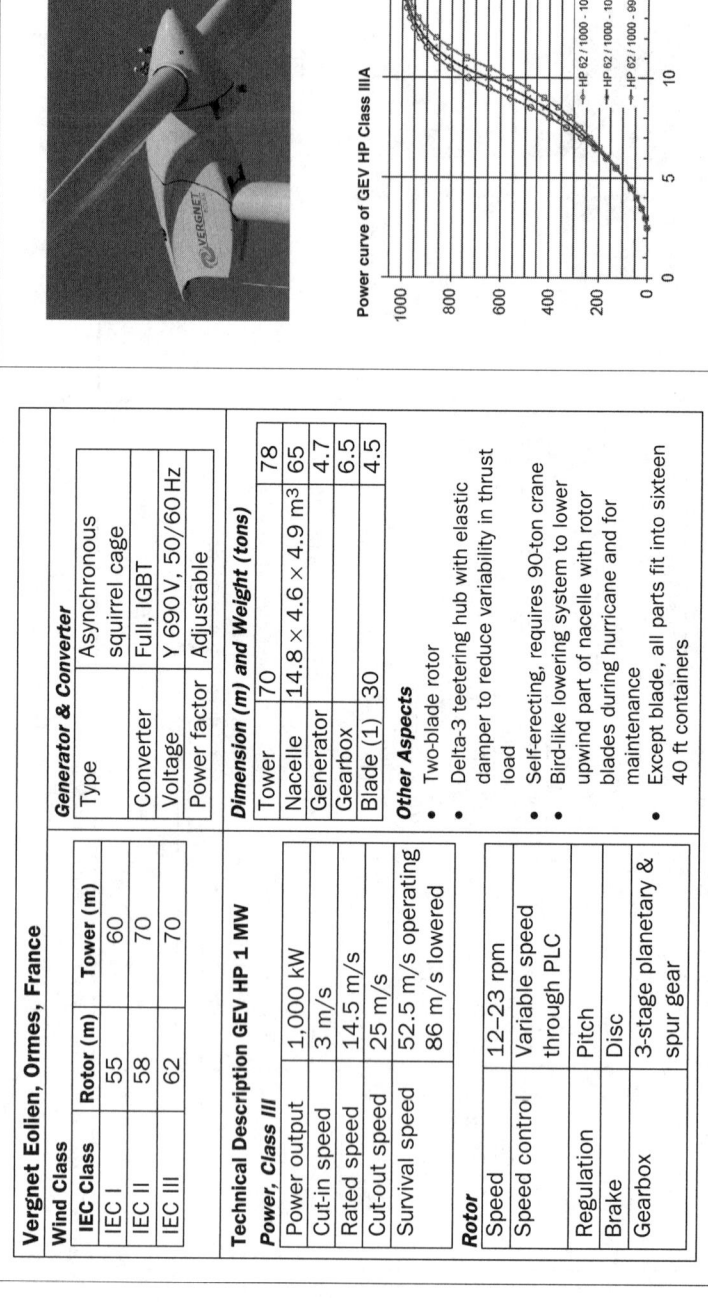

Source: With permission from Vergnet.

TABLE 9-6 Detailed Specifications of Vergnet 1-MW Turbine

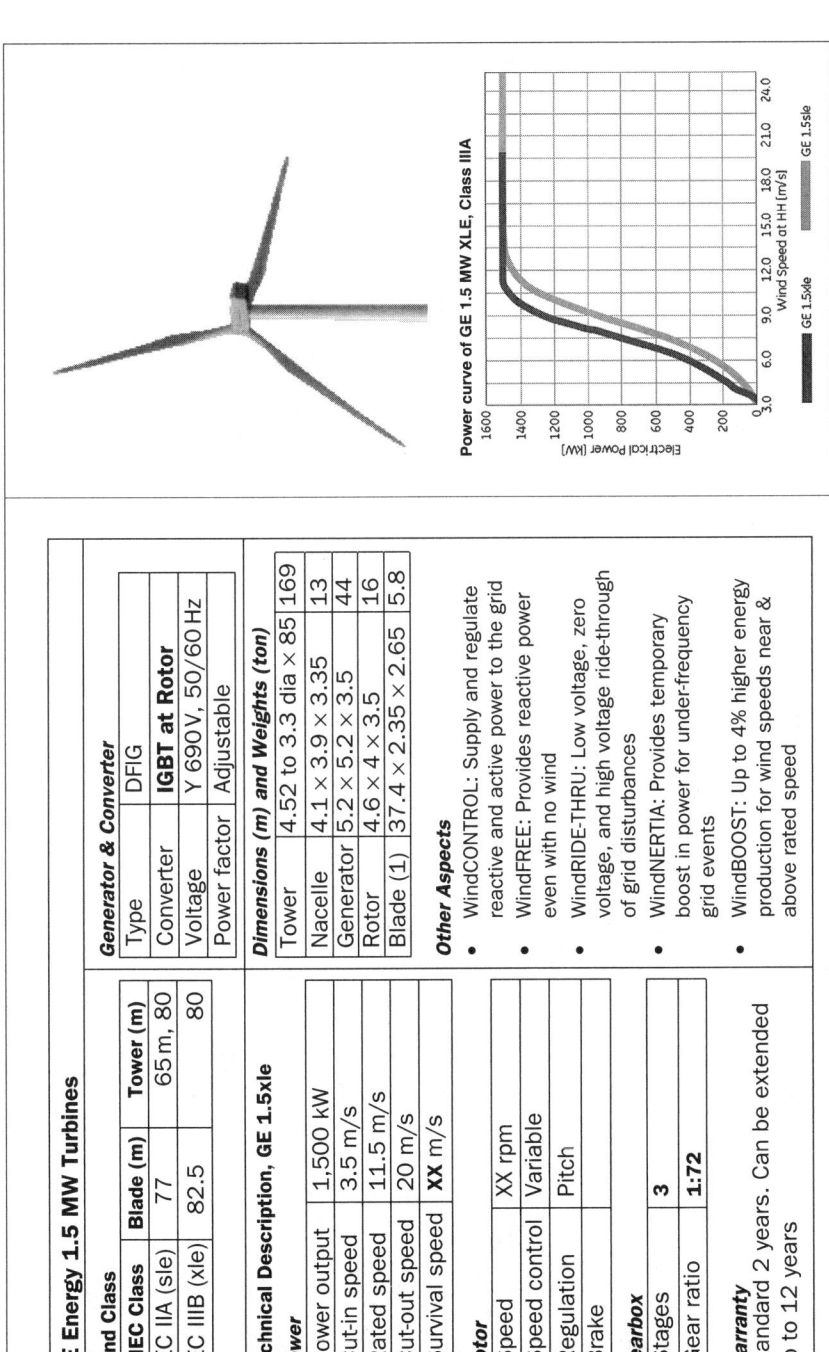

GE Energy 1.5 MW Turbines

Wind Class

IEC Class	Blade (m)	Tower (m)
IEC IIA (sle)	77	65 m, 80
IEC IIIB (xle)	82.5	80

Technical Description, GE 1.5xle

Power

Power output	1,500 kW
Cut-in speed	3.5 m/s
Rated speed	11.5 m/s
Cut-out speed	20 m/s
Survival speed	**XX m/s**

Rotor

Speed	XX rpm
Speed control	Variable
Regulation	Pitch
Brake	

Gearbox

Stages	3
Gear ratio	**1:72**

Warranty

Standard 2 years. Can be extended up to 12 years

Generator & Converter

Type	DFIG
Converter	**IGBT at Rotor**
Voltage	Y 690V, 50/60 Hz
Power factor	Adjustable

Dimensions (m) and Weights (ton)

Tower	4.52 to 3.3 dia × 85	169
Nacelle	4.1 × 3.9 × 3.35	13
Generator	5.2 × 5.2 × 3.5	44
Rotor	4.6 × 4 × 3.5	16
Blade (1)	37.4 × 2.35 × 2.65	5.8

Other Aspects

- WindCONTROL: Supply and regulate reactive and active power to the grid
- WindFREE: Provides reactive power even with no wind
- WindRIDE-THRU: Low voltage, zero voltage, and high voltage ride-through of grid disturbances
- WindNERTIA: Provides temporary boost in power for under-frequency grid events
- WindBOOST: Up to 4% higher energy production for wind speeds near & above rated speed

Power curve of GE 1.5 MW XLE, Class IIIA

TABLE 9-7 Detailed Specifications of a GE 1.5-MW Turbine

Vestas V90-1.8 MW

Wind Class

IEC Class	Blade (m)	Tower (m)
IEC IIA	90	80, 95

Technical Description

Power

Power output	1,800 kW
Cut-in speed	4 m/s
Rated speed	12 m/s
Cut-out speed	25 m/s
Survival speed	m/s

Rotor

Speed	14.5 rpm
Speed control	Variable
Regulation	Pitch
Brake	Full blade feathering with 3 pitch cylinders

Gearbox

Stages	Three-stage planetary/ helical
Gear ratio	

Generator & Converter

Type	Six-pole asynchronous with variable speed
Converter	
Voltage	Y 690 V, 50/60 Hz
Power factor	Adjustable

Dimensions (m) and Weights (t)

Tower	80	155
Nacelle	4 × 3.4 × 10.4	70
Generator		
Rotor	3.3 × 4 × 4.2	18
Blade (1)	44 × 3.5	6.7

Other Aspects

- Vestas Converter Unity System for constant and consistent output to grid
- Complete variable speed capability
- LVRT

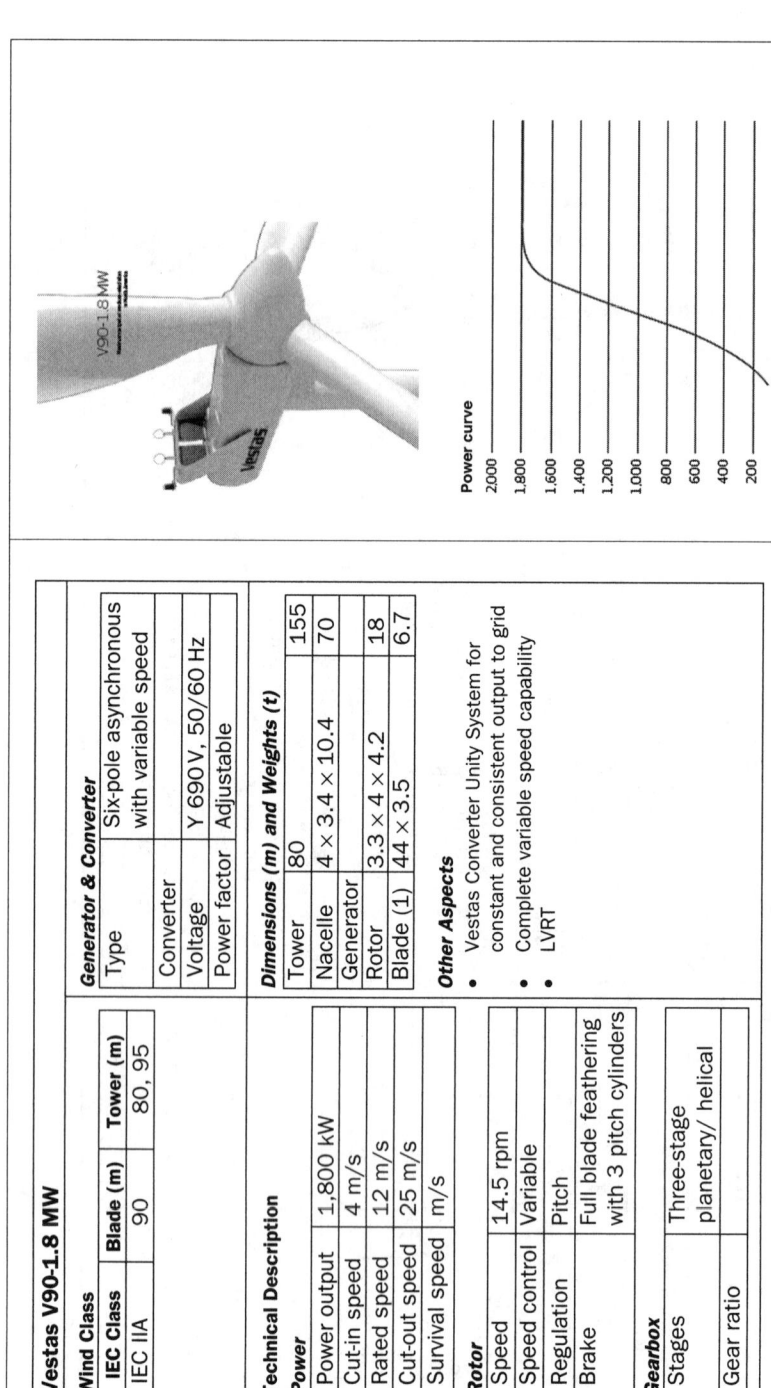

Power curve

TABLE 9-8 Detailed Specifications of a Vestas V90 1.8-MW Turbine

References

1. Scheuer, P., and Langemeier, C. Big challenges: The role of resins in wind turbine rotor blade development. *Reinforced Plastics.com.* February 1, 2010.
2. Ashwill, T. Blades: Trends and Research Update [Online] 5 12, 2008. http://www.sandia.gov/wind/2008BladeWorkshop/PDFs/Mon-02-Ashwill.pdf.
3. Red, C. "Offshore wind: How big will blades get?" *Composites Technology.* 2009, May.
4. Schlereth, A. "Maintenance." *Wind Systems.* 2010, Febr.
5. Morgan, K., and Ntambakwa, E., *Wind Turbine Foundation Behavior and Design Considerations,* AWEA WindPower Conference, Houston, TX, 2008.
6. Earth Systems Southwest. Patrick & Henderson Tensionless Pier [Online] http://earthsys.com/Library/.../The-Patrick-Henderson-Tensionless-Pier.pdf.
7. International Electrotechnical Commission. *Wind Turbine Part 1 Design Requirements.* Geneva, 2005. IEC 61400-1 Edition 3.
8. How 2 HAWC2, the user's manual. *Risø National Laboratory DTU* [Online] December 2007. http://www.risoe.dtu.dk/da/knowledge_base/publications/reports/ris-r-1597.aspx?sc_lang=en.
9. GH Bladed: Wind Turbine Design Software. *GH Bladed* [Online] Garrad Hassan, UK. http://www.garradhassan.co.uk/products/ghbladed/index.php.
10. DNV. DNV-Type Certification [Online] 2010. http://www.dnv.com/industry/energy/segments/wind_wave_tidal/wind_turbine_type_certification/.
11. NREL. What is Wind Turbine Certification? [Online] NREL, 2010. http://wind.nrel.gov/cert_stds/Certification/certification/index.html# Design Evaluation.

CHAPTER 10

Basics of Electricity and Generators

I shall make electricity so cheap that only the rich can afford to burn candles

—Thomas Alva Edison

Introduction

In this chapter, the generator side of wind turbines is described. The aerodynamics chapters alluded to the notion that the type of generator has a significant impact on the efficiency of the turbine rotor. These aspects are covered in this chapter. The chapter starts with basic principles of electromagnetism followed by basic principles of alternating current and basic principles of electrical machines. This is followed by descriptions of synchronous, permanent magnet, and asynchronous generators. The chapter concludes with a comparison of most commonly used types of generators in wind turbines.

Basic Principles of Electromagnetism

A current carrying straight conductor creates a magnetic field (lines of force) that is circular and in a plane that is perpendicular to the conductor.

A current carrying circular conductor creates a magnetic field that is perpendicular to the plane of the circular conductor.

A current carrying circular conductor that is cylindrically wound in the form of a coil produces a magnetic field that is parallel to the axis of the cylinder. When a core of ferromagnetic material is placed inside the coil, then the magnetic field is magnified and this is called

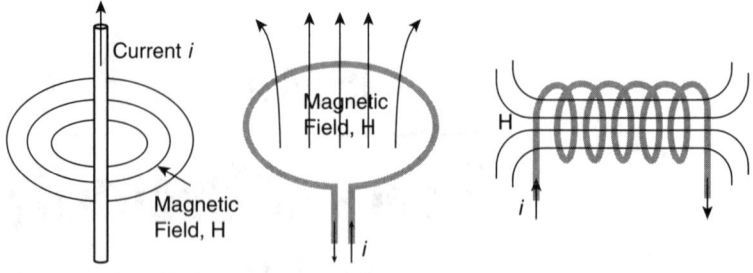

FIGURE 10-1   Illustration of magnetic field created by current in a conductor.

an electromagnet. In electrical machines cylindrically wound coils are used in the stator and sometimes in the rotor.

These three basic principles of electromagnetism are illustrated in Fig. 10-1 and are applicable to generators.[1]

Faraday's Law of Induction

The voltage induced in a conductor is proportional to the rate of change of lines of magnetic field that pass through the conductor. It is important to note that for voltage to be induced there must be cutting of magnetic field. There are several ways to cut magnetic field. The obvious one is to move the conductor in such a manner that magnetic field is cut. A second method is to fix the conductor, but to generate a magnetic field that changes over time; the changing magnetic field provides the relative movement that leads to cutting of magnetic field. A third method is to fix the conductor and move the magnet or electromagnet, thereby creating the relative movement.

Lenz Law

Induced voltage from Faraday's law will cause a current to flow in a closed circuit in such a direction that the magnetic field, which is caused by the current, will oppose the change of magnetic field that induced the voltage.

Lorenz Law or Biot-Savart Law

When a current carrying conductor is placed in a magnetic field in such a manner that the current is perpendicular to the magnetic field, then a force is exerted on the conductor. The magnitude of the force is product of flux density, current, and length of conductor. The direction of the force is determined by the left-hand rule.

Basic Principles of Alternating Current

An alternating current is a sinusoidal variation in amplitude of current.

$$i = i_m \sin \omega t \tag{10-1}$$

where i is alternating current, i_m is the peak current, ω is the frequency, and t is time. If the frequency is 60 Hz, then $\omega = 2\pi.60 = 120\pi$ radians per second.

A method to get an alternating current is to apply an alternating electromotive force (same as voltage) across a pure resistor R.

$$i = \frac{e}{R} = \frac{e_m \sin \omega t}{R} \tag{10-2}$$

The power in such a circuit is the product of voltage and current.

$$P = e.i = e_m i_m \sin^2 \omega t = \frac{1}{2} e_m i_m (1 - \cos 2\omega t) \tag{10-3}$$

$$P_{av} = \frac{1}{2} e_m i_m \tag{10-4}$$

In this circuit, current and voltage are in phase. Next, consider a pure inductive circuit with no resistance in which the current and voltage are at 90° phase difference. A pure inductor is closely approximated by winding a copper coil on a laminated iron core. When alternating current is passed through such a circuit, an alternating magnetic field is created. This magnetic field cuts the copper coil and, therefore, creates a self-induced electro-motive force (EMF). The self-induced EMF opposes the applied EMF. If $i = i_m \sin \omega t$ is the current, then the self-induced EMF e' is:

$$e' = -L \frac{di}{dt} = -L \omega i_m \cos \omega t = -e \tag{10-5}$$

where e is the applied voltage. Note the current and voltage are 90° out of phase. The power consumed in this circuit is:

$$P = e.i = e_m i_m \sin \omega t \cos \omega t = \frac{1}{2} e_m i_m \sin(2\omega t) \tag{10-6}$$

$$P_{av} = 0 \tag{10-7}$$

The power alternates between positive and negative at twice the frequency. When the current is rising ($\omega t \in (0, \frac{\pi}{2})$), circuit supplies power to create a magnetic field, when the current is falling ($\omega t \in (\frac{\pi}{2}, \pi)$), magnetic field decreases. This decreasing magnetic field supplies power to the circuit.

Consider a circuit with resistance (R) and inductance (L) in series. Assume the potential drop is $e_m \sin \omega t$ and the current is $i = i_m \sin(\omega t - \varphi)$, then

$$e_m \sin \omega t = i_m R \sin(\omega t - \varphi) + i_m L\omega \cos(\omega t - \varphi)$$
$$= i_m(R(\sin \omega t \cos \varphi - \sin \varphi \cos \omega t) + L\omega(\cos \omega t \cos \varphi + \sin \omega t \sin \varphi))$$
$$= \sin \omega t \, [i_m R \cos \varphi + i_m L\omega \sin \varphi] + \cos \omega t \, [-i_m R \sin \varphi + i_m L\omega \cos \varphi] \tag{10-8}$$

Equating the $\sin \omega t$ and $\cos \omega t$ terms yields:

$$0 = -i_m R \sin \varphi + i_m L\omega \cos \varphi \implies \tan \varphi = \frac{L\omega}{R} \tag{10-9}$$

$$e_m = i_m R \cos \varphi + i_m L\omega \sin \varphi \implies |e_m| = i_m \sqrt{R^2 + (L\omega)^2} \tag{10-10}$$

$$P = e_m i_m \sin \omega t \sin(\omega t - \varphi) = \frac{1}{2} e_m i_m(\cos \varphi - \cos(2\omega t - \varphi)) \tag{10-11}$$

$$P_{av} = \frac{1}{2} e_m i_m \cos \varphi \tag{10-12}$$

The power delivered to the circuit is lower than a pure resistance circuit. This is because the voltage and current are not in phase. Therefore, the sum of the product of the two quantities over time is lower. $\cos \varphi$ is called the power factor: the ratio of the true power and the apparent power.

Three phase alternating current is current in three conductors with a phase difference of $120°$.

$$i_1 = i_m \sin \omega t, \, i_2 = i_m \sin(\omega t + 120), \, i_3 = i_m \sin(\omega t + 240) \tag{10-13}$$

Basic Principles of Electrical Machines

Electricity generators are rotating electrical machines that have a stator and a rotor. Stator is the stationary part of the generator. This is the outer cylindrical casing of the generator that contains three, six, or other multiple of three numbers of coils. In rare configurations, the stator may be the inner cylinder. To simplify the exposition only three coils will be considered as shown in Fig. 10-2; a-a', b-b', and c-c' are the three coils. The coils are connected to three-phase AC.

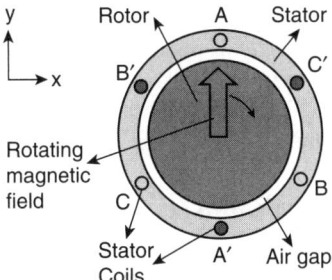

FIGURE 10-2 Schematic of a generator. The stator has three coils arranged 120° apart: AA′, BB′, CC′. Rotors of synchronous generators contain permanent magnets or windings that are excited by DC. Rotors of induction generators contain copper bars or copper windings.

The magnetic field created by the three coils is:

$$h_1 = H \sin \omega t$$

$$h_2 = H \sin(\omega t + 120)$$

$$h_3 = H \sin(\omega t + 240) \tag{10-14}$$

$$h_x = h_1 + h_2 \cos 120 + h_3 \cos 240 = H \sin \omega t \left[1 + \frac{1}{4} + \frac{1}{4}\right] = \frac{3}{2} H \sin \omega t \tag{10-15}$$

$$h_y = h_2 \sin 120 + h_3 \sin 240 = H \cos \omega t \left[\frac{3}{4} + \frac{3}{4}\right] = \frac{3}{2} H \cos \omega t \tag{10-16}$$

Resultant magnetic field is:

$$h_r = \sqrt{h_x^2 + h_y^2} = \frac{3}{2} H \tag{10-17}$$

The direction of the magnetic field is:

$$\tan \theta = \frac{\sin \omega t}{\cos \omega t} = \tan \omega t => \theta = \omega t \tag{10-18}$$

Therefore, the resultant magnetic field has constant magnitude and its direction changes with time with an angular velocity of ω (see Fig. 10-2). This property is crucial to the operation of generators. Stators contain three coils that are arranged such that the axis of the coils are 120° apart. When a three-phase alternating current is fed to the coils, then a rotating magnetic field is generated. The magnetic field created by the stator is akin to a spinning permanent magnet placed on the axis of the stator with N-S pointing radially and the axis of the spin coincides with the axis of the stator, as shown in Fig. 10-2.

The rotor is the moving part of the generator. This is connected to the turbine rotor to which the blades are attached. The different types of rotors will be explained later in the chapter.

Conversion of Mechanical to Electrical Power

A wind turbine captures linear kinetic energy of wind and converts it into rotational kinetic energy. This rotational kinetic energy from a turbine is transferred to an electrical generator either directly or through a gearbox. The electrical generator then converts the mechanical energy into electrical energy.

Consider a fixed-speed generator with a steady angular velocity of ω; let E be the amount of kinetic energy per second and ε be the overall efficiency of power transfer from wind to the generator. The torque transferred to the rotor of generator is:

$$\tau_g = \varepsilon \frac{E}{\omega} \qquad (10\text{-}19)$$

Next, consider the stator of the generator that is connected to a grid with terminal AC voltage of V_T with a frequency of ω. Power in an electrical circuit is $V_T i \cos \varphi$, where i is the current flowing through the circuit and φ is the phase difference between the voltage and current.

$$\tau_g \omega = V_T i \cos \varphi \qquad (10\text{-}20)$$

Simplistically, in the above equation, ω and V_T are constant. Therefore, torque governs the amount of current and phase angle. The remaining question is: How is the mechanical torque that is applied to the rotor of the generator balanced by an opposing torque? If the torque is not balanced, the rotor will accelerate out of control. The answer is: The balancing torque is provided by two mechanisms described below. Remember that the stator creates a rotating magnetic field with angular speed ω and the rotor is rotating with an angular speed of ω.

- In a synchronous generator, the magnetic pull between the rotor and stator provides the balancing torque. This magnetic pull is the tangential force applied when the opposite poles of the stator and rotor are not radially aligned.

- In an asynchronous generator, balancing torque is provided by the tangential force experienced by current carrying conductors in the rotor in the presence of magnetic field created by the stator. For the torque to be generated, the speed of rotor has to be slightly higher than the speed of the stator's magnetic field.

In summary, assuming 100% efficiency in conversion of mechanical power from turbine rotor to electrical generator and direct connection between the turbine rotor and generator rotor, when a steady mechanical power of $\tau_g \omega$ is used to turn the generator, then electrical power of an equal amount is produced when current flows through the circuit.

Synchronous Generator

Rotor of a synchronous generator is either permanent magnet or DC-excited electromagnet. The simplest generator is a two-pole (one north pole and one south pole) rotor with the stator connected to the grid. The grid supplies the voltage and frequency, meaning the voltage and frequency of the stator circuit are fixed. The grid also supplies the current to energize the stator. The stator produces a rotating magnetic field of angular speed ω, as described above. If a compass were placed at the center, it would rotate with the magnetic field. Instead of a compass, the rotor has a permanent or electro magnet that moves with the stator magnetic field with angular speed ω. At zero load, the speed of rotation of the rotor is the same (no relative speed between rotor and stator) and the phase difference is zero (north pole of stator-generated magnetic field is aligned with the south pole of the rotor). There is zero torque in this scenario and, therefore, zero power.

From electric circuit standpoint, the grid supplies the terminal voltage of V_T to the stator and current i. The rotating magnetic field of the rotor (which is synchronized with the rotating magnetic field of the stator) cuts conductor in the stator and, therefore, induces EMF in the stator coils; this generated EMF will be called E_g. When the stator and rotor poles are perfectly lined up, E_g and V_T are 180° apart. Perfect alignment of stator and rotor poles means that the angular speed of stator's magnetic field is the same as the angular speed of rotation of the rotor and the opposite poles of stator and rotor face each other (radial alignment). The magnitude of E_g is determined by the strength of the permanent magnet or the DC excitation current of the electromagnet. Assuming zero resistance in stator coils, the only load in the stator circuit is pure inductance. Therefore, current flowing through the stator coils is 90° from $(V_T - E_g)$.

Note that even when no power is delivered to the grid by the generator, the grid is still supplying current to energize the magnetic field of the stator. In this case, grid current drawn (by the stator of the generator) is out of phase with grid voltage. Therefore, no "real" power is drawn from the grid. However, the grid delivers "reactive" power to the stator of the generator—power is delivered to the stator from the grid during one-half of the cycle and power is returned to the grid during the other half of the cycle. Therefore, over one complete cycle, the net power delivered to or drawn from the grid is zero. For 60-Hz frequency, one cycle is 1/60 of second.

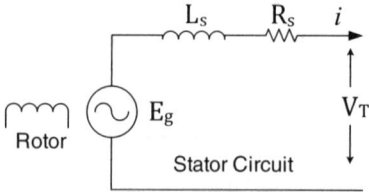

FIGURE 10-3   The stator circuit is energized by the magnetic field of the rotor. This induces EMF of E_g, and current i flows through the coils of the stator with inductance L_s and resistance R_s. The grid is supplied current i at terminal grid voltage V_T.

A simple circuit of a synchronous generator is seen in Fig. 10-3. In the circuit, the rotor produces a magnetic field either through a permanent magnet or a DC-excited wound rotor. This induces EMF E_g in the coils of the stator.

The equation for voltages are:

$$-\overline{E_g} + i\overline{Z} + \overline{V_T} = 0 \tag{10-21}$$

where

$$Z = R_s + j\omega L_s \tag{10-22}$$

Assuming that the stator coil has negligible resistance, $Z = j\omega L_s$. Assuming $R_s = 0$ and the phase difference between $\overline{E_g}$ and $\overline{V_T}$ is δ:

$$\begin{aligned}
\overline{E_g} &= E_g \lfloor \delta \\
\overline{V_T} &= V_T \lfloor 0 \\
\overline{Z} &= Z \lfloor 90° \\
i &= \frac{E_f \lfloor \delta - V_t \lfloor 0}{Z \lfloor 90°}
\end{aligned} \tag{10-23}$$

For reference, an angle of V_T is chosen to be zero in the above equation. When $\delta = 0$, then current is orthogonal to the terminal voltage. Therefore, it delivers no torque or power (see Fig. 10-4).

In generator mode, mechanical power (P) is applied to the rotor of the generator from a turbine. This torque ($\tau = P/\omega$) causes the rotor to accelerate temporarily from its state of zero power and constant angular velocity. As the rotor accelerates, its poles move ahead of the rotating magnetic field of the stator (see Fig. 10-4b). This causes the stator's magnetic field to apply a tangential force on the rotor. The force is magnetic in nature: the north pole of the stator pulls on the south pole of the rotor that has moved ahead. Simultaneously, the phase difference between E_g and V_T changes from 180 to 180-δ degrees as rotor's

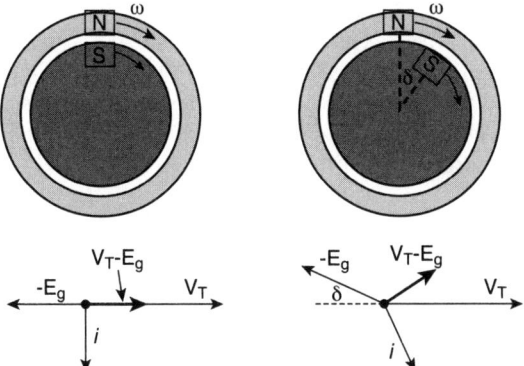

FIGURE 10-4 Stator's magnetic field rotates with angular speed ω, as does the rotor. When the opposite poles of stator's magnetic field and rotor's pole are aligned, then $\delta = 0$ and the voltage vectors $(\overline{E_g}, \overline{V_T})$ are opposite and current is orthogonal to the terminal voltage. When load is applied, rotor's rotating pole then moves ahead of the stator, causing the induced EMF to be at an angle δ to the terminal voltage. This changes the current to have a component along the terminal voltage, thereby delivering power.

pole moves δ degrees ahead of stator's pole. This causes the angle between V_T and current i to be not 90°, which means electrical power is generated and delivered to loads on the grid. That is, mechanical power is converted to electrical power.

Analysis of Synchronous Generator

The relationship between rotational speed, grid frequency, and poles is:

$$f = \frac{p}{2}\frac{\omega}{60} \qquad (10\text{-}24)$$

where f is the frequency in Hz, p is the number of poles, and ω is the rotational speed of generator in revolutions per minute.

There are two types of synchronous generators: Fixed speed and variable speed. Fixed-speed generators have a constant angular speed of the rotor that is governed by the grid frequency. The stator of this type of generator is connected to the grid, from which it derives frequency and terminal voltage. For a grid-connected generator with 60 Hz AC power and two poles, the speed of rotor must be 3600 rpm or 60 revolutions per second (*rps*). For 50 Hz AC power, the speed of rotor is 50 rps.

Variable speed generators are not connected to the grid directly, but instead are connected to a power convertor, which converts variable frequency alternating current (AC) into direct current (DC), and then converts DC to AC at grid frequency. Variable-speed generators are typically connected to direct-drive turbines with no gearbox or a simple gearbox. The speed of rotors of turbine and generator depend on wind speed and are not fixed. This will be described in more detail later in the chapter.

The equation for total power in all three phases is:

$$P_T = 3\overline{V_T} . \overline{i} \tag{10-25}$$

Working in the complex domain, total power is:

$$P_T = P + jQ = 3\frac{V_T}{Z}(E_g(90 - \delta) - V_T\,90) \tag{10-26}$$

$$= 3\frac{V_T}{Z}(E_g\cos(90 - \delta) + E_g\,j\sin(90 - \delta)) - V_T\,j$$

$$\text{Real power, } P = 3\frac{V_T}{Z}(E_g\sin(\delta)) = P_{max}\sin\delta \tag{10-27}$$

$$\text{Reactive power, } Q = 3\frac{V_T}{Z}(E_g\cos(\delta) - V_T) \tag{10-28}$$

$$\tau = \frac{P}{\omega} = \frac{P_{max}\sin\delta}{\omega} = 3\frac{V_T}{Z\omega}(E_g\sin(\delta)) = \tau_{max}\sin\delta \tag{10-29}$$

For the purposes of this simplified analysis, assume a fixed-speed synchronous generator, therefore, the following are constants: f, E_g, V_T, ω, Z. The grid mandates fixed values of terminal voltage and frequency. Fixed frequency means fixed ω. Thus, the real power P and torque τ are functions of the power angle δ. A plot of power versus load angle is in Fig. 10-5. The maximum power that may be generated and the maximum torque that it can withstand are given by P_{max} and τ_{max}, respectively. When the applied torque is greater than τ_{max}, then the power output reduces, the rotor accelerates, and the generator pulls out of synchronism. Note the power and torque curves are both sinusoidal. When $\delta < 90$, larger torque is required to increase δ, which makes the system stable. However, when $\delta \geq 90$, lesser torque is required, meaning in this condition, the rotor will start accelerating because the resisting torque is smaller than the applied torque and will never to able to resist the additional torque. When this condition is achieved, the generator will be destroyed. Therefore, in order to avoid this condition, adequate protection must be put on the turbine or drive train side, such as aerodynamic or mechanical brakes.

Note, in a synchronous generator the external torque is resisted by the tangential force between poles of a magnet. The tangential force exists as long as the there is angular displacement between the poles. The same machine can, therefore, serve as both a generator and a motor. When the angular displacement is negative, the electrical

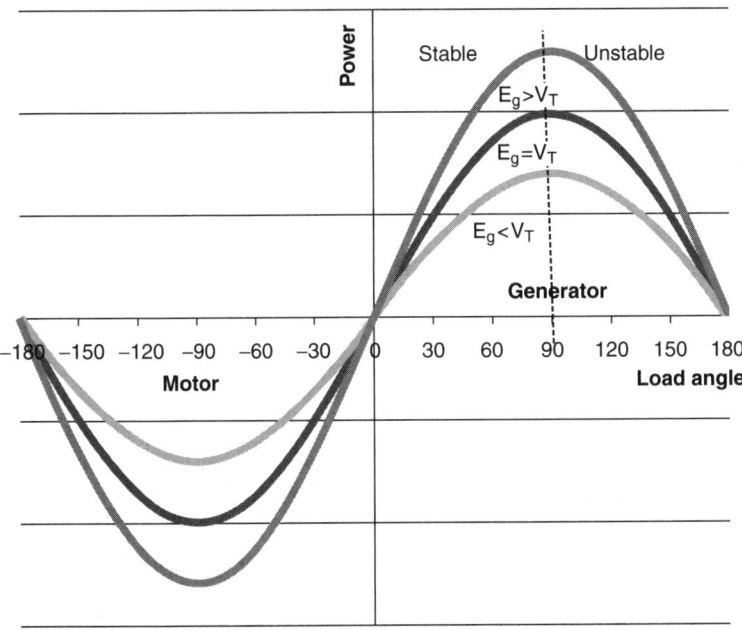

Figure 10-5 Power characteristics of a synchronous generator in a power versus load angle curves. Positive load angles makes this machine a generator. The power produced depends on the relationship between EMF generated (E_g) in the stator versus the terminal voltage (V_T).

energy provided to the stator coils is used to pull the lagging rotor. The lagging rotor in this case is connected to a mechanical load. If the steady load is more than the maximum allowable mechanical load, then the rotor stops.

Thus far, a grid connected synchronous machine was described, regardless of the nature of the rotor: Wound rotor or permanent magnet rotor. Next, a characteristic of a wound rotor is described, ability to control amount of reactive power (see schematic in Fig. 10-6). As indicated in Eq. (10-28), a synchronous generator produces reactive power. The amount of reactive power may be controlled by changing the DC excitation current in the rotor winding, thereby changing the value of E_g. Three modes of operation are possible: Underexcited mode $E_g < V_T$, with lagging power factor, which is current lags voltage; $E_g = V_T$, unity power factor; and overexcited mode $E_g > V_T$ with leading power factor, which is current leads voltage.

A grid connected synchronous machine in Fig. 10-6 that has its stator directly connected to the grid is forced to run at a fixed rotational speed. Next, synchronous machines that are not directly connected to the grid are described. In these machines, the stator power goes

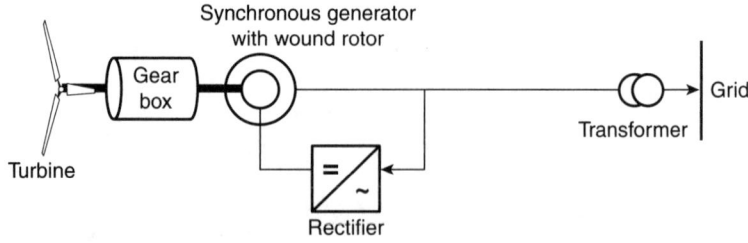

Figure 10-6 Schematic of a grid-connected synchronous generator with wound rotor. Power factor control is achieved by controlling the DC excitation of the wound rotor through the rectifier.

through a power converter (AC to DC and then back to AC) before delivering to the grid.

Variable-Speed Permanent Magnet Synchronous Generators

The inherent nature of wind energy demands variable speed generators. The reason may be seen in the torque-rotor speed curves for different wind speeds. A constant-speed generator is able to capture energy most efficiently only for single wind speed. In Fig. 10-9 (see later), the constant-speed generator captures the maximum power only for wind speed of 6 m/s; at other wind speeds, this type of generator does not operate at peak power. A variable speed turbine that can change rotor speed such that the turbine is operating at peak power for all wind speeds (below the rated wind speed) will yield higher energy output. This is the motivation for the movement toward variable speed generators.

A special case of permanent magnet synchronous generator (PMSG) is discussed in which the speed of the rotor is variable. With p as the number of poles in the rotor, and rotor turning at speed of ω_r, the EMF generated in the stator is:

$$E_g = K_{pm} \frac{\omega_r \, p}{2} \tag{10-30}$$

where K_{pm} is a constant that depends on the strength of the magnet, ω_r is the speed of the generator, and p is the number of poles.

Circuit and power equations are the same as Eqs. (10-25) to (10-29). For a grid-connected synchronous generator, V_T is known. However, for a variable-speed generator there is an intermediate variable frequency voltage (V_g) that is delivered to the rectifier.

$$-\overline{E_g} + i_g \overline{Z} + \overline{V_g} = 0 \tag{10-31}$$

Voltage, power, and torque are given by:[2]

$$V_g = E_g \cos \delta \tag{10-32}$$

$$P_g = \frac{3}{Z} V_g E_g \sin\delta = \frac{3}{2} \frac{E_g^2}{Z} \sin 2\delta = \frac{3}{8Z}(K_{pm}\omega_r\, p)^2 \sin 2\delta = \frac{3i_g^2 Z}{\tan\delta}$$

(10-33)

$$\tau_g = \frac{3}{8Z}(K_{pm}\, p)^2 \omega_r \sin 2\delta = \frac{3}{4L} K_{pm}^2\, p \sin 2\delta = T_{max} \sin 2\delta$$

$$= \frac{3i_g^2 Z}{\omega_r \tan\delta} = \frac{3i_g^2 Lp}{2\tan\delta}$$

(10-34)

where $Z = \omega L$ is the inductance of the stator. The angular frequency of generator current is $\omega = p\omega_r/2$, where ω_r is the angular speed of the rotor.

Variable-speed generators will generate voltage and current that vary in magnitude and frequency, unlike the constant voltage and frequency of the grid. Therefore, the voltage and frequency from these generators is conditioned to be compatible with the grid by converting from variable AC to DC and then back to constant frequency and voltage AC (see Figs. 10-7 and 10-8). The first part of conversion is done using rectifier and the second half is done using an inverter. In sequel, a pulse width modulated (PWM) type of inverter is discussed. The DC voltage and current after the AC output of PMSG is rectified are:

$$V_{DC} = \frac{V_g 3\sqrt{6}}{\pi}, i_{DC} = \frac{i_g \pi}{\sqrt{6}}$$

(10-35)

As expected, the DC power is $P_{DC} = V_{DC} i_{DC} = 3V_g i_g$.

Turning attention to the wind side of rotor, in a variable-speed turbine, the turbine rotor is directly connected to the generator rotor. One of the advantages of the variable-speed turbine is the ability to stay on optimal performance by changing the rotor speed as the wind speed changes (see Fig. 10-9). Next, a simplistic control mechanism is described that achieves this objective of keeping the rotation speed to an optimal value. This involves controlling how much current (i_{DC}) the PWM inverter draws (see Fig. 10-7). That is, imagine a controller

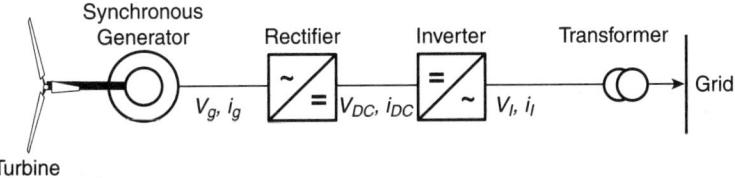

FIGURE 10-7 Schematic of a variable-speed synchronous generator with permanent magnet rotor.

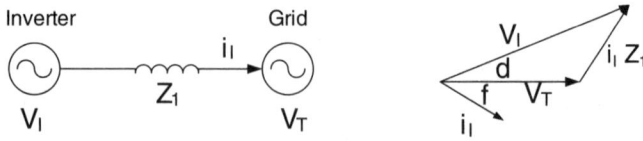

FIGURE 10-8 Schematic of power electronics and control of real and reactive power with a permanent magnet generator.

at the inverter that regulates how much i_{DC} is drawn by the inverter. Consider the two cases below:

- Lesser i_{DC} is drawn => lesser i_g flows through the stator coils => Lesser torque is produced to oppose the mechanical torque from wind => ω the angular speed of rotor increases

- Higher i_{DC} is drawn => higher i_g flows through the stator coils => higher torque is produced to oppose the mechanical torque from wind => ω the angular speed of rotor decreases.

In summary, the PWM inverter is controlled in such a manner that optimal values of i_{DC} is drawn from the rectifier. This optimal value determines the optimal value of i_g, which determines the optimal value of torque τ_g (Eq. 10-34), which determines the optimum value of

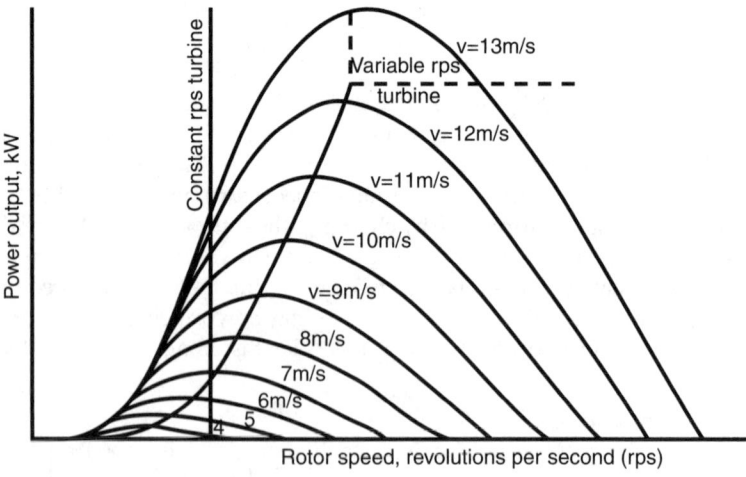

FIGURE 10-9 Power versus rotor speed characteristics of a wind turbine for different wind speeds.[3] A constant-speed turbine cannot extract peak power at different wind speeds. A variable-speed turbine that tracks peak power at different wind speeds will deliver higher overall power output. At wind speeds greater than the rated wind speed of turbine, different control strategies are deployed: Controlling frequency through power electronics to limit the power of the inverter; and RPS and power may also be regulated by changing pitch.

ω. A control mechanism that achieves this objective can be designed to adjust the turbine rotor speed to accomplish optimal power for different wind speeds.

In addition, such a system can deliver reactive power to the grid; the amount of reactive power is controlled by using an inductor and controlling the inverter output voltage (V_I) and its phase angle (δ_1) relative to the grid.[2] If V_I and V_T are output voltage of inverter and terminal grid voltage, then real and reactive power delivered to the grid is:

$$P_T = 3V_T i_T \cos \phi = \frac{3V_I V_T \sin \delta_1}{Z_1} \qquad (10\text{-}36)$$

$$Q_T = 3V_T i_T \sin \phi = \frac{3V_I V_T \cos \delta_1}{Z_1} - \frac{3V_T^2}{Z_1} \qquad (10\text{-}37)$$

Direct-Drive Synchronous Generator (DDSG)

Enercon popularized this type of generator. Enercon's E66 1.5 MW generator[4] has 72 poles and produces six-phase power, which is rectified to DC. This DC current is fed to the rotor and to an inverter. In order to accommodate the large number of poles, the rotor is large. As an illustration, consider a DDSG turbine with rotation speed of 10 to 22 rpm. Compared to a 1500 rpm induction generator, the torque is about 70 times for the same amount of power. The rated torque of an electrical machine is proportional to the volume of the machine. Therefore, the volume of DDSG generators is large, which is usually achieved by increasing the radius. A control mechanism, similar to the one described for PMSG, can be used to capture optimal power by changing the angular speed of the rotor. A schematic of the DDSG is in Fig. 10-10.

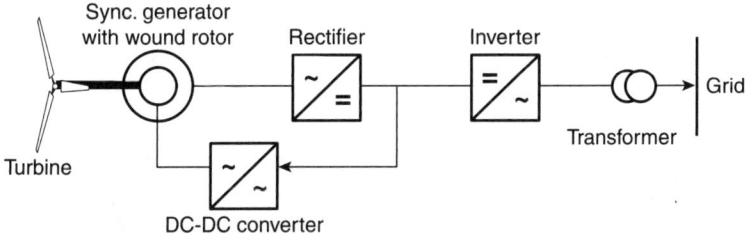

FIGURE 10-10 Schematic of a direct drive synchronous generator. Output of generator is variable frequency power that is rectified and then inverted to grid frequency.

Asynchronous Generators

Asynchronous generators are also called induction generators (Fig. 10-11). The principles are similar to a transformer in which the primary coils around one side of the core generate magnetic field. This changing magnetic field produced EMF in the secondary coil on the other side of the core. Here, energy is transferred from one level of AC voltage to a different level of AC voltage at the same frequency.

In an induction generator, instead of a solid core, there is an air gap through which the magnetic field travels. The primary core is in the stator and the secondary core is in the rotor. In a squirrel cage-induction generator, the rotor circuit is shorted, so, in an idealized case, the resistance is zero and, therefore, no electrical load. Like synchronous generator, a rotating magnetic field is created by connecting the stator to the grid. When the rotational speed of the rotor (ω_0) is the same as the speed of the rotating magnetic field of the stator, then the relative speed is zero. In this situation, there is no induced EMF in the rotor because relative to the rotor there is no change in magnetic field. There is no current in the rotor, and, hence, no force, no torque, and no power.

When the wind energy is delivering power to the generator, a torque is delivered to the rotor. Magnitude of the torque is: P/ω. The torque causes the rotor to accelerate. As the speed of the rotor becomes $\omega_1 (> \omega_0)$, the conductor in the rotor has a relative velocity ($= \omega_1 - \omega_0$) with respect to a rotating magnetic field of the stator. This causes EMF to be induced in the rotor and current starts to flow. The frequency of the current in the rotor conductor is $\frac{(\omega_0 - \omega_1)}{\omega_0} f$, where f is the grid frequency. Since the current is flowing in the rotor conductor, which is immersed in stator's magnetic field, a force is exerted on the rotor conductor. This force resists the external torque. The magnitude of the force is proportional to the current, which is proportional to the

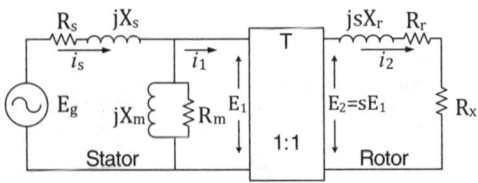

Figure 10-11   Equivalent single-phase circuit of an induction generator[5] E_g is the source/grid voltage on the stator; R_s, X_s, R_r, X_r are stator side resistance, reactance, and rotor side resistance and reactance. R_m, X_m are resistance because of losses and magnetizing reactance; R_x is the external resistance connected to the rotor to control performance for wound rotors only. s is the slip. The stator circuit is at frequency of f, where as the rotor circuit is at frequency sf.

relative speed. The rotor continues to accelerate until the external torque has been fully resisted.

Slip is defined as:

$$s = \frac{\omega_0 - \omega_1}{\omega_0} \tag{10-38}$$

For most generators, the slip is in the range of 1.5 to 4%. The rotor speed is, therefore, very close to the synchronous speed.

A circuit of an induction generator with a variable resistor in the rotor is in Fig. 10-11. The current in the rotor is given by:[5,7]

$$i_r = K_i \frac{s}{\sqrt{R_{rx}^2 + s^2 X_r^2}} \tag{10-39}$$

where K_i is a constant, R_{rx} and X_r are total resistance $(R_r + R_x)$ and reactance of the rotor. R_x is the variable resistance in the rotor; this will be described later in this section. Torque is proportional to $i_r^2 R_{rx}/s$:

$$\tau = K_\tau \frac{s R_{rx}}{R_{rx}^2 + s^2 X_r^2} \tag{10-40}$$

Power is, therefore:

$$P = \tau \omega = K_\tau \frac{s R_{rx} \omega}{R_{rx}^2 + s^2 X_r^2} \tag{10-41}$$

where ω is the speed of the rotor.

The relationship between the torque and the slip is given by the curve in Fig. 10-12. When resistors (R_x) are added to the rotor, then the curve shifts to the right. At higher level of wind energy, this allows the generators to work at larger value of absolute slip, that is, it allows generators to work at higher revolutions per minute while keeping a steady torque.

Table 10-1 below compares the properties of a synchronous and asynchronous generator.

Several improvements to the basic asynchronous generator are described next.

- During startup, there is a high demand for current from the grid, which is expensive and, over time, damaging to the equipment. A soft starter that controls the flow of current is, therefore, part of an induction generator. This is illustrated in Fig. 10-13.

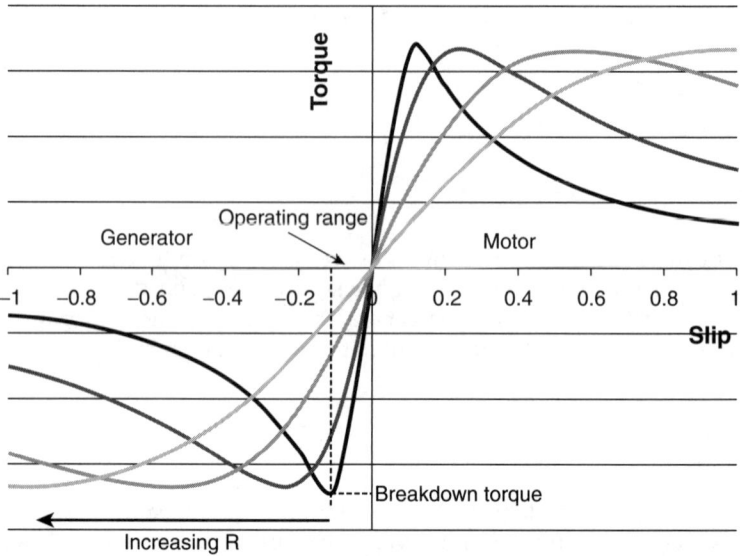

Figure 10-12   The operating characteristics of an asynchronous generator. When stator and rotor are in sync, the slip is zero; negative slip corresponds to generator mode; for small slip, torque increases linearly until it reaches the breakdown point. Higher slips can be supported by increasing the external resistance R_x. Note the rotor speed increases from right to left.

- During normal operations, an induction generator requires about 30% reactive power from the grid, and during low load operations, the reactive power demand gets higher. Procuring reactive power from the grid adds cost; therefore, capacitor banks are deployed to reduce the need for reactive power. This is illustrated in Fig. 10-13.

- Traditional induction generators operate with a very small range of slip, that is, the rotor speed is almost constant. As illustrated earlier, variable rotor speed allows a turbine to extract additional power (Fig. 10-9). The first scheme to increase the slip in order to make the generator variable speed is, variable resistance on the rotor. It is widely deployed in Vestas

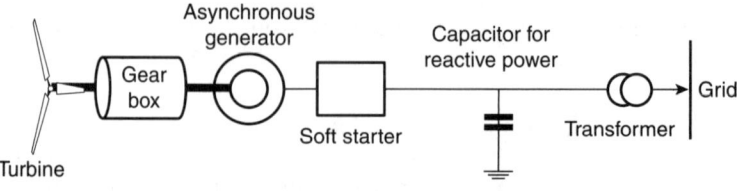

Figure 10-13   Schematic of connection of a fixed-speed system with an asynchronous generator connected directly to the grid.

Asynchronous Generator	Synchronous Generator
Simple, inexpensive, rugged and reliable	
Requires reactive power. Capacitors may be required to manage the reactive power.	Requires no reactive power; can provide reactive power on demand
Traditional asynchronous generator is unable to run at variable speed.	A constant rotor speed synchronous generator is not able to capture wind energy efficiently.
Opti Slip generator or DFIG is able to capture wind energy more efficiently.	A variable speed synchronous generator captures wind energy in the most efficient manner.
During startup, a current surge occurs. Soft starter circuit that utilizes phase-controlled antiparallel thyristors is required to manage the inflow of large current during startup.	During startup time, there is low demand on the grid for current
During operation, it demands reactive power, which may be of the order of 30% of the kVA that is delivered to the grid*	During operation, it can supply both reactive and active power
Gusts do not cause synchronization problems. With variable external resistance, larger negative slips are allowed, thus higher power fluctuations can be cushioned.	Wind gusts can cause stability issues for constant speed generator, as it can go out of synchronization

*Bhadra, S. N., Kastha, D., Banerjee, S. *Wind Electrical Systems,* Oxford University Press, New Delhi, 2005.)

TABLE 10-1 Comparison of Properties of a Synchronous and Asynchronous Generator

turbines under the label "Opti Slip". Figure 10-14 contains a schematic of such a generator.

• A second scheme to achieve variable speed is a doubly-fed induction generator (DFIG), a variation of an induction generator with a wound rotor. Note that although this is an asynchronous generator, the relative speeds of magnetic field is zero. This is because the slip (difference is speed of magnetic field of stator and the rotational speed of the rotor) is compensated by the speed of the frequency of the current in the rotor.

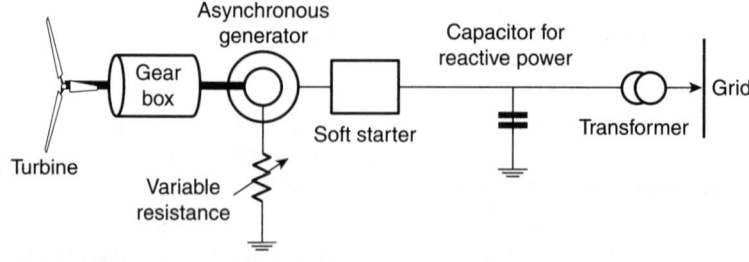

Figure 10-14   Schematic of an asynchronous generator in which slip is managed with a variable resistor on the rotor.

Traditional induction generators work on a small amount of slip. However, the DFIG works with a slip in the range of 60 to 110%.[2] During the 60 to 100% slip zone, energy is provided to the rotor from the power electronics at a frequency that ensures zero relative speed; note that the stator is still producing energy. During the 100 to 110% slip zone, energy is extracted from the rotor and fed to the grid. Above 110%, the blade pitch mechanism is triggered to limit the energy that is delivered to the rotor. The current in the rotor is at the slip frequency; hence, it must be rectified and then inverted before delivering to the grid. Figure 10-15 is a schematic of DFIG.

Variable Speed

Variable-speed generators are more efficient at capturing wind energy over a wider range of wind speeds. Therefore, the utility-scale wind turbine market has moved to this type of generator. Three types of generators have become popular in the recent years: Double-fed induction generator (DFIG), direct-drive synchronous generator (DDSG) with DC excitation, and direct-drive permanent magnet (DDPM) generator. Comparison of the features of the three generators is seen in Table 10-2.

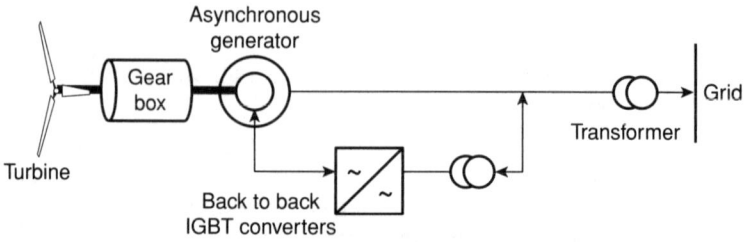

Figure 10-15   Schematic of a double-fed induction generator. When slip is negative (rotor is faster than stator's frequency), then rotor also provides power to the grid; when slip is positive, then rotor pulls power from the grid.

DFIG and Other Induction Generators	DDSG	DDPM
Asynchronous generator works in a wide range of slip conditions. Wound rotor either delivers or is fed energy, depending on rotor speed.	Rotor is fed DC excitation current. Variable-speed operation with large number of poles yields variable frequency power that is conditioned before delivering to grid.	Rotor has permanent magnets. Variable-speed operation yields variable frequency power that is conditioned before delivering to grid.
Power factor can be regulated	Power factor can be regulated	Power factor can be regulated. Additional circuits required to manage power factor
Stator is directly connected to the grid. Grid disturbances can affect the generator.	Generator is isolated from the grid. Better ability to manage voltage ride-through	Generator is isolated from the grid. Better ability to manage voltage ride-through
Output voltage is typically less than 1 kV	Output voltage is high, can be order of tens of kilovolts. No transformer required if connected to distribution lines*	Output voltage is high, can be order of tens of kilovolts. No transformer required if connected to distribution lines
Gearbox required	No gearbox. Cost, vibration and noise associated with gearbox is eliminated	No gearbox
Compact generators	Multiple pole generators are large	Compact generators
GE 1.5, Vestas (Opti Slip) and Siemens turbines use this generator	Enercon and Emergya use this type of generator	GE2.5, Clipper, Vensys, and NorthWind use this type of generator
Slip rings and brushes on rotor winding, which leads to higher maintenance and lower reliability	No slip rings or brushes.	Same as DDSG
Operating wind speed range is narrower	Operating wind speed range is higher	Same as DDSG

TABLE 10-2 Comparison of Three Variable-Speed Generators (*Continued*)

DFIG and Other Induction Generators	DDSG	DDPM
Ability to handle wind gusts is limited. Gusts cause significant increase in torque.	Gusts lead to increased speed while keeping torque within limit. Energy from gusts is converted to electric energy.	Same as DDSG
rpm of generator is large 1800/3600 rpm for 60 Hz or 1500/3000 rpm for 50 Hz; so torque is smaller	rpm is in the range of 15 to 30 rpm; so torque is large	Same as DDSG
Small power electronics converters are required to manage rotor circuit	Large rectifiers and inverters are required to convert all power to DC and then to AC at grid frequency. High frequency harmonics are produced which may be mitigated by filters.	Same as DDSG
Compact generator, but complicated power electronics	Large generator requires precision manufacturing and assembly to maintain uniformly small air gap	Material cost of magnets is high, and assembly is complicated by requirement for small air gap

*Hau, E., von Renouard, H. *Wind Turbines: Fundamentals, Technologies, Application, Economics,* New York, Springer, 2005.

TABLE 10-2   Comparison of Three Variable-Speed Generators (*Conitnued*)

References

1. Richardson, D. *Rotating Electric Machinery and Transformer Technology,* Reston Publ. Co., Reston, VA, 1978.
2. Bhadra, S. N., Kastha, D., and Banerjee, S. *Wind Electrical Systems,* Oxford University Press, New Delhi, 2005.
3. Quaschning, V. *Understanding Renewable Energy Systems,* EarthScan, London, 2005.
4. Freris, L., and Infield, D. *Renewable Energy in Power Systems,* Wiley, New York, 2008.

5. Wildi, T. *Electrical Machines, Drives and Power Systems,* Prentice Hall, Upper Saddle River, NJ, 2006.
6. Hau, E., and von Renouard, H. *Wind Turbines: Fundamentals, Technologies, Application, Economics,* Springer, New York, 2005.
7. Salam, M. A. *Fundamentals of Electrical Machines,* Alpha Science International Ltd., Oxford, 2005.

CHAPTER 11

Deploying Wind Turbines in Grid

We speak of electrical energy as "current": it exists only while it runs away; we use it only by delaying its escape.
—Wendell Berry, "The Use of Energy,"
The Unsettling of America, 1977

Introduction

In this chapter, the focus is on the interactions between the grid and a wind farm. The first section discusses how the variability of wind energy is handled on the grid and how wind energy is integrated into the grid. The next section is an introduction to the various electrical components that are between the turbine and the grid, followed by introduction to transmission and distribution, and conductors that carry current in a wind farm. The standards for interconnection are discussed next. This is followed by description of alternate wind farm topologies.

Electrical protection systems that protect the various elements of a wind farm, along with lightning protection, is described in this chapter. This is followed by interconnection studies conducted by utilities in the United States to approve a project. The final section of the chapter is about Supervisory control and data acquisition (SCADA) systems for data collection, monitoring, and control.

What Happens on a Grid When There Is No Wind?

Despite all the advantages of wind energy, there is a serious disadvantage: Variability of the wind resource, the raw material. When wind speeds are low, there is no or low energy production. This section will

221

explain the place of wind energy in a grid and how variability in wind resource is managed in a grid.

Generation capacity on a grid can be classified into three types: Base-load generators, spinning reserves, and nonspinning reserves. Base-load generators are large thermal or nuclear power plants that run at fixed capacity 7 days a week, 24 hours a day. These plants run at capacity factors of 90-plus%, because they are most efficient at high-capacity factors. These plants supply energy to cover the "base load." Base load is the minimum amount of hourly energy consumption across the entire grid on any day. Since the base load amount of demand will exist on the grid at all times, base-load generators meet this demand. Among renewable energy sources, geothermal can supply to base load, because it can be designed to produce energy at a constant rate 24 × 7. Spinning reserves is the second type of generation capacity comprising of natural gas or diesel-powered generators. These generators, as the name suggest, are spinning or running at low capacity all the time and can react very quickly to increase in demand. These generators meet the variable portion of demand during the day. Nonspinning reserves are generators that have a reaction of time of 10 min to 1 h. As the name suggests, these reserves are not running and need time to warm up and start producing energy. The goal of these reserves is to fill in for known changes to supply and demand, for example, scheduled maintenance of a base-load generator. In a highly reliable grid, the capacity of the three types of generators is planned to meet the variability in demand.

Wind and solar plants supply energy to offset the production of spinning reserves. The amount of wind and solar energy is becoming larger and is likely to be a significant fraction of total electricity production in a grid. With wind penetration of up to 20% of system total peak demand, additional spinning reserves will be required. It has been found that this results in additional cost of up to 10% of the wholesale value of wind energy,[1] which is a modest increase in cost.

Wind resource varies second-to-second, hour-by-hour, and has diurnal and seasonal patterns. As wind energy is aggregated from multiple turbines in a wind farm, the short-term variations in wind resources is attenuated as a percentage of the overall power output. When wind energy from multiple wind farms is aggregated, minute-to-hour variation in energy is attenuated and an hour-to-day variation in wind energy production and delivery to the grid is observed. Grid system operators manage such variations using spinning reserves.

Consider another case when wind energy penetration into a grid is high and there is insufficient load on the grid. This may occur during high wind periods. Because wind energy is considered an intermittent resource, wind plants are asked to curtail energy production, after all the spinning reserves have attained the minimum energy output level. Note the base-load generators typically do not curtail production.

Overall, connecting Wind power to an existing grid does not pose significant operational issues as long as: (i) wind penetration levels into the grid are low, and (ii) grid has sufficient spinning reserves with high response rates. As the size of wind plants increases to 500-MW+ and approaches the size of traditional fossil-based power plants and wind energy penetration is higher than 20%, then integration with grid and management of the variability of wind resource becomes a bigger challenge.

The US Department of Energy released a technical report in 2008 "20% Wind Energy by 2030"[2] that explores one scenario for reaching the goal of 20% wind energy by 2030 and compares it to a scenario of no new US wind power capacity. Its findings were:

- To achieve 20% wind energy on the grid, 300 GW of wind generation would be required
- Cost to integrate this amount of wind energy is modest
- 2% additional investment would be required for 20% wind scenario
- 50 cents increase per month, on average, household electricity bill
- 50% reduction in natural gas consumption by electric utilities
- Avoid construction of 80 GW of coal power plants
- Reduction in CO_2 emissions of 825 million metric tons annually by 2030
- Cut water consumption by electricity sector by 17%, which is 450 billion gallons annually by 2030
- Additional transmission would be required; transmission is a big bottleneck in remote wind-rich areas
- United States has affordable and accessible wind resource to achieve 20% wind scenario

Another study[3] done for the Southwest Power Pool by the Charles River Associates concludes: "The analytical results of the study show that there are no significant technical barriers to integrating wind generation to a 20% penetration level into the SPP system, provided that sufficient transmission is built to support it."

"Scheduling" and Dispatch of Wind Resources

Electricity grid operators use a method of day-ahead scheduling of generation resources in order to do short-term planning of matching supply (generation) and demand (load). Simplistically, at any given point of time, there cannot be imbalances on the grid, that is, supply must match demand. Because there is no storage on most networks, excess supply or excess demand lead to imbalances. On a second-by-second basis, supply and demand are kept in balance with load—following spinning reserves generators. On a day-to-day basis, a grid operator plans grid operations by seeking day-ahead generation plans from each generation entity on the grid. The day-ahead plan, or forecast, is in the form of an hour-by-hour generation output that will be made available to the grid. This is called generator's schedule. In the United States, if a generation entity is unable to supply to the hour-by-hour schedule (of the day-ahead plan), then there are penalties. These penalties are called imbalance charges; the intent is to encourage accurate forecasting such that reliability of the system can be enhanced. This mechanism works well for fossil fuel-based generators (base-load or spinning reserves) because there is no variability in availability of raw material. The mechanism does not work well for renewable energy sources like solar and wind.

An alternate mechanism that favors renewable energy has been developed by California Independent System Operator (CAISO), which removes penalties based on hour-by-hour schedule and allows for netting of schedules at the end of the month. In order to avail of this netting, the wind plant operator must participate in the centralized wind output forecasting program through which CAISO provides forecasting and scheduling. Other approaches that have been floated, not necessarily implemented, are to allow wind plant operators to provide 4-h ahead forecasts[3] for wind resource and for system operators to balance supply and demand at this granularity.

Single-Line Diagram

A wind project of a reasonable size requires a diagram to represent the various electrical components of the wind project and interfaces with the grid. A simple method to achieve this is with a single-line diagram. The basic components in a circuit for wind project are seen in Fig. 11-1.

Most turbines produce AC power at 480 V (smaller turbines) or 690 V (utility-scale turbines). In order to minimize losses, this power is transformed into distribution-level voltage of 11 to 34.5 kV using a transformer at the base of the tower. The circuit on the higher voltage side of the transformer is called the medium voltage (MV) circuit. The MV circuit consists of switches to disconnect power and underground cables to carry the power to a substation. At the substation, the power is transformed again and the voltage is lifted to the grid voltage. A circuit breaker in the substation monitors the grid parameters and compares to the power parameters that are on the output side of the substation. The circuit breaker trips when it detects a fault and disconnects the entire wind farm from the grid. Switchgear is a combination of fuses, circuit breakers, and others. In this section, the two terms circuit breaker and switchgear will be used interchangeably.

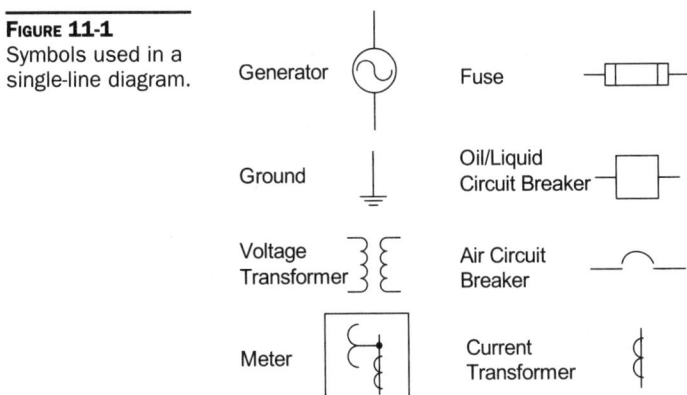

FIGURE 11-1
Symbols used in a
single-line diagram.

Figure 11-2 contains a single-line diagram of a typical low-voltage circuit; the following components are illustrated:

1. Generator

2. "Gen-Specific" box contains one or few of the following components depending on the type of generator and turbine manufacturer. These components are usually part of the turbine's electrical system.

   • Generator earthing

   • External resistors for slip control, for an induction generator with variable resistors to control slip

   • Capacitors for phase compensation, for induction generator

   • Power electronics converter for AC to DC to AC conversion for permanent magnet and variable-speed synchronous generators

   • Soft starter to control inrush current, for an induction generator

3. Current transformer, intelligent electronic device (IED), and circuit breaker are part of a programmable protection system that is described later in this chapter. The purpose is to monitor

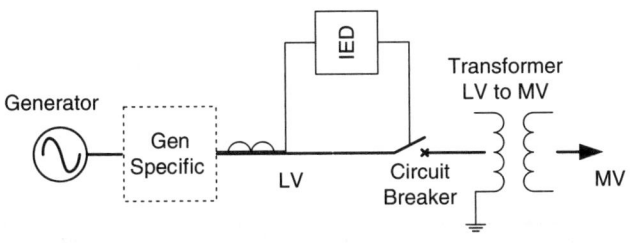

FIGURE 11-2   Single-line diagram of the components from generator to transformer at the base of the tower.

various parameters on the line like voltage, current, and frequency in order to connect or disconnect the left-hand side components from the transformer.

4. Three-phase transformer to raise voltage from low (LV) to medium voltage (MV). Typically, this transformer will increase the voltage from 480 or 690 V (depending on generator) to 11 or 69 kV (depending on design of MV line inside the wind farm). The capacity of the transformer is specified in terms of MVA, which is the product of the maximum voltage and maximum current from the generator. In simple terms, the MVA number is 15–25% higher than the rating of the generator. If the turbine is rated 2 MW, then a transformer rating from 2.3 to 2.5 MVA may be chosen. These ratings also apply to almost all components like the circuit breaker, fuses, and metering equipment.

Another example of a single-line diagram that covers the entire wind farm is in seen in Fig. 11-3. In this diagram, the following components are illustrated:

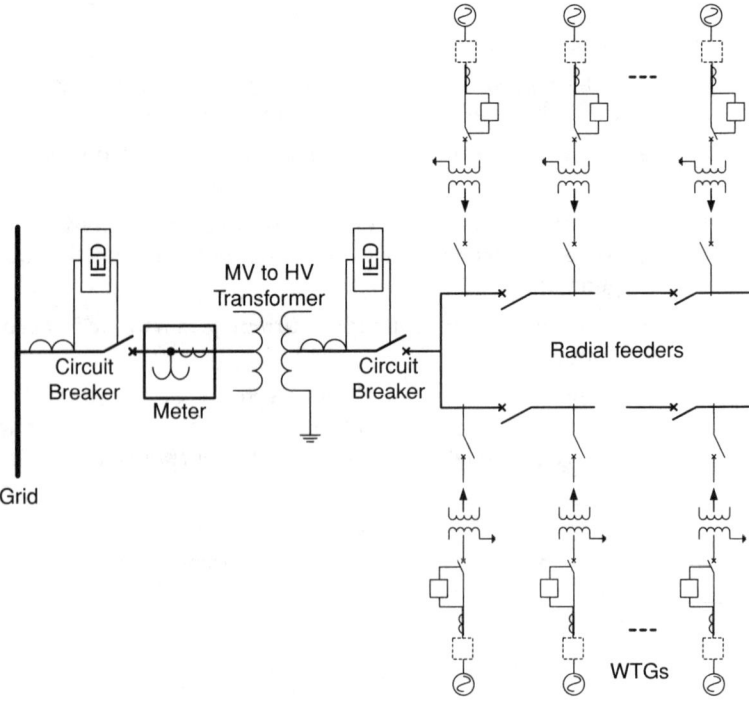

Figure 11-3 Single-line diagram of a wind farm in a radial topology. Individual turbines are connected to a radial feeder.

1. Wind farm with multiple turbines connected to feeders. Each turbine with switch gear and transformer is connected to a switch and then to the feeder. The switch is used to disconnect a single turbine before any work is done on it.

2. The two feeders collect energy from turbines and put it on a common bar. This is then stepped up to the grid voltage using a transformer. This transformer must be rated to step-up the entire power output of the wind farm. The bus-bar and transformer are separated by a circuit breaker.

3. A utility-grade meter is installed to measure the amount of energy that is delivered to the grid.

4. The last item is a circuit breaker that monitors grid voltage, current, frequency, phase difference, and other parameters. Depending on the data and the processing logic programmed into the IED, the circuit breaker connects/disconnects the wind farm from the grid.

Transmission and Distribution

Wind turbines are connected to three types of electricity networks: Transmission, distribution, and directly to the delivery point. The distinction between the three is based on the line voltage (see Table 11-1 for voltage and power-carrying capacity). The current-carrying capacity depends on the size of the conductor. The table contains sample current capacity of conductors. The power is carried in three conductors, one for each phase. There is usually a fourth smaller conductor, which is the neutral conductor. The power carrying capacity of three-phase AC is:

$$P = \sqrt{3}Vi \qquad (11\text{-}1)$$

The other factors that influence power-carrying capacity include reactive power, sagging of power line, and temperature.

The primary limitation on the capacity of a transmission line is the thermal capability of the line. Thermal capability is determined by:

Electricity Network	Voltage	Current, amps	Power
Transmission	110–750 kV	~500–1000	60–1300 MW
Distribution	11–69 kV	~500	10–60 MW
Secondary	400–11 kV	5–200	0.8 kW–3.6 MW

TABLE 11-1   Guide to the Amount of Power Carried in the Three Types of Power Networks

Conductor Type	Area, mm²	Normal Ampacity, Amps	Emergency Ampacity, Amps	Comments*
ACSR 26/7	282	726		A
ACSR 30/19	400	918		A
ACSR 54/19	805	1359		A
ACCR 26/7	205	1059	1136	B
ACCR 24/19	418	1692	1820	B
ACCR 54/19	808	2590	2795	B

ACSR (Aluminum Conductor Steel Reinforced) [Online] American Wire Group. http://buyawg.thomasnet.com/viewitems/all-categories/acsr-aluminum-conductor-steel-reinforced?&sortid=1172&measuresortid=1005&pagesize=100.
3M. Proven to Deliver More Amps. *3M ACCR* [Online] 3M. http://solutions.3m.com/wps/portal/3M/en_US/EMD_ACCR/ACCR_Home/Proven_Benefits/More_Amps/.
*A, Normal condition is 25°C ambient and 75°C conductor; Emergency condition is 90°C at which temperature the total exposure must be limited to 100 h. B, Normal conditions is 210°C; emergency condition is 240°C at which temperature the total exposure must be limited to 1000 h.

TABLE 11-2 Overhead Transmission Conductor Sizes and the Current-Carrying Capacity (Ampacity)

- *Amount of current flowing through the conductor.* The losses in transmission because of resistance is $i^2 R_t$; this loss is converted into heat, where R_t is resistance of transmission.

- *Amount of sag allowed on the conductor.* As the current increases, the temperature increases causing sag to increase

- *Ambient temperature and wind.* At higher ambient temperature, the current-carrying capacity is reduced because amount of heat dissipated is lower. Wind assists in carrying the heat away from the cable. In most cases, a worst-case scenario is chosen to specify the current-carrying capacity of the transmission line.

An overhead cable is a collection of conductors. The most popular overhead cable used to transport electricity is the aluminum conductor steel reinforced (ACSR). An ACSR 26/7 cable contains 26 aluminum conductors that surround a core of seven steel wires for reinforcement. Aluminum conductors carry the current. Aluminum conductor composite reinforced (ACCR) is an advancement over ACSR with an aluminum core and an aluminum zirconium outer strand for reinforcement. It is able to carry much higher currents at higher temperatures. The current carrying capacity of the two types of conductors and the temperature limits are specified in Table 11-2.

As a rule of thumb, any wind project that is less than 10 MW will connect to distribution lines, and projects over 10 MW will connect to

Voltage	Power-Carrying Capacity	Power Length
Secondary network, low voltage	200 kW	34 kW-km
Distribution network, medium voltage	10 MW	85 MW-km

TABLE 11-3   Power-Carrying Capacity and Power Length of 150 mm² Aluminum Underground Cable

transmission lines. The exact criteria depend on the current load on the lines—the amount of energy that is currently being carried on the lines.

In most cases, within a wind farm, power is transmitted using underground cables. All transmission within wind farm from turbines to substation is called a collection system. Underground cables used in collection systems are three-phase cables bundled into one cable with appropriate shielding. The most common cable is the 150-mm² aluminum conductor.[6] The amount of power it can transmit as a function of voltage and "power length" is shown in Table 11-3. Power length divided by the power yields the distance that power can be transmitted to yield a loss of under 5%.

Standards for Interconnection

Power Factor and Reactive Power

Inductance on transmission lines and inductive loads can cause the power factor to be different from unity. The general equation for the real power carried in three-phase AC lines is:

$$P = \sqrt{3}Vi \cos \varphi \qquad (11\text{-}2)$$

where $\cos \varphi$ is the power factor. The goal of any electricity network is to obtain a power factor that is as close to unity as possible. The causes of poor power factor are: Magnetizing loads like transformers, motors, generators, and induction furnaces; long transmission lines; inverter producing poor current wave form. These types of loads cause the voltage and current to be out of phase.

As an illustration of the impact of low power factor, consider a power factor of 70%, which implies a phase difference of 45° between voltage and current. Generators, transmission lines, and transformers are rated in terms of the product of voltage and current with units of kilo-volt-amps (kVA) or mega-volt-amps (MVA). This is called the total apparent power. Power factor is also the ratio of real power to the total apparent power. For simplicity, consider a system that is designed for 10 MVA operating with a power factor is 70%. When the power factor is 70%, although the generators, transmission, and other components are running at full capacity (10 MVA), the system is delivering only 7 MW of real power; the rest is reactive power or "wattless power" that performs no work. All the components are at full capacity or fully loaded in this system because the maximum amount of current is flowing through the system causing losses of $i^2 R_t$ with an appropriate amount of temperature rise. In this situation, despite current flow at full capacity, only 70% of real power that is expected at the consumption point is, in fact, delivered. Suppose a load in this system requires 9.5 MW of real power. In order to satisfy this load, $9.5/0.7 = 13.5$ MVA of total apparent power is required. This will overload the 10 MVA system. The solution to low power factor is to install VAR compensators at appropriate points in the system in order to deliver higher amount of power to loads.

An induction generator demands reactive power of the order of 30%[7] of its rated kVA from the grid. See Fig. 11-4 for an illustration of a power factor as a function of slip for a squirrel cage induction generator. In normal operating conditions, the generator operates at a slip of about 5%, where the power factor is about 80 to 90%. At partial loads, slip is smaller and the power factor may be 50% or lower. This demands 50% or more reactive power. One method to improve the power factor is to use Static VAR compensator[8]. It is a capacitor in parallel with an adjustable inductor. With a 10 MVAR capacitor and 10 MVAR inductor, reactive power ranging from 0 to 10 MVAR can be provided.

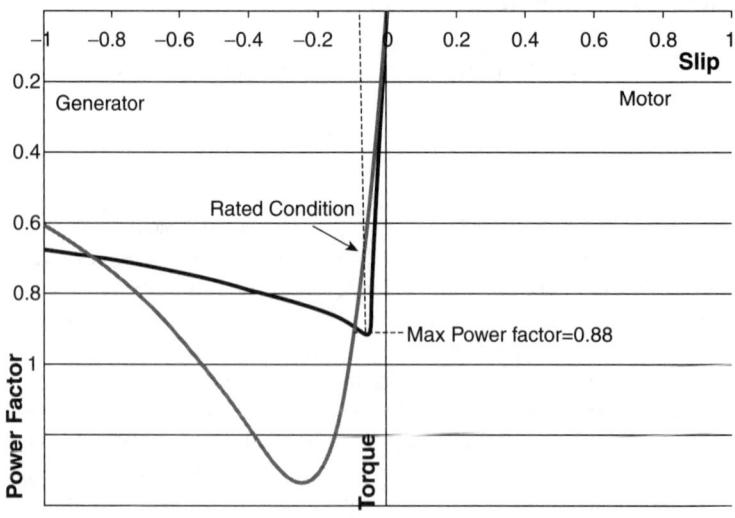

Figure 11-4 Power factor and torque curves for a squirrel-cage induction generator. In this illustration, the generator is rated at 5% slip and a power factor of 88%. At lower slip and lower torque the power factor is 50% or below.[10]

The old induction generators are becoming a rarity; doubly-fed induction generators, synchronous generators, and permanent magnet generators are able to control the reactive power demand.

Utilities around the world have started requiring wind plants to maintain a power factor of 0.95 or higher. In 2005, the US Federal Energy Regulatory Commission (FERC)[9] proposed a requirement that "large wind plants maintain a power factor in the range of 0.95 lagging to 0.95 leading to be measured on the high voltage side of the wind plant substation transformer." The method of providing reactive power was left flexible: Power electronics or switched capacitors.

Low-Voltage Ride-Through

Thus far, the discussions of connection of a wind farm to a grid have involved disconnect or islanding of the wind plant in case of a fault on the grid. Specifically, sudden change in voltage on the grid is one of the conditions to cause a circuit breaker to trip, thereby disconnecting the wind farm from the grid. Often, the fault is temporary (for duration of less than about 3 s) because of lightning strikes, equipment failure on the grid, or downed power line. Taking a large wind plant offline in such situations can cause further instability in the grid by causing other cascading tripping of generators and loads. Starting in 2005, FERC[9] recommended that a large wind plant stay online and connected to the grid during temporary voltage drops. FERC has proposed a voltage profile that serves as low-voltage ride through criteria (see Fig. 11-5): A wind plant should stay online as long as the grid

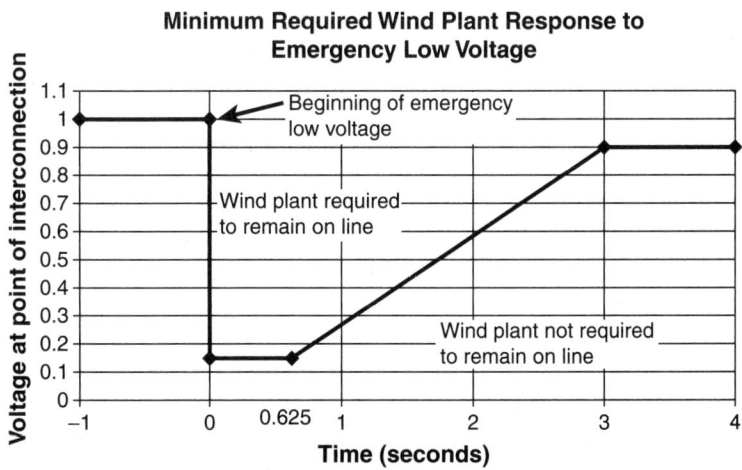

Figure 11-5 Low-voltage ride-through specifications by US FERC[9] *y*-axis units are ratio of actual to nominal voltage.

voltage is at least 15% of the normal voltage for less than 0.625 s and voltage recovers linearly to 90% of normal voltage within 3 s. This helps to improve the reliability of the grid. LVRT is accomplished by supplying reactive power to the grid during the low-voltage condition.

Power Quality: Flicker, and Harmonics

Flicker, as the name suggests, is a short-term variation in voltage that causes a light bulb to noticeable change in intensity. The source of flicker is (i) change in torque as blade passes the tower during normal operations, if a single turbine were supplying energy and (ii) startup and disconnect of turbine(s) from the grid.

The next power quality issue is harmonics. Harmonics are introduced when power converters are used, for example, in variable-speed turbines. The source of harmonics is because of the imperfect sinusoidal wave (sinusoidal is approximated by a step-function) generated by power converter during the DC–AC stage. Pulse-width modulation, a method employed by most modern turbines produces harmonics with a frequency of about 2000 Hz and small amplitude. Such high-frequency and low-amplitude harmonics can be easily removed by high-frequency filters. Higher-frequency harmonics, if not filtered, can cause reduction in the impedance of capacitors. Lower-frequency harmonics seen with thyristor-based power converters is lower frequency and higher amplitude, which causes higher current heating of capacitors and transformers.

IEC-61400-21[11] specifies parameters that are used to characterize power quality of a grid-connected wind turbine, procedures for measuring the parameters, and procedures for assessing compliance with power-quality requirements. The IEEE 1547-2003,[12] Standard for Interconnecting Distributed Resources with Electric Power Systems provides technical specifications for power quality and other aspects that projects with generation capacity of 10 MW or less must meet when connecting to the grid. Each utility uses these standards as a basis for developing their own interconnection requirements for wind projects that are 10 MW or less.

Short-Circuit Power

Strength of a grid at the point of common coupling (PCC) is an important measure of how much power can be injected into the grid. Short-circuit power, S_{cc}, is the amount of kVA or MVA at the PCC under maximum load. The rated grid voltage is assumed to be constant at V_g, voltage at PCC is V_{cc}, line impedance is Z_{cc}. Short-circuit power is computed as if short-circuit were created at PCC:

$$S_{cc} = V_{cc}^2 / Z_{cc} \qquad (11\text{-}3)$$

$$\rho_{cc} = S_{cc} / P \qquad (11\text{-}4)$$

where ρ_{cc} is the ratio of short-circuit power to the amount of planned injected power (P) at PCC by a wind farm (or any other source); note S_{cc} and P must have the same units of kVA and kW or MVA and MW. ρ_{cc} is used during sizing of a wind farm.

- *Weak grid.* When ρ_{cc} is 8 or lower, a grid is likely to experience disturbing voltage variations. In such situations, a grid is considered a weak grid with respect to the amount of power P that is being injected into it.

- *Strong grid.* In design situations when ρ_{cc} is 20 or higher, the grid is considered to be strong with respect to the amount of power that will be injected to the grid at the PCC. Under strong conditions, the variations in voltage are minor and predictable.

As the distance of the interconnection point increases from the nearest substation, the short-circuit power decreases at the point of interconnection. Values of S_{cc} and V_{cc} are determined by power-flow or load-flow studies under normal operations and short-circuit conditions. The grid operator normally does these studies using network analysis and simulation software.

Wind Farm Topologies

The layout of a wind farm is a function of wind speeds and wake effects. Once a layout of turbines has been determined, an intrawind farm collector system is designed. Considerations include:

- Topology of the collector system include the following, options: radial, feeder-subfeeder, and looped feeder.[13] Most wind farms operate on a feeder-subfeeder topology.
- Voltage level for the collector system
- Mix of underground and overhead cables in the wind farm
- Location of substation
- Design of the neutral and grounding systems

Consider a wind farm with 20 turbines. A simple radial topology in Fig. 11-6 that strings the 20 turbines is low cost, but high-risk topology with no redundancy. If there is a break in the feeder link, all

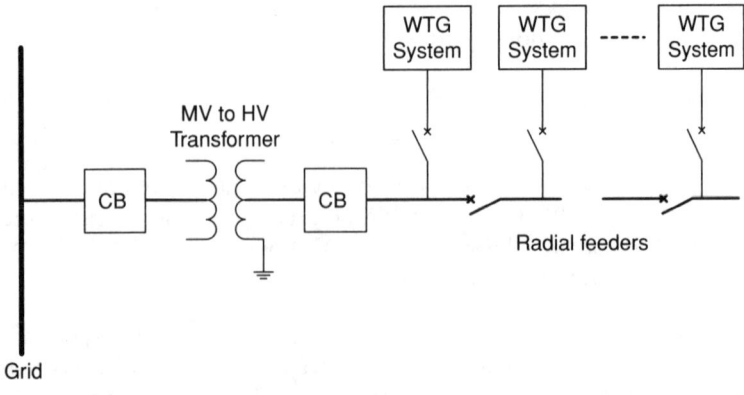

FIGURE 11-6 Radial topology with circuit breaker.

turbines to the right will be offline. A radial system has single circuit breaker between the transformer and the feeder and all the turbines are connected in sequence. Note that a fault in the main circuit breaker can cause the entire wind farm to be offline.

Three wind farm topologies are described next.[13] The associated figure have been modified from Reichard[13] paper.

- Figure 11-7 is a bifurcated radial topology with slight improvement over radial topology. Improvement is because a break in feeder link causes fewer turbines to be offline. In a wind farm, this topology is more likely compared to radial because turbines are unlikely to be arranged geographically in one direction. Normally, the reliability of the circuit breaker, bus,

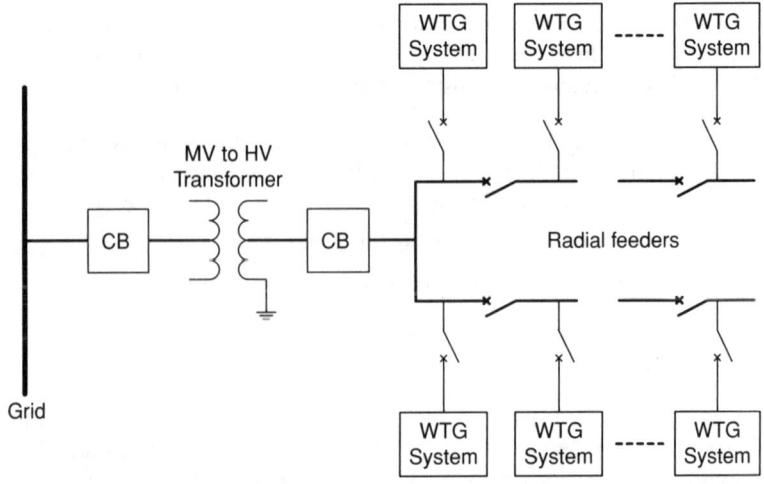

FIGURE 11-7 Bifurcated radial topology with circuit breaker after common bus.

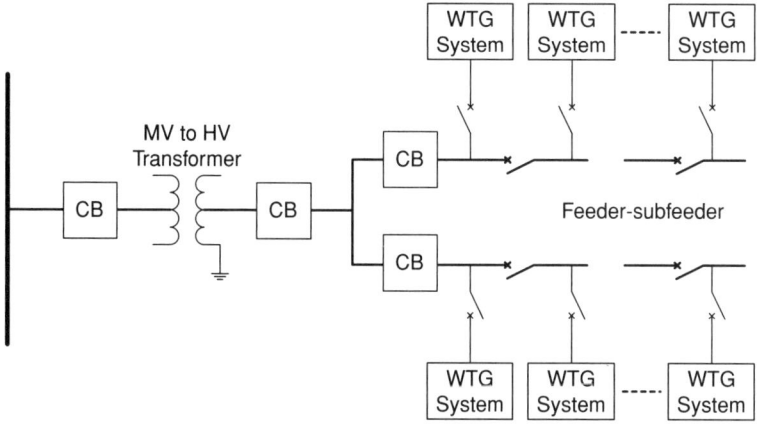

FIGURE 11-8   Feeder–subfeeder topology with circuit breaker at the end of each feeder.

and MV cables is sufficiently high that redundant circuits or redundant circuit breakers may not be warranted.

- Figure 11-8 is a feeder–subfeeder topology with subfeeders that collect energy and supply to the feeder. Note each subfeeder has a circuit breaker. For large wind farms with large number of turbines relying on single circuit breaker increases the risk sufficiently to warrant circuit breakers on each subfeeder.

- Figure 11-9 is a looped feeder with the highest reliability because failure of a circuit breaker does not cause a large number of turbines to be offline. This configuration has redundant paths so that if there is any fault in the MV transmission

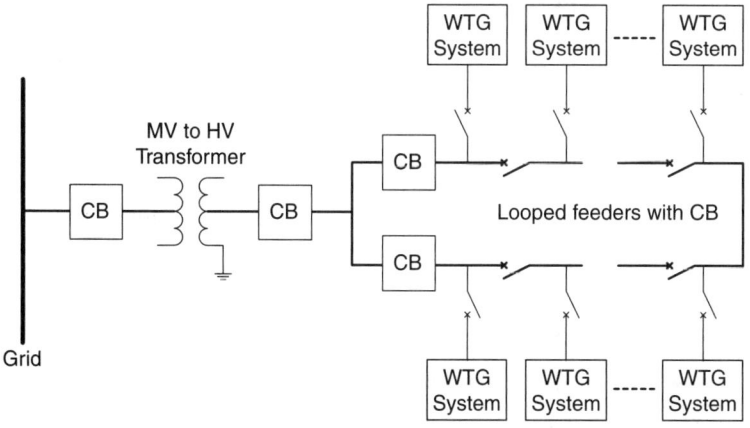

FIGURE 11-9   Looped feeder for redundancy.

circuit, power can still flow from all turbines, in most cases. Such redundancy may be warranted when using overhead cables for intrawind farm MV circuit because in strong wind areas, overhead cables reduce reliability of the MV circuit.

Protection Systems

One of the key elements of electrical design is a protection system. There are three main components of this type of system:

- Switch or circuit breaker that is operated electromechanically or electronically that can open or close connections in a circuit to protect the generator, transformer, grid, and other systems. The speed of switches is very important; the time of operation is of the order of 0.1 s after a fault is detected.

- Monitoring system that measures quantities like current, voltage, frequency, and others. Quantities are measured in each of the three phases and may be measured at the input and output of the generator, input and output of transformer, input to the grid, and flow in the grid. The measured quantities are used in the logic described below.

- Logic system that compares measured quantities against preset threshold levels, compares two measured quantities and performs other computations following which the logic system computes the action required. The signal about the action is sent to the switch, which executes the action. In a programmable protection system, the logic, the threshold levels, and other parameters are programmed.

Newer wind farms also have a fourth component, which is a sophisticated communication system that connects the different protection units in a wind farm using a fast fiber-based Ethernet to each other, a centralized controller, and SCADA system. Such protection systems, with the three or four components, are referred to in the industry by the generic name of intelligent electronic device (IED). IED is defined as[14] "device with versatile electrical protection functions, advanced local control intelligence, monitoring abilities, and the capability of extensive communications directly to a SCADA system."

Several quantities are monitored by protection systems:

- Over and under voltage
- Over current and direction of current
- Over and under frequency
- Directional and nondirectional earth fault protection

For the purposes of illustrating the stringent time requirements, consider the following example. A protection system has detected a drop in frequency on the grid of 6% and, in such situations, the utility requires that the wind farm be taken offline. The sequence of events is: (a) Sensor will detect the situation, (b) controller will analyze data and specify the action, (c) action is sent to the circuit breaker, and (d) circuit breaker will "throw the switch." There are two time periods of a protection system: Relay time to detect and analyze the condition, and circuit-breaker time to "throw the switch." Relay times are in the range of 15 to 30 ms and circuit breaker times are in the range of 30 to 70 ms.[15] Therefore, the total time from detection of fault on the grid to isolation of system is of the order of 50 to 100 ms. On a 60-Hz system, this is three or six cycles.

Readers may ask why not use other means to take the wind farm offline, like pitch the blades to stop energy production? The answer is, mechanical components do not react as swiftly as electronic components. The mechanical components can take minutes to react: The process of turning gears to change the pitch from operating to feather position, and the process of losing all the inertia of a rotating machine.

Grounding for Overvoltage and Lightning Protection

Grounding or earthing design is crucial for safe operation of a wind project. Most turbine manufacturers provide a grounding system and supervise its installation. A common grounding system consists of a ring installed in or around the foundation of the turbine with long copper rods driven into the earth.[16] The ring is bonded to the reinforced concrete foundation and then connected to a neighboring turbine using conductors that emanate out radially from the two sides of the ring (see Fig. 11-10). The radial earthing conductor is placed in the same trench that carries power or fiber for SCADA communication. The same grounding system is used for grounding the wind turbine and all its components, and grounding lightning strikes. Grounding system must provide: (a) Low impedance path for current to flow and (b) equipotential bonding, which is to ensure that all the metal components like tower, electrical cabinets, and foundation are at the same potential.

The length of the electrodes depends on the resistivity of soil (see IEC 61400-24).[16] In place of a ring, reinforcements of the concrete foundation may be used if the foundation is of 15 m or larger diameter.

Design of grounding systems is a very detailed process that will not be described here. It must be emphasized that good grounding is necessary for safe operation of a wind plant. It is not only the grounding of individual components, but good grounding also ensures that ground fault relays are triggered promptly in case of ground fault.

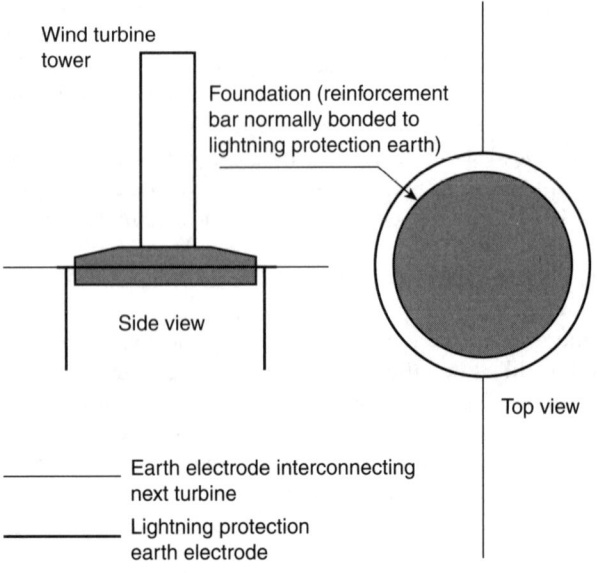

FIGURE 11-10 Representation of a grounding system around the foundation of a turbine.[16] (With permission from IEC 61400-24.)

Lightning Protection

IEC61400-24[16] provides guidance on design of wind turbine for lightning protection. In order to design a system that is safe during lightning strikes, pathways must be designed for the lightning current to be safely conducted down to the earth without causing unacceptable damage or disturbances to the systems. Traditional tall structures like buildings and transmission lines use conductors that are put above and around the structure to provide lightning attachments points. In a flat area, an attachment point of height h protects an area of radius $3h$. Wind turbines do not have separate conductors, other than blades, that provide attachment points. Since most modern blades are made of composites, the blades are not good conductors. In order to conduct lightning currents, blades have embedded conductors with attachment points starting a few centimeters from the tip of the blade and placed all along the length of the blade. Steel wires of 12 mm or more are used to conduct the lightning current safely to ground and without damage to the blade. Serious damage to blade occurs when arcs are formed inside the blade as the lightning current jumps from the conductor to air chamber or between layers of composite material. The internal arcs cause pressure shock waves that can cause severe damage, including explosion of blade. From the blade, lightning may pass through pitch bearings, hub, main shaft, main shaft bearings, gears, generator bearings, bedplate, yaw bearing, and tower. As the lightning enters the main shaft, it finds a low impedance path. Most modern turbines have carbon brushes to accomplish this. It avoids passing lightning through the main bearings, which can cause severe pitting in the rolling elements of the bearing. Since bearings are expensive to replace, alternate path are designed through the carbon brushes. The height and shape of a turbine is such that not all lightning strikes start at the end of blades; side strikes to the blade and strike to the tower are not uncommon. Although the blades are non-conducting, water on the surface makes them conducting.

The second most common damage because of lightning is to the control systems. This is primarily because the control systems are designed for low voltages. High voltages and high currents enter the control systems through the mechanisms described below; methods to mitigate the damage are also described below:[16]

- Conduction primarily across improper insulation and small air gaps. This can be minimized by designing a low-impedance path for lightning current, proper insulation, and good bonding methods. Good bonding ensures that all the components and metal fixtures are at the same potential, which eliminates jumping of lightning current.

- Capacitive coupling occurs when lightning current carrying conductors are in proximity to other conductors. This can be minimized by separating the grounding cables from others and shielding through metal enclosures.

- Magnetic coupling is caused by high rate of change of current flowing through conductors that cause strong and changing magnetic field. This can induce damaging voltages into loops of conductors. Metal enclosures, twisted cables, and avoiding large loops can reduce magnetic coupling.

Transformers for Wind Applications

Step-up transformers in wind applications are different compared to standard off-the-shelf transformers used for power applications. Significant numbers of failures indicate that attention must be paid to the uniqueness of the conditions in which wind application transformers operate. The following are some of the differences:[17]

- Wind turbine step-up transformers (WTST) are subject to loads that correspond to the average capacity factor of the wind farm. That is, the transformer operates at 35 to 45% of its rated capacity, on average. Traditional transformers are subject to loads at the rated capacity or slightly higher. The light loading leads to higher core losses and lower winding losses.

- Loading on WTST is considerably more cyclical than traditional transformers. The thermal cycling caused by varying load can cause damage to insulation and to internal and external connectors.

- Imperfect sinusoidal waveform because of power electronics converters cause additional loading. If high-frequency filters are not used, then electrostatic shielding may be required to prevent transfer of harmonic frequencies to the higher voltage side.

- Ability to withstand low-voltage ride-through (LVRT) requires that transformer be able to handle full short-circuit current for at least 0.625 s. During such a transient period, the transformer experiences high mechanical forces.

Wind-Plant Interconnection and Transmission Study

Wind farms that deliver power into the utility grid are required to submit to an interconnection study that is performed, in most countries, either by the local utility or a regional transmission organization (RTO) or a regulatory agency. The description of the various studies and agreements in this section is specific to the United States, although most countries have similar processes. Furthermore, the intent of this section is not to be comprehensive, since there are variations to the process from one country to another and within a country from one region to another.

An interconnection agreement only provides the right to connect to the transmission line; it does not provide the right to transmit power. A separate transmission agreement is required. Interconnection and transmission agreements do not guarantee that a utility will buy the energy; a power purchase agreement is required with a buyer to sell the energy produced in the wind farm. The interconnection study, as practiced in the United States, is separated into two types:

- Large generator interconnection request (LGIR) for 20 MW or more

- Small generator interconnection request (SGIR) for less than 20 MW with a request for energy resource service

LGIR is a three-step process. Although the following process is specific to United States, most other countries have a variation to the following process:[18,19,22]

- *Feasibility study*: The primary purpose is to check if the system can accept power injection at the interconnect point chosen by the wind farm developer, and to provide high-level cost estimates for interconnection. It includes a power flow and short-circuit analysis. In the United States, a deposit of $10,000 is required to initiate the study, which is performed by the RTO or its consultants. It must be completed in 45 days after a completed application has been received; RTO may seek an extension by providing a reason to the customer.

- *System impact study*: This is a detailed transmission planning study to understand the impact of the proposed wind farm on existing transmission, generation, and customers. The analysis performed includes, impact on short-circuit power, dynamic and stability issues, protection systems, safety and reliability of transmission, and any other adverse impacts on the system. The deliverable is a written report on improvements to the transmission system, and estimated cost. In the United States, a deposit of $50,000 is required; the study is performed by the RTO. This study must be completed in 90 days; however, if a cluster or group study is performed, then an additional 90 days is available.

- *Facility study*: This is a detailed engineering study to design the interconnection substation and transmission and to provide estimates for the cost and time to implement the upgrades. In addition, it delineates the responsibilities of the wind farm developer and the local transmission company to upgrade transmission. In the United States, the study is normally performed by the local utility for the RTO; it requires a deposit of $100,000. This study should be completed in 90 or 180 days depending on accuracy of cost estimates requested, ±20 or ±10%, respectively.

In all three studies, a refund is issued or additional charges are billed based on the actual cost of the study. An interconnection request is inserted into a queue and time to process the request varies from region to region. The timeframes specified above are based on FERC guidelines. Prerequisite to an interconnection application are access to land either through ownership or lease agreements. The interconnection will require a developer to choose between two types of services: Energy resource interconnection service (ERIS) and network resource interconnection service (NRIS). An ERIS has a much lower bar because it does not require any transmission system upgrades. It allows a generation to interconnect in order to deliver output using existing firm or nonfirm transmission capacity on an "as available" basis. NRIS requires a complete network analysis, while treating the proposed wind farm as a network resource comparable to any other large generation facility. In this analysis, the system is analyzed at peak load and under a variety of stressed conditions to ensure that the proposed generation facility operating at full capacity along with all other generation facilities on the network will be able to deliver power to the loads.

For small wind projects that are less than 20 MW, SGIR with ERIS requires only the feasibility study for a fee of $1,000.

Irrespective of the size of a project, the cost of transmitting energy from the wind farm to the interconnect point is the responsibility of the entity requesting the interconnection, which is usually the wind farm developer. The cost of upgrades to the transmission system beyond the interconnect point is allocated to the stakeholders, based on a variety of laws. If a project is designated as a network resource service, then the entity requesting interconnection service is reimbursed 50% of the cost of network upgrades. ERIS projects are responsible for the entire cost. In most cases, the network upgrades are performed by the interconnecting utility.

Transmission Bottlenecks

Areas with high wind energy are not areas where people like to live. So, after wind energy is harvested, it must be transported/transmitted to population centers where most energy is used. Since few people live in high-wind areas, the electricity grid is weak in those areas and the predominant flow of electricity has to be reversed from flowing into the region to flowing out of the region. For this and other reasons, significant upgrades to the transmission infrastructure are required in most countries to increase penetration of wind energy to significant levels. In the United States, several studies have indicated[2,3] that wind energy penetration levels of 20% into the electricity grid are feasible without adversely impacting system reliability. In order to achieve this level of penetration, transmission upgrades, voltage control devices and dynamic voltage support will be required. Since transmission upgrades are longer lead time projects, the planning and implementation of the upgrades must precede wind projects.

SCADA Systems

Supervisory control and data acquisition (SCADA) system is the nerve center for a wind plant that collects data from the wind farm and controls the wind farm locally and remotely. In a typical wind farm, all turbines, switchgears, meters, met-towers, and all other systems that collect data and can be remotely controlled are connected using fiber-optic cables to a central SCADA computer onsite. The SCADA computer can be accessed from a remote computer. Most turbine manufacturers provide a SCADA system that is best suited to work with their turbines. However, manufacturer-independent SCADA systems are common on large wind farms with multiple types of turbines.[20]

IEC 61400-25[21] provides a standard for the information model, information exchange, and mapping to communication profile.

The main functions of a SCADA system are: Data acquisition, reporting, and control.

Data Acquisition

The SCADA system is the main repository of all the data. All this real-time data is collected and stored in a database and, as expected, the volume of this data is enormous. For convenience, the data types are classified into four types:

Energy Production Data

Data about energy production by each turbine and energy availability at critical collection points and substation are measured and sent to the SCADA system.

Events and Availability Data

Events are change in status of turbines, switchgears, met-towers, substation, and SCADA system itself. These events along with grid events are classified, time stamped, and recorded.

Machine Condition Data

Turbines report condition data that includes temperature of fluids, acoustic emission, vibration, strain gauge readings, and any other form of data that the turbine manufacturer uses to monitor the condition and predict impending failure. The trend is toward use of embedded sensors in structural elements, like blades and towers, in order to monitor changes in structural dynamics of these components. This data is used by O&M for routine and unscheduled maintenance.

Wind Measurement Data

Each turbine is equipped with meteorological sensors. In addition, permanent met-towers are installed to independently measure wind data and other weather parameters. This data is used to compare actual versus predicted power curve, actual versus predicted energy generation, and others. The SCADA system collects and records this data.

Reporting

The real-time data collected by the data acquisition system is impossible to examine and analyze because of the volume of data. Therefore, SCADA provides various aggregates of data and report templates to support various entities like operations, maintenance, and business departments. Other features include online and offline reports for current and historical data, and daily, weekly, monthly, and annual reports that are electronically dispatched to all the interested parties, including project owner, turbine manufacturer, and system operator of the utility.

The most commonly used reports are energy production, meteorological, availability, and power curve comparison. Because of the volume of data and demand from different constituencies, reporting can get very complex.

Control

The most active role of SCADA system is operational control of a wind power plant. The SCADA system[20] can issue the following command to an individual turbine or group of turbines: Release to run, stop, or reset. The SCADA control system also implements a variety of functions required by the grid operator, like curtailment of energy production, reactive power output, and voltage and frequency control. Examples of other control logic that can be implemented pertain to environmental constraints:

- Shutdown or slowdown of turbines during bird migratory seasons or when birds and bats are detected in close proximity to turbines
- Slowdown of turbine during certain hours of a day to reduce noise level, if the turbines are in proximity to noise-sensitive areas like residential areas

A SCADA system is, therefore, a crucial tool for:

- Monitoring production
- Monitoring health of each turbine
- Monitoring reason for poor performance and warranty claims. It provides data that can provide a basis for determining the cause of poor performance: Poor wind, poor wind farm layout, poor turbine performance, and others
- Adjusting resource estimation methodologies for future wind farms.

References

1. Smith, J. C., Milligan, M. R., DeMeo, E. A., and Parsons, B. "Utility Wind Integration and Operating Impact State of the Art," *IEEE Transactions on Power Systems* 22: 8, 2007. http://www.nrel.gov/docs/fy07osti/41329.pdf.
2. Energy Efficiency and Renewable Energy, US Department of Energy. *20% Wind Energy by 2030*. US Department of Energy, 2008. www.nrel.gov/docs/fy08osti/41869.pdf. DOE/GO-102008-2567.

3. Charles River Associates. *SPP WITF Wind Integration Study.* Charles River Associates, Boston, MA, 2010. http://www.crai.com/News/listingdetails.aspx?id=12090.
4. ACSR Aluminum Conductor Steel Reinforced [Online] American Wire Group. http://buyawg.thomasnet.com/viewitems/all-categories/acsr-aluminum-conductor-steel-reinforced?&sortid=1172&measuresortid=1005&pagesize=100.
5. 3M. Proven to Deliver More Amps. *3M ACCR* [Online] 3M. http://solutions.3m.com/wps/portal/3M/en_US/EMD_ACCR/ACCR_Home/Proven_Benefits/More_Amps/.
6. Hau, E., and von Renouard, H. *Wind Turbines: Fundamentals, Technologies, Application, Economics* New York, Springer, 2005.
7. Bhadra, S. N., Kastha, D., and Banerjee, S. *Wind Electrical Systems,* Oxford University Press, New Delhi, 2005.
8. Muljadi, E., Butterfield, C. P. Yinger, R., and Romanowitz, H. *Energy Storage and Reactive Power Compensator in a Large Wind Farm,* National Renewable Energy Laboratory, Golden, CO, 2003. http://www.nrel.gov/docs/fy04osti/34701.pdf. NREL/CP-500-34701.
9. USA Federal Energy Regulation Commission, *Interconnection for Wind Energy,* June 2005. Docket No. RM05-4-000 – Order No. 661, 18 CFR Part 35.
10. Nasar, S.A. *Electric Machines and Power Systems, Vol. I.* McGraw-Hill, New York, 1995.
11. International Electrotechnical Commission (IEC). *Wind turbines–Part 21: Measurement and assessment of power quality characteristics of grid connected wind turbines.* International Electrotechnical Commission, Geneva, 2008. IEC 61400-21 Edition 2.0.
12. IEEE. *1547 IEEE Standard for Inteconnection of Distributed Resources with Power Systems.* The Institute of Electrical and Electronics Engineers, Inc, New York, 2003.
13. Reichard, M., Finney, D., and Garrity, J. *Windfarm System Protection Using Peer-to-Peer Communications,* 60 Annual Conference for Protective Relay Engineers, 2007. www.gedigitalenergy.com/smartgrid/Aug07/windfarm.pdf.
14. Strauss, C. *Practical Electrical Network Automation and Communication Systems,* Newnes, New South Wales, 2003.
15. Kundur, P. *Power System Stability and Control,* McGraw Hill, New York, 1994.
16. International Electrotechnical Commission. *Wind Turbine Generator Systems - Part 24 Lightning protection,* International Electrotechnical Commission, Geneva, 2002. IEC 61400-24.
17. Dickinson, M. *The Unique Role of Wind Turbine Step-Up (WTSU) Transformers* [Online] GoArticles.com, 2009 11, GoArticles.com. http://www.goarticles.com/cgi-bin/showa.cgi?C=2155656.
18. Vosberg, R. M. *Utility Wind Integration,* Southeast Regional Offshore Wind Power Symposium. http://www.clemson.edu/scies/wind/Presentation-Vosberg.pdf.
19. Xcel Energy. Frequently Asked Questions [Online] 11 2008. http://www.class4winds.org/faq.pdf.
20. Smith, G. *Development of a Generic Wind Farm SCADA System,* Garrad Hassan and Partners. 2001. http://www.berr.gov.uk/files/file17854.pdf. ETSU W/45/00526/REP.

21. International Electrotechnical Commission. *Wind turbine - Part 25-1: Communications for monitoring and control of wind power plants - Overall description of principles and models.* International Electrotechnical Commission, Geneva, 2006. IEC 61400-25-1 First Edition.

22. Windustry. Community Wind Toolbox, Chapter 14: Interconnection – Getting Energy to Market. *Windustry* [Online] http://www.windustry.org/ your-wind-project/community-wind/community-wind-toolbox/ chapter-14-interconnection/community-wind-to.

CHAPTER 12

Environmental Impact of Wind Projects

Don't blow it—good planets are hard to find.
—Quoted in *Time Magazine*

Introduction

Wind energy has significant environmental benefits, including no emissions into the environment during energy production, no thermal pollution of river or lake water, and no exclusive land use. Since no fuels are used, the energy production causes no emissions at all. This is unlike coal, diesel, and natural gas-fired power plants that are large contributors to greenhouse gases and to a variety of other pollutants into the atmosphere and land. Mining of coal and extraction and processing of crude and natural gas also cause pollution. In coal and nuclear power plants, a large amount of water is required either to produce steam or to cool the fuel rods. Any other fuel that releases energy during combustion has to release part of the heat to the environment causing thermal pollution. Although wind farms use a lot of land, this land is not exclusively used by the wind plant. If the land is agricultural land, agriculture can continue on over 90% of the land; if it is forestland or wilderness land, then forests and land-based animals can continue to thrive.

Wind energy does create adverse impacts on the environment including: Noise, aesthetics, killing of wildlife to a limited extent, and others. In addition, wind projects may interfere with aviation airspace, microwave communications, and long-range radar.

This chapter will start with a framework for analyzing environmental impact. This is followed by description of impact on wildlife and methodology to assess the impact. The next three sections describe the noise, shadow flicker, and aesthetic impact. Subsequent sections describe the potentials of hazard to aviation, microwave interference, and other electromagnetic interference. In each section, the impact, ways to analyze the impact, and methods to mitigate the impact are described.

Framework for Analyzing Environmental Impact

Environmental impact assessment (EIA) came out of legislation in the United States called the National Environmental Policy Act (NEPA) of 1969. It required all federal agencies to fully analyze environmental effects of their programs and actions. The EIA has become a standard practice for wind energy projects in the European Union countries. However, in the United States, a national-level EIA does not exist for private wind projects on nonfederal lands. It is largely up to the local and state planning regulatory agencies to require varying degrees of environmental assessments. Since offshore projects are on federal lands, the Mineral Management Service (MMS) of the Department of Interior has to conduct an EIA before allocating areas for offshore wind farm development.

A comprehensive framework for analyzing the environmental impact of a wind project does not exist in the United States. The National Research Council[1] has proposed guidelines for an analytical framework, but it is not a definitive framework. The reasons are lack of data and analytical methods to fully understand the full impact of wind farm development on the environment: (i) Relative to other methods of generation, (ii) for multiple temporal and spatial scales, and (iii) as a result of cumulative actions. These three considerations must be taken into account in any analytical framework.[1]

Context of Environmental Impact

Environmental impact must be evaluated relative to the impact of other human activities; two examples will illustrate this. First, consider the noise generated by a turbine. Wind turbine that is 100 m away generates 55 dBA of noise; a wind turbine 350 m away generates 35 to 45 dBA of noise. However, a busy office has a noise level of 60 dBA and a busy road 5 km away generates 35 to 45 dBA of sound. When the noise level is put in context, a more realistic picture emerges. The second example is that wind projects kill tens of thousands of birds each year in the United States. However, buildings and windows kill 500-plus million birds annually, transmission lines kill 100-plus million birds annually, and domestic cats kill hundreds of millions of birds. As the examples illustrate, putting the environmental impact of wind farm in context will help all stakeholders make better informed decisions.

Temporal and Spatial Scale

Environmental impact of a wind farm is easier to assess locally and for a short duration. For instance, the impact of a wind farm over a few years into the future and over tens of kilometers from the wind farm is routinely done. However, wind farms have impact over a much larger timeframe, for example, loss of birds and bats that have long life span and do not reproduce frequently can have a significant impact over a longer timeframe. As a second example, consider the spatial scale, fossil fuel-based energy-generation plants have a global impact because of climate change caused by greenhouse gas emissions. In the final example, consider the impact of wind turbines on a larger spatial scale: Turbines placed in the path of migratory birds can have impacts on bird populations that live thousands of kilometers away. As the examples illustrate, an EIA must be performed over several temporal and spatial scales in order to obtain a more comprehensive understanding of the impact.

Cumulative Effects

Understanding the impact of single actions is an easier task than understanding the impact of past, current, and future actions. For example, the impact of installing a few turbines on a wildlife habitat may not be large, but past or subsequent wind developments in the same region may fragment the habitat to an extent that the survival of a species in that region may be at risk. Another example is development of wind farms along the migratory path of birds that cover hundreds to thousands of miles. In regions where migratory bird paths coincide with high-wind areas, a single wind farm may be acceptable, but the cumulative effect of wind farms dotted along the migratory bird path may cause harm that is an order of magnitude higher that predicted by single installation. As the two examples illustrate, an EIA must examine the combined impact of multiple wind projects on the environment.

Quick Comparison of Wind Versus Fossil Fuel–Based Electricity Production

In the United States, the amount of greenhouse gases (CO_2, NO_x, and SO_2) emitted by coal and natural gas power plants is presented in Table 12-1.

Greenhouse Gas Emissions, kg/MWh			
Type of fuel	CO_2	NO_x	SO_2
Coal	992	2.2	4.9
Natural gas	538	0.8	0.1
System average	631	1.4	2.7

Source: From *Environmental Impacts of Wind Energy Projects,* The National Academies Press, Washington, DC, 2007.

TABLE 12-1   Emission of Greenhouse Gases for Electrical Generating Units in Year 2000

In addition to the greenhouse gases, the following are produced:[2] 34 kg/MWh of flying ash, 5.1 kg/MWh of bottom ash, and 8.9 kg/MWh of gypsum.

These pollutants have significant impact on wildlife and humans. Other environmental impacts that are common to fossil fuel-based and wind-power plants are:[1]

- Clearing of vegetation to construct plant and access roads. Both types of plant impact the environment; the relative difference is difficult to quantify.

- Aesthetics of scenic areas or visibility from public areas like highways, public parks, and others. Both types of plants impact the aesthetics. Wind plants occupy a larger area of land and, therefore, larger areas are impacted.

Impact of Wind Farms on Wildlife

There are two primary impacts on wildlife: (a) Bird and bat fatalities, and (b) impact on land-based animals because of forest clearing and changes in forest structure.

It was believed that wind turbines cause fatalities in birds and bats primarily through impact with rotating blades. Recently, a study revealed that fatalities in bats happen through barotrauma when bats enter zones of rapid decrease in air pressure near blades.[3] According to the report, "Barotrauma involves tissue damage to air-containing structures caused by rapid or excessive pressure change; pulmonary barotraumas is lung damage due to expansion of air in the lungs that is not accommodated by exhalation."

In contrast to bats, birds are vulnerable to collision with turbine blades. The following information is from the NRDC report:[1] 6% of the reported fatalities are of raptors; other species with the highest rate of reported fatalities are the nocturnally migrating passerines. Species that are abundant around turbines, but have rare cases of fatalities, include, crows, ravens, and vultures. Three factors play a significant role in bird and bat fatalities: Abundance of species, behavior of species, and wind project site-related factors. Wind project site-related factors that lead to higher fatalities include landscapes that are conducive to nesting, feeding, preying, and flying.

Endangered species and their habitats may be impacted by clearing of land and operation of wind turbines. Since all agencies of the US government are required to participate in conservation of endangered species, an early determination of the potential existence of endangered species is important.

Impact on wildlife is managed by a variety of governmental agencies. In the United States, the US Fish and Wildlife Service (USFWS) is the lead agency that must be contacted to assess the impact of wind turbines on migratory birds, threatened and endangered species, and their habitats. Although USFWS will be in the lead, other agencies are likely to be involved. Nonmigratory birds are the domain of state governments. The US Army Corp of Engineers is the lead agency if the site contains Wetlands.

In 2007, the USFWS Wind turbine guidelines advisory committee[4] was formed to develop "effective measures to avoid or minimize impact to wildlife and their habitats related to land-based wind energy facilities." It has released a 6.1 draft recommendation to the Secretary of the Interior.[5] In it, a tiered approach is proposed for assessing the adverse impact on wildlife and their habitat. In contrast to a one-size-fits-all, the tiered approach tailors amount of required evaluation based on location of project, scale of project, and the potential for significant adverse impact. Figure 12-1 is a flowchart of the assessment process. This is a three-step process:

1. The first step in each tier is a list of questions that a project must answer in the tier. The guidelines contain a list of suggested questions for each tier, and the project chooses questions that are appropriate for the project.

2. The second step in each tier is to: (a) Define methods to collect data in order to answer the questions and (b) define metrics to make decisions.

3. In the third step, methods and metrics are used to make decisions with three possible outcomes:
   a. Abandon the project because it poses high degree of risk for adverse impact to wildlife.
   b. Proceed with the project because a determination has been made that there is a low probability of adverse impact to wildlife from the wind project.

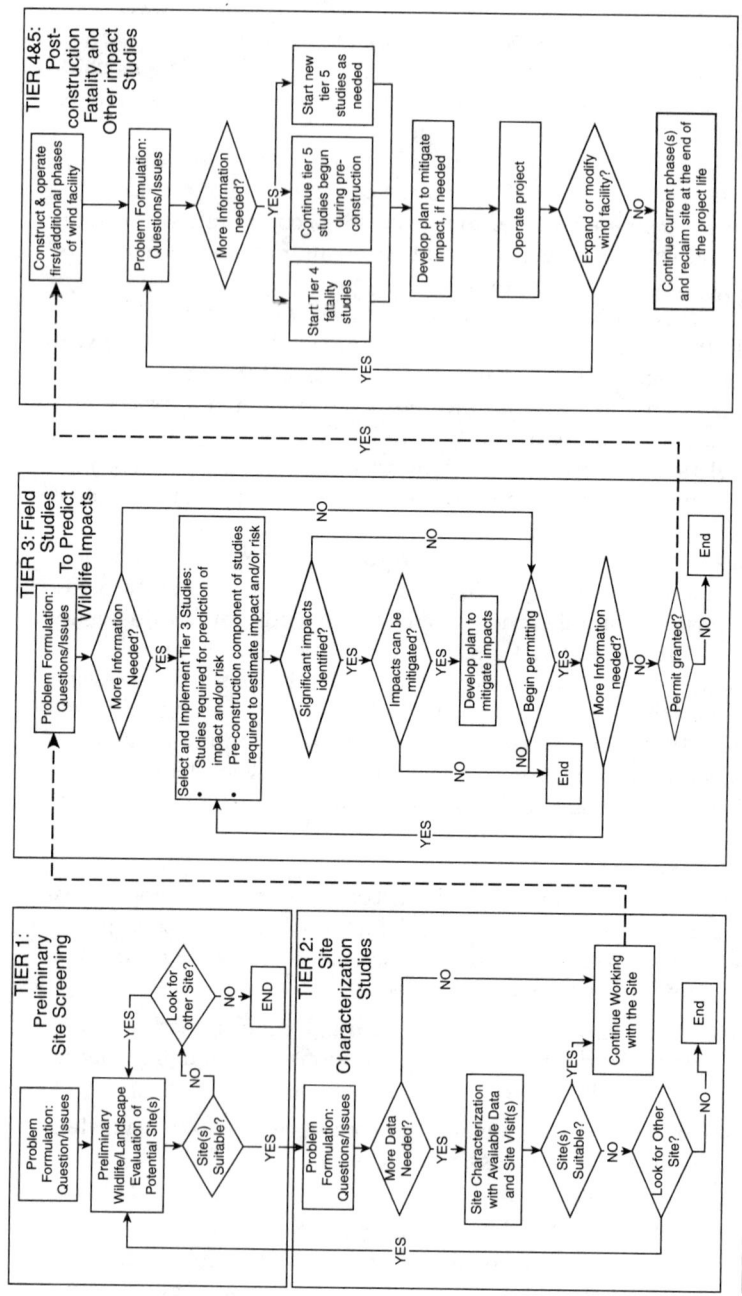

FIGURE 12-1 Process for wildlife assessment[4].

c. Move to the next tier because a determination cannot be made in the current tier. In the subsequent tier the following are performed: Additional data collection and analysis; redesign of the project for mitigation of impact; and design and execution of postconstruction monitoring.

The five tiers are[5] listed below. The first three tiers are during the preconstruction evaluation phase, while the last two are during the postconstruction monitoring phase.

Tier 1: Preliminary evaluation or screening of potential sites. Akin to determining if a site has sufficient wind with publicly available data, this tier is a first look at screening sites in terms of sensitivity to wildlife. Questions that must be answered are:

a. Does the site contain species of concern or their habitat?

b. Does site contain areas where development is precluded by law, or areas designated as sensitive by scientifically credible information?

c. Does site contain areas of wildlife congregation?

d. Does site contain sensitive species that need contiguous blocks of habitat? If the answer is yes to any of the above questions, then the site should be precluded unless there is good plan to significantly reduce the adverse impact to wildlife.

Tier 2: Site characterization studies. This involves site visit and a more in-depth review of literature, data, and information from local organizations to answer the questions posed in tier 1. In addition, specific species that are at risk in the proposed site are identified. If the answer is *yes* to any of the tier 2 questions, then the site should be abandoned because there is a high probability of adverse impact. If the answer is *no* to all the questions, based on sufficient information, then the wind project may proceed with no additional tiers. Finally, if answer is inconclusive for at least one question, because of insufficient data or the uncertainty is high, then proceed to tier 3.

Tier 3: Field studies to document site wildlife conditions and predict project impacts. In this tier, rigorous quantitative scientific field studies are conducted to assess the impact. Similar set of questions as in tier 1 and 2 are answered, but with scientifically rigorous onsite data collection related to abundance, behavior, and site use by species of interest. The data is analyzed and risk of impact is computed. If there is sufficient data to determine, with high degree of certainty, that the impact on wildlife is not significant, then the permitting processes is begun. If significant impacts are identified, then the question that is addressed is: Can the impacts be mitigated? If the answer is *yes*, then plans are developed to mitigate and the permitting process is begun. After the permit is obtained, construction can begin. A list of best management practices for site development and construction are presented in the guidelines. These can ensure that the potential impact is minimized during the construction phase.

Tier 4: Postconstruction fatality studies. This tier is performed if required by any of the tier 1–3 studies. The activities performed in this tier include search for bird and bat carcasses, classify finding based on species, compare actual fatalities with predicted, and analyze variation of fatalities based on turbine location within the project site.

Tier 5: Other postconstruction studies. This tier is performed in cases when (a) the actual fatalities are higher than predicted in tier 3 and/or the mitigation strategies implemented are ineffective, and (b) the fatalities measured in tier 4 are likely to cause a significant impact on the population. This tier is costly and should be performed in rare cases. The intent of the recommended guideline is to ensure that tier 1–3 are done with enough rigor that a project is not in a situation in which tier 5 is required.

In order to perform preliminary wildlife assessments, few states in the United States provide online tools that provide an inventory of species. The State of Illinois Department of Natural Resource has an online application, Ecological Compliance Assessment Tool (EcoCAT)[6], that provides a list of protected natural resources in and around an area of interest.

Noise from Wind Turbines

Noise is a form of pollution generated by turbines. However, the impact of sound is limited to a few hundred meters from the base of a turbine. Noise is generated in a turbine from two primary sources:

1. Aerodynamic interactions between the blades and wind. This is the persistent "whoosh" sound as the blades slice the wind. This is the dominant noise from a turbine.

2. Mechanical noise from different parts of the turbine, like gearbox and generator.

Sound is a pressure wave. A unit of measure for sound is decibels. Sound level is quantified as ten times the logarithm of the ratio of

Sound Pressure Level, dBA	Noise Source
140	Jet engine at 25 m
120	Rock concert
100	Jackhammer at 1 m
80	Heavy truck traffic
60	Conversational speech & TV
50	Library
40	Bedroom
30	Secluded wood
20	Whisper

Source: From *A Guide to Noise Control in Minnesota,* Minnesota Pollution Control Agency, St. Paul, MN, 2008.

TABLE 12-2 dBA of Common Noise Sources

energy contained in a sound wave with respect to a reference wave of 0 decibels. For a given frequency, the energy is proportional to the square of the amplitude of the pressure in the pressure wave.

$$\text{Sound level in decibels} = 10 \log_{10}\left(\frac{SE}{RE}\right) = 20 \log_{10}\left(\frac{SP}{RP}\right) \qquad (12\text{-}1)$$

where SE, SP is the measure of sound energy, sound pressure, and RE, RP is the reference energy, reference pressure. Reference pressure = 2×10^{-5} Pa.

dB(A) is the A-weighted decibels for sensitivity to human ear. Because the human ear does not have a flat spectral response, sound pressure levels are weighted based on audible frequencies, which are the higher harmonics of middle A (between 2 and 4 kHz). In the rest of the document, dBA is used as the unit of measure of sound power and sound pressure.

Sound pressure describes the effect on a receptor (a person listening) and is specified as $X\ dBA$ at a distance of Y meters. Sound power, on the other hand, describes the power of an emitter, for instance, a wind turbine emits 105 dBA. Table 12-2 contains decibel levels of common noise sources. Another important aspect is ambient noise or background noise. Wind is a major source of ambient noise. In wind projects, the relevant sound level is the level above the ambient noise.

Since sound is a compression wave, the decibel level drops quickly as distance from the sound source is increased. For a point source of sound, doubling of distance reduces pressure by one-half, which reduces decibel level by about 6 dBA $\left(= 20 \log_{10}\left(\frac{SP}{RP}\right) - 20 \log_{10}\left(\frac{0.5SP}{RP}\right)\right)$. For a line source of sound, doubling of distance reduces decibel level by 3 dBA. Considering the geometry of a turbine relative to a person on the ground, the loudest sound is heard when the blade is the closest to the ground as it whooshes through the air, and the source of sound can be approximated to be a line source.

Since decibel level is logarithmic, simple addition or subtraction of decibel levels is incorrect. As an example, consider an ambient decibel level of 60 dBA for wind speed of 8 m/s. Next, consider a spinning turbine subject to wind speed of 8 m/s and decibel level of 63 dBA, which includes ambient and turbine noise. The 3 dBA increase suggests that the energy level is double and the pressure amplitude ratio is $\sqrt{2}$. This implies that the decibel level because of turbine alone, without the ambient component, is 60 dBA. Continuing with this example, consider a person who is standing equidistant from two turbines.

There are two aspects to a noise analysis: (a) Level of noise generated by a specific turbine and attenuation of noise as a function of distance, and (b) acceptable noise level.

Wind analysis software like WindPRO and WindFarmer provide noise computation tools. These tools implement multiple noise computation methods like ISO 9613-2 and a variety of country-specific standards. Turbine manufacturers provide sound measurement data in terms of sound power at the hub that is subject to 8 m/s wind speed; a few manufacturers also provide data for different tonal frequencies. The noise computation tools use the sound power at the hub combined with terrain, air density, and location of turbines in a wind farm to compute iso-decibel lines. The iso-decibel lines are then used to color code regions based on decibel level.

The United States does not have a national standard for computing sound levels and does not have national standard for acceptable levels of sound level from a turbine. The acceptable levels are managed by state or local agencies through sound ordinances. Most sound ordinances distinguish between day and night time. As an example, consider the noise ordinance in the state of Minnesota[7] for Class 1 land-use areas that include housing communities, farmhouses, and educational organizations. In Table 12-3, the noise levels are specified in terms of percentage of time in an hour; L50 is 50% of the time in an hour and L10 is 10% of the time in an hour.

Mitigation of Noise

The amount of noise is directly related to the maximum tip speed of the turbine, which depends on the angular speed of the turbine in revolutions per minute (rpm), length of blade, and wind speed. Strategies to lower the noise level include:

Daytime		Nighttime	
L50	**L10**	**L50**	**L10**
60 dBA	65 dBA	50 dBA	55 dBA

Source: From *A Guide to Noise Control in Minnesota,* Minnesota Pollution Control Agency, St. Paul, MN, 2008.

TABLE **12-3**   Noise Ordinance in the State of Minnesota

- Use of smaller length blades when installed in noise-sensitive areas
- Setting of turbine control to lower rotor's revolutions per minute. Most utility scale turbines possess such capabilities.
- Use of direct-drive turbines, which are quieter because they have no gearbox.

During the planning phases of a wind project, noise computation tools in WindPRO or WindFarmer are used to analyze noise levels. In addition to the turbine layout and terrain model of the area, noise-sensitive areas (residences, barns, etc.) are identified in these tools. The outcome of the analysis is: a) Noise level at the noise-sensitive locations; b) if necessary, relocation of wind turbines to ensure that all the noise-sensitive areas are below the permitted decibel level.

Low-Frequency Noise

In addition to the above audible frequencies, turbines produce low-frequency noise in the range of 20 to 100 Hz. These frequencies are not adequately represented in the A-weighted decibel level; instead, a C-weighted decibel level captures lower frequency sound. IEC 61400-11 suggests that if the C-weighted decibel level is higher by 20 db compared to A-weighted decibel level, then there is a significant low-frequency noise component from the turbine and must be modeled.

Myths about harmful effects of low-frequency sound exist in the literature. However, an independent scientific panel concluded that:[14]

- There is no evidence that the audible or subaudible sounds emitted by wind turbines have any direct adverse physiological effects.
- The ground-borne vibrations from wind turbines are too weak to be detected by, or to affect, humans.
- The sounds emitted by wind turbines are not unique. There is no reason to believe, based on the levels and frequencies of the sounds and the panel's experience with sound exposures in occupational settings, that the sounds from wind turbines could plausibly have direct adverse health consequences.

Shadow Flicker

During sunrise and sunset, wind turbine blade casts a shadow that alternates as blades cover the sun for a short period. This is called shadow flicker and is relevant only if there are houses or other buildings with windows on which a shadow falls. The shadow flicker is pronounced on buildings that are less than 1,000 ft from a turbine.[1] The shadow flicker happens only during sunrise and sunset for a few minutes each day. Wind assessment applications, like WindPRO and WindFarmer, have tools to assess the impact of shadow flicker. In addition to summary numbers like 20 h of flicker annually, they also provide detailed location-by-location and day-by-day assessment of start time and end time of shadow flicker.

Only Germany has guidelines for calculating shadow flicker and an allowable limit on the amount of shadow flicker. The calculation guidelines are:[8] (a) The angle of the sun over the horizon must be at least 3°; (b) the blade of the wind WTG must cover at least 20% of the sun.

The maximum shadow impact for a neighbor to a wind farm according to the German guidelines is: (a) Maximum 30 h per year of astronomical maximum shadow (worst case); (b) maximum 30 min worst day of astronomical maximum shadow (worst case); (c) if automatic regulation is used, the real shadow impact must be limited to 8 h per year.

The known impact of shadow flicker is to epilepsy patients. According to the Epilepsy Foundation, frequencies above 10 Hz are likely to cause epileptic seizures.[1] For most utility-scale turbines, the frequency of rotation is 0.6 to 1 Hz, so the frequency of flicker for a three-blade turbine is 1.8–3 Hz.

Aesthetic Impact

Visual impact is the most difficult to quantify and is often the primary reason for opposing a wind project. Although there is widespread support for wind energy projects, in general, the support drops when the project is close to one's community because of the not-in-my-backyard (NIMBY) syndrome. Most projects do not adequately address visual impact issues and even when visual assessments are done, subjective approaches are adopted to assess visual impact. In this section, a few established methods will be covered for visual assessment.

NRC has outlined a detailed assessment process for evaluating the visual impact of wind projects. It is a six-step process:[1]

1. Project description. All relevant wind project data is collected in this step including: Turbine details like size, color,

lighting, foundation area; meteorological towers; substation; access roads; transmission lines, and others.

2. Project visibility, appearance, and landscape context. Photo-montages and 3-D simulation of walk-through of an area that is within 10 miles of the project area can provide an objective view of the visual impact on the landscape. In this step, the following facets are included: Detailed topographical maps with digital elevation; all features in the modeled area like trees, buildings, roads, and water bodies, etc.; digital pictures of views of interest. 3-D simulation walk-through software can take all this information along with visual information from the project description and create a simulation in which a user can walk-through or fly-through the areas of interest. A run-time copy of the 3-D simulation can be provided to residents and other stakeholders. This provides concerned citizens a means to learn about the visual impact of the project from different vantage points.

3. Scenic resource values and sensitivity levels. The US Forestry Service and the US Bureau of Land Management have developed methods to determine scenic resource values for visually sensitive landscapes. In most cases, the scenic resource values have already been determined and recorded in public planning documents pertaining to previously proposed transmission, power plants, or other projects. Scenic values are determined based on distinctive features that are characterized by unique combinations of land form, patterns of vegetation, water bodies, and others.

4. Assessment of visual impacts. Visual impact is assessed based on visual context and project characteristics. Visual context is determined based on the following factors: Distance of viewers from the project, view duration, angle of view, panoramic versus narrow view, scenic quality of view, focal point within view, number of observers, viewer expectations, documented scenic resources, visibility, and weather conditions. Project characteristics that have visual impact are: Scale of the project relative to the surroundings, number of turbines in view and visual clutter, visibility of project infrastructure, noise, and lighting.

5. Mitigation techniques. Some of the mitigation techniques that may be applied to wind projects are:
   a. Appropriate siting so as not to overwhelm sensitive scenic resources
   b. Downsizing, relocation, or adjusting turbine layout pattern to minimize impact

c. Minimize lighting, use nonreflective materials, choose appropriate colors, commitment to maintain turbines so they are functioning and have no visible blemishes like oil leaks, and commitment to decommission nonoperating turbines.

d. Design infrastructure like substation and transmission in a manner that hides it from view

e. Minimize vegetation removal, screening, and noise.

6. Determination of unacceptable or undue aesthetic impacts. The final aspect of the assessment is to make a determination if the proposed project will pose an unacceptable visual impact based on the information collected in the five steps above.

Hazard to Aviation

Although hazard to aviation is not strictly an environmental impact, it is part of the permitting process. In this section, the process used in the United States is described as an example. Projects must consult local aviation jurisdiction for specific guidance, processes, and rules.

The US Federal Aviation Administration (FAA) administers the efficient use of navigable airspace under Title 14 of Code of Federal Regulations Part 77. To accomplish this, FAA performs an Obstruction Evaluation/Airport Airspace Analysis (OE/AAA) through Form 7460-1, Notice of Proposed Construction or Alteration. The application for evaluation may be filled out online at https://oeaaa.faa.gov/oeaaa/external/portal.jsp. The outcome of this process is a determination by the FAA that the proposed structure is or is not a hazard to aviation. In addition, FAA provides guidance on Obstruction Marking and Lighting, which describes the standards for lighting and marking required on the structure.

Notification to FAA is required for:[9]

1. Any construction or alteration of more than 200 ft in height above the ground level

2. Any construction or alteration:
   i. Within 20,000 ft of a public use or military airport that exceeds a 100:1 surface from any point on the runway of each airport with at least one runway more than 3,200 ft
   ii. Within 10,000 ft of a public use or military airport that exceeds a 50:1 surface from any point on the runway of each airport with its longest runway no more than 3,200 ft
   iii. Within 5,000 ft of a public use heliport that exceeds a 25:1 surface

In the above specification, 25:1 surface is a plane that starts from the edge of the airport and has a gradient of 1-ft elevation from ground for every 25 ft of radial distance from the airport edge. Note the above specifies that any structure that is 200 ft or higher in height, regardless of its proximity to an airport, requires FAA notification. Notification is also required for temporary structures like met-towers that exceed 200 ft. in height above the ground level.

The criteria used by the FAA to deem a structure to be an obstruction to the navigable airspace is defined in http://edocket.access. gpo.gov/cfr_2004/janqtr/pdf/14cfr77.23.pdf.

After the FAA completes the OE/AAA study, there are three possible outcomes:

1. *No objection.*

2. *Conditional determination.* The proposed project is acceptable as long as the lighting and marking standards are implemented.

3. *Objectionable.* The proposed project is determined to be a hazard.

An objectionable ruling is not necessarily the death of a project. The FAA will work with a project sponsor to determine how the impact of structures can be mitigated. In the case of wind farm, this may involve relocating a few turbines.

Electromagnetic Interference

Telecommunications towers use radio waves to transmit and receive microwave, TV, radio, and radar signals. If a telecommunications tower is in close vicinity of a wind turbine, then there is likely to be interference with microwave antenna. A large wind turbine can change the impedance of high-gain antenna that is used by microwave signal. This interference is significant within a radius of 200 m. Outside of this radius, interference is because of obstruction of the direct signal or scattering of signal, which is described next.

Microwave

Microwave transmissions (300 MHz–300 GHz) are used for a variety of purposes, including backhaul in cellular networks, microwave radio relay links for TV and telephone, and communication with satellites. It is a line-of-sight transmission of radio signals in the microwave frequency range between a transmitting tower and receiving tower. Microwave antennas are parabolic dishes in tall towers. In the United States, the Federal Communications Commission (FCC) issues licenses to operate microwave links and is a repository of all the information pertaining to locations of microwave towers.

Since the transmission is point-to-point and line-of-sight, interference by moving blades of a turbine or any other part of the structure of a turbine, can cause attenuation or interruption of the signal. The path of the microwave link forms a volume around the line connecting the transmitter and receiver, and this volume is called the Fresnel zone. The first Fresnel zone must be free of obstacles. The width of the first Fresnel zone can be approximated using:[10]

$$d_1 = \sqrt{\lambda(1{,}000d_t)(1 - d_t/d_{AB})} = 17.3\sqrt{d_t(1 - d_t/d_{AB})/f}$$

where λ is the wavelength in meters, f is the frequency in GHz, $\lambda = 0.3/f$, d_t is the distance between the turbine and the closest tower in km, and d_{AB} is the distance between towers A and B in km. As an example, consider a 6 GHz microwave signal being transmitted between two towers that are 25 km apart; the Fresnel zone will be the widest at the middle. The diameter of the Fresnel zone in the middle is:

$$d_1 = \sqrt{(300/6{,}000)12{,}500(1 - 0.5)} = \sqrt{312.5} = 17\,m$$

In addition to d_1, the height of the microwave path must be compared with the height of the obstacle. The height of the microwave anten- nas on towers is available from the FCC database. Figure 12-2 is an illustration of the microwave analysis. In case, w is less than 100 m, that is a turbine is less than 100 m away from an active or proposed microwave link, then a more detailed analysis must be performed regardless of interference. The detailed analysis should include

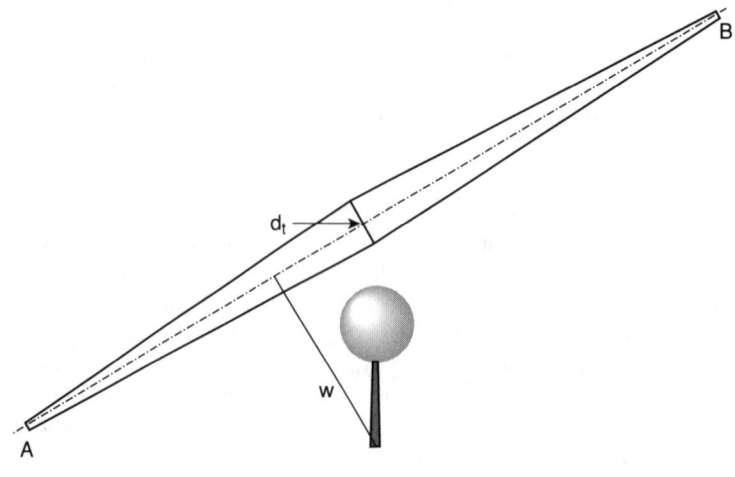

Figure 12-2 A and B are locations of microwave towers; w is the distance from turbine to the closest point on the path between A and B. d_t is the width of the Fresnel zone around the microwave link. Diagram is not to scale.

validating the exact locations and height of microwave antennas by contacting the licensee. The reason is that FCC database may not have precise locations and heights of microwave transmitters or receivers either because of measurement error or because the installation was updated without informing the FCC. In case microwave interference is found, then the turbine must be relocated.

TV and Radio Transmissions

Similar to microwave signal, TV signal are susceptible to interference. Here, in addition to direct obstruction of the line-of-sight between transmitter and receiver, a second mechanism causes signal interference. This mechanism is scattering of signal, which causes the receiver to get two signals: Direct signal and scattered signal that is delayed, which impact the picture quality by producing ghost images. Sound quality is typically not affected. TV reception can normally be fixed with improvements to antennas and changes to setup of the antenna.

Radar

Wind turbines interfere with radars installed for the purpose of monitoring weather, and detecting and tracking incoming aircraft and missiles. In the United States, radar operators include National Weather Service, Department of Defense, Homeland Security, FAA, and others. Rotation of wind turbine blades can create clutter interference and Doppler interference with radar signals. Examples of impact of wind farms on radar include:

1. In the United States, the National Weather Service (NWS) has 159 Next Generation Weather Radar (NEXRAD). When wind farms are located in the radar line of sight (RLOS) then there are three types of impact:[11]
   - Wind turbines reflect energy whose signal can be confused with a storm. This contaminated reflectivity data is called clutter.
   - If turbines are within a few kilometers, then the turbine can block a significant portion of the radar signal that can cause the radar to not "see" events behind the wind farm. Figure 12-3 is a chart of impact from the NWS.
   - The velocity and spectrum width data can be impacted, which may result in issues with algorithms to detect certain storm characteristics, such as tornadoes, storm motion, and turbulence.

2. As shown in Fig. 12-3, the impact is the greatest when the turbine is within a few kilometers. According to Vogt et al.,[11] turbines within 200 m of radar are not safe; turbines within

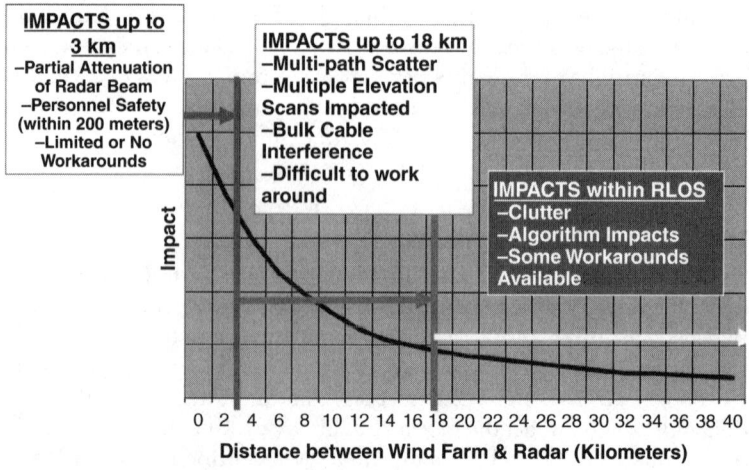

IMPACTS up to 3 km
- Partial Attenuation of Radar Beam
- Personnel Safety (within 200 meters)
- Limited or No Workarounds

IMPACTS up to 18 km
- Multi-path Scatter
- Multiple Elevation Scans Impacted
- Bulk Cable Interference
- Difficult to work around

IMPACTS within RLOS
- Clutter
- Algorithm Impacts
- Some Workarounds Available

Impact

0 2 4 6 8 10 12 14 16 18 20 22 24 26 28 30 32 34 36 38 40

Distance between Wind Farm & Radar (Kilometers)

FIGURE 12-3 This graph shows how wind turbine clutter impacts on weather forecast office (WFO) operations vary with distance (approximate) from the Weather Surveillance Radar-1988, Doppler (radar). The impacts (blue line) curve upward (increase), as turbines are sited closer to the radar. Impacts increase quickly very close to the radar. Two key distances are 11 and 2 miles. These distances delineate where additional impacts generally begin (assuming a level terrain). The distances of these impacts can vary by many miles depending on terrain and weather conditions. (Courtesy of the WSR-88D Radar Operations Center).[11]

1 km prevent forming of proper radar beam leading to significant degradation of radar performance; turbines within 18 km can cause clutter.

3. Air traffic control software could temporarily lose track of an aircraft that is flying over a wind farm. [12] Wind farms create a vertical "cone of silence" and when an aircraft flies into it, the radar system has difficulty detecting the plane. On the civilian side, this is an issue only with smaller planes that do not have transponders.

4. Normally, an air traffic control radar system will filter out large returns from stationary targets, but the Doppler shifts introduced by the WTG blades are interpreted as moving aircraft that can confuse radar operators and compromise safety[12]

Conventional radar's clutter filters are ineffective in removing clutter signal reflected from moving blades of a fixed turbine. A study by MITRE[12] finds that there are no fundamental technical constraints that prevent detection of clutter caused by wind turbines. The problem is the aging long-range radar infrastructure that was not designed to mitigate clutter from wind turbines. Several solutions have been proposed to mitigate this problem including[13]:

- Changing the Doppler signature of blade by making them less reflective to radar signals
- Changing digital processing software to recognize wind turbines
- Curtailing production during significant weather events such that uncluttered data can be obtained

In the United States, the National Telecommunications Information Administration (NTIA) is the central clearinghouse for all

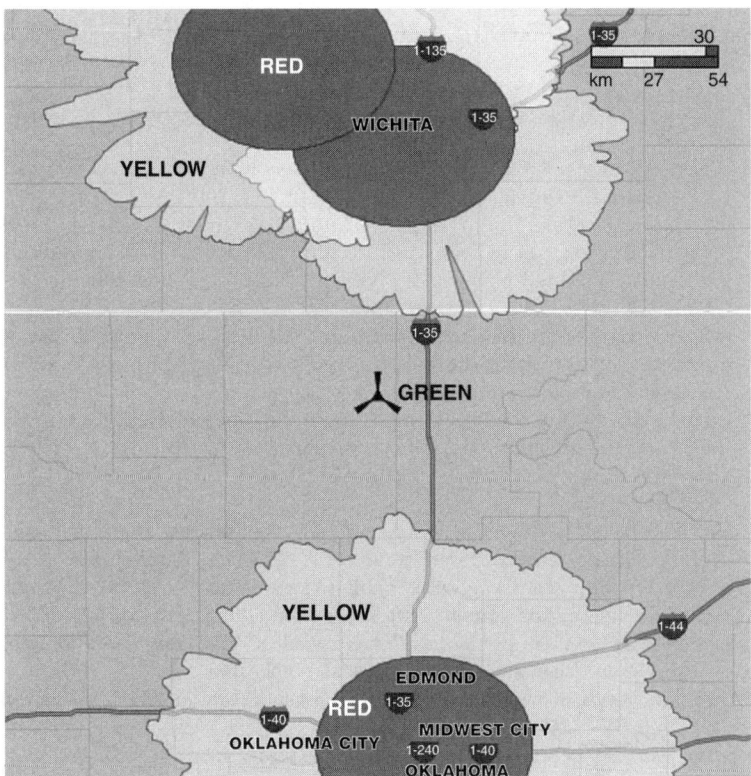

Figure 12-4 Output of FAA tool for preliminary analysis of potential impact to long-range radar for Air Defense and Homeland Security. Legend: RED: Impact highly likely to Air Defense and Homeland Security radars. Aeronautical study required. YELLOW: Impact likely to Air Defense and Homeland Security radars. Aeronautical study required. GREEN: No anticipated impact to Air Defense and Homeland Security radars. Aeronautical study required.

government agencies to receive information about proposed wind farms and to respond with a determination of radar interference. Thus far, submission of wind farm data to NTIA is voluntary; NTIA collects objections from the government agencies and responds in about 45 days.

The FAA also provides an online evaluation tool for prescreening of sites with respect to impact on long-range and weather radars https://oeaaa.faa.gov/oeaaa/external/gisTools/gisAction.jsp?action =showLongRangeRadarToolForm. Its scope is: (a) Air Defense and Homeland Security Radar; (b) weather surveillance radar; and (c) military operations.

The outcome of the analysis is a map around the user-specified location with three colored regions that describe the impact to air defense and homeland security radars: Green (no impact), yellow (impact likely), and red (impact highly likely). An example of such an analysis for a location in the state of Oklahoma is seen in Figure 12-4.

References

1. National Research Council. *Environmental Impacts of Wind Energy Projects*, The National Academies Press, Washington, DC , 2007.
2. Sokka, L., Koskela, S., and Seppälä, J. Life cycle inventory analysis of electricity production from hard coal-Including data on Polish coal mines and a Finnish coal-fired power plant [Online] 2005. http://www.energia.ee/OSELCA/files/Hard_coal_overview.pdf.
3. Baerwald, E. F., D'Amours, G. H., Klug, B. J., and Barclay, R. M. R. "Barotrauma is a significant cause of bat fatalities at wind turbines."*Current Biology, Cell Biology*, 18: 695–696, August, 2008.
4. US Fish & Wildlife Service. Wind turbine guidelines advisory committee. *USFWS Habitat and resource conservation* [Online] 2010. http://www.fws.gov/habitatconservation/windpower/wind_turbine_advisory_committee.html.
5. USFWS Wind Turbine Guidelines Advisory Committee Synthesis Workgroup. Draft recommended guidelines. *USFWS* [Online] December 16, 2009. http://www.fws.gov/habitatconservation/windpower/Wind_FAC_Synthesis_Workgroup_Draft_Recommendations_v6.1.pdf.
6. Illinois Department of Natural Resources. EcoCAT. *Illinois DNR* [Online] http://www.dnr.state.il.us/orep/ecocat/srchcnsltns.asp.
7. *A Guide to Noise Control in Minnesota*, Minnesota Pollution Control Agency, St. Paul, MN, 2008.
8. Nielsen, Per. *WindPRO 2.5 Users Guide*. EMD International, Aalborg, Denmark, 2006.
9. Federal Aviation Administration (FAA). PART 77–OBJECTS AFFECTING NAVIGABLE AIRSPACE. *2004 CFR Title 14, Volume 2* [Online] National Archives and Records Administration, 2004. http://www.access.gpo.gov/nara/cfr/waisidx_04/14cfr77_04.html.
10. Sengupta, D. L., and Senior, T. "Electromagnetic Interference from Wind Turbines," *Wind Technology*, D. Spera,Chapter 9, ASME Press, New York, NY, 1994.

11. Vogt, R. J., Crum, T. D., Sandifer, J. B., Ciardi, E. J., and Guenther, R. A way forward, Wind farm-weather radar coexistence. *NOAA National Weather Service* [Online] 2009. www.roc.noaa.gov/WSR88D/Publicdocs/WindPower2009_Final.pdf.
12. Brenner, M. *Wind Farms and Radar.* The MITRE Corporation, McLean, VA, 2008. http://fas.org/irp/agency/dod/jason/wind.pdf. JSR-08-125.
13. Tenant, A., and Chambers, B. Signature management of radar returns from wind turbine generators, *Smart Materials and Structures* 15: 2006.
14. Colby, W. D. et al., Wind turbine sound and health effects, an expert panel review, December 2009. http://www.awea.org/newsroom/releases/AWEA_CanWEA_SoundWhitePaper_12-11-09.pdf.

CHAPTER 13

Financial Modeling
of Wind Projects

Failure to prepare is preparing to fail!
—John Wooden

Introduction

Financial models for wind projects are unique in many respects. The uniqueness is in the fact that (i) wind energy projects are capital intensive, with large amount of upfront investment, (ii) no raw material costs, (iii) relatively small operating costs, and (iv) significant tax-related incentives.

In this chapter, various aspects of a financial model for wind projects are covered. The chapter starts with detailed description of the revenue, capital costs, and recurring costs. The next section deals with depreciation and tax issues. Financial statements and the performance metrics are covered next. The final two sections deal with financing and organizational structure of wind projects, and examples of scenarios and alternatives that are evaluated to optimize the performance of a wind project.

Financial Model

The three major components of a financial model of a wind project are: Revenue model, capital costs, and recurring costs.

Revenue Model

Revenue generated from a wind project is from the sale of energy. The sale price of energy is either: (a) Negotiated through a power purchase

269

agreement (PPA) with a buyer (typically, a utility), (b) market-based pricing, (c) feed-in tariff, or (d) retail rate through net-metering. In the rest of this section, the laws and policy drivers that determine pricing in the United States are discussed.

In the United States, according to the Public Utility Regulatory Policy Act (PURPA), a utility is required to buy power from an independent power producer at the "avoided cost" of production. The responsibility of setting the avoided cost rests with individual states. In the 1990s, the electricity market was deregulated and a framework for wholesale energy generation market was created. Utilities were required to provide to independent power producers the same transmission service that is provided to their own generators. The service must be provided through nondiscriminatory open access tariffs.

In the United States, renewable portfolio standards (RPS) is another policy tool that influences power prices paid to wind projects. RPS is a state policy that requires utilities to produce or purchase a minimum percentage of energy from renewable energy sources within a certain timeframe. As of 2010, RPS has been enacted in 29 states and District of Columbia. Of the 29 states, six states have set goals; the rest of the states mandate the requirement. For example, RPS of the state of Illinois mandates[1] that all utilities get 25% of electricity from renewable sources by 2025.

Power Purchase Agreement

The PPA is the most widely used model to set prices and sell electrical energy for large wind projects in the United States. It is a negotiated price for sale of power in a "take or pay" structure, which means the utility is obligated to take all the energy that is produced, regardless of demand. The floor price of the PPA is the avoided cost; in the United States one could argue that if the price is the "avoided cost" then the project owner does not need a PPA because this price is guaranteed by the law. Therefore, the PPA price is higher than the avoided cost. Figure 13-1 is a chart of the past prices of wind energy in the United States. Utilities in states that have RPS mandates usually have higher PPA price compared to states without RPS, because the utility has a mandate to purchase a fixed percentage of renewable energy. In most states, utilities are far below the mandated percentage and, therefore, are eager to buy renewable energy.

PPA is typically a long legal document that is negotiated between a buyer and seller of energy ahead of commissioning of a project. The PPA typically covers the entire life of a project, which is typically 20 years. In addition to sales price of energy and duration of agreement, PPA contains a variety of negotiated agreements about: Exit clause, the commissioning process, curtailment agreements, transmission issues, milestones and defaults, credit, insurance, and ownership of environmental attributes or credits. Although PPA provides the

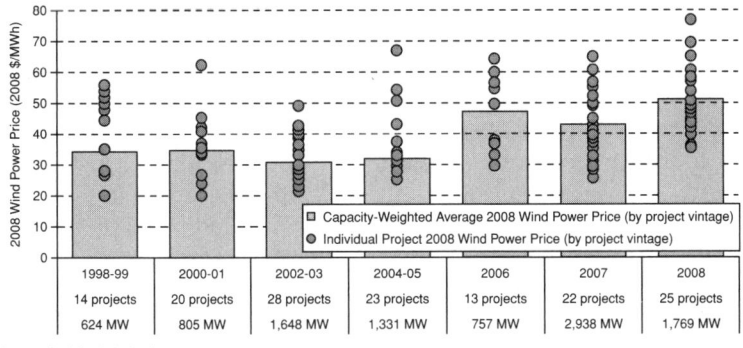

Source: Berkeley Lab database

FIGURE 13-1   Sale price of wind energy in United States.[2]

price of energy sales in order to compute revenue, other items in the PPA that can materially affect revenue are:

> *Price profile.* PPA sales price of energy may be fixed or contain provisions for price escalation, or, in the case of Minnesota's Community Based Energy Development (C-BED), PPA is required to provide a higher price in the first 10 years of the project. The negotiated price is close in value to the wholesale price of electricity in the utility's region. In the United States, a utility may pay more for the electricity if the State has a renewable portfolio standard (RPS). The negotiated price in a PPA is a guarded secret and not released publically. As a rule of thumb, PPA rate is about 50–75% of the local retail electricity rate. In the United States, PPA rates range from 4.5 to 6.5 cents per kWh. In the Caribbean, PPA rates are 10-plus cents per kWh, because most of the energy is generated using diesel or heavy oil and the retail cost of energy is 20-plus cents per kWh.
>
> *Curtailment clauses.* Curtailment of energy delivery may occur due to insufficient demand on the grid, insufficient transmission capacity, or unscheduled maintenance. Under such conditions, wind farm will be asked to curtail production or will voluntarily curtail production. PPA will specify the conditions under which a wind farm will be compensated.
>
> *Penalties for delivering less energy than scheduled.* As mentioned in Chapter 11, wind farms are required to provide day-ahead plans that states the amount of hourly delivery of energy. When the deliveries are not met, penalties may apply per the PPA.
>
> *Penalties for delay in commissioning of project.* PPA may contain clauses that specify penalties for missing milestones. Interim and final milestones are usually defined; examples of

milestones are obtaining permits, delivery of turbine, construction contract, start of construction, and commercial operation date. These penalties should be taken into account as either lost revenue or increase in cost.

Lower prices for delivery of energy that is in excess of planned amount. PPA may have a clause for lower prices, which must be applied to the revenue.

Market-Based Pricing

Market-based pricing is available as an option to wind project owners that do not wish to sign long-term fixed price PPA. Because of the unpredictability of the spot price of energy, investors will not fund wind projects that rely on selling 100% of energy in this market. This option is rarely exercised in isolation; some projects choose to sell a small percentage of electricity on the spot market to avail of market-based pricing.

Feed-in Tariff (FiT)

FIT is a common pricing model in Europe, Canada, and several other countries. This provides the most simple and predictable model for pricing from the point of view of renewable energy investors. In this model, the price of energy is transparent, fixed, and applies to all producers of renewable electricity. Utilities are mandated by legislation to provide access and a fixed price for electricity that is normally based on cost plus normal profit margin. The feed-in tariff is specific to the type of renewable resource, for instance, solar usually has a higher feed-in tariff compared to wind; in some cases, it may depend on the size of the project. Feed-in tariff has proved to be the best mechanism to promote renewable energy. An example of FiT in Germany in 2005 is:[3]

- For onshore wind projects, pricing is 8.7 eurocents per kWh for the first 5 years and then 5.5 eurocents per kWh from 6 to 20 years

- For offshore wind projects, pricing is 9.1 eurocents per kWh for the first 12 years and then 6.2 eurocents per kWh from 13 to 20 years

As of 2009, FiT for wind energy is essentially nonexistent in the United States, although several states are considering enacting FiT legislation.

Retail rate through net-metering   Net-metering is a popular pricing mechanism for small wind projects, where the renewable energy source, the demand center (house, factory, or office), and the grid are all connected to the same meter. The meter runs backward when the energy production is greater than demand, that is, renewable energy is injected to the grid. The meter runs forward when the renewable energy production is less than demand, that is, energy is used from the grid. In the United States, several states have legislated net-metering laws. The laws are state-specific and the most updated information may be found at DSIRE, http://www.dsireusa.org/. The net-metering scheme is characterized by the following:

a. *Capacity limit of the power plant.* In majority of the cases, net-metering applies to small generators of size 100 kW or less. For instance, as of 2009, Nebraska has a net-metering limit of 25 kW, Oklahoma is 100 kW (or 25 MWh per year), Illinois is 40 kW, and Massachusetts is 2 MW.

b. *Netting period.* The meter is netted at the end of the specified period. For example, if the netting period is 1 month, then, at the end of each month, the excess energy supplied to the grid is paid at the excess energy price. In Massachusetts, there is no netting period, that is, credits can be transferred from one month to another indefinitely. In Illinois, the netting period is a year after which the credits expire, meaning that net excess generation can be transferred month-to-month at full retail value until the twelfth month; at the end of 12 months, the credits expire. In Oklahoma, the netting period is 1 month.

c. *Payment of excess generation.* In some cases, the utility receiving the energy pays for the excess energy produced by the renewable source at a discounted rate. In Massachusetts, the net excess generation is monetized at a rate that is slightly less than the full retail rate (goes hand-in-hand with no netting period). In Oklahoma, utilities are not required to pay for net excess generation; however, if a utility wishes to, it can then pay the avoided cost rate.

In addition to paying for energy usage, which is a charge per kilowatt-hour, nonresidential utility customers pay a variety of other charges: Demand charge, reactive power charge, and other charges. Demand charge is computed based on peak power usage (not energy). Peak is computed by finding the maximum power usage in each 15-min or 1-h interval for the billing cycle of 1 month; the interval used depends on the utility. The demand charge is high because it reflects the cost of infrastructure—generators, transmission, and all the equipment—required by the utility to deliver peak power to customers, in addition to the extra cost of turning on the most expensive generation capacity. It is not uncommon to see higher kilowatt demand charges compared to the kilowatt-hour energy charges. Customers are, therefore, rewarded handsomely to reduce peak demand.

In order to understand the real savings of a wind energy project under the net-metering scenario, factors like reduction in demand charge and reactive power charge should also be taken into account. This is particularly true for industrial and business customers that may have a demand charge that is greater than the charge for energy (kilowatt-hour).

Renewable Energy Credits and Carbon Credits

In addition to revenue from selling electricity, there is revenue from selling renewable energy credits, which can take the form of renewable energy certificates (REC), carbon credits, and other tradable certificates. In the United States, REC is a proof of 1 MWh of renewable energy generation. It is also called green tags or tradable renewable certificates. In states with renewable portfolio standards (RPS), RECs are the method of accounting for RPS obligations and verifying if the RPS is achieved.[4] RECs have become a tradable commodity and when trading occurs in the RPS context, it is called the REC compliance market. The REC compliance market is state-specific. Voluntary REC market is where RECs trade in the green power market on a voluntary basis and the voluntary buyers are businesses and households. Renewable energy generators in non-RPS states sell RECs into this market. There are two voluntary markets: National and West.

In the eastern states, in January 2009, the RECs traded[2] in the compliance market for about $14 in New Jersey, about $28 in Connecticut, about $33 in Massachusetts, and about $45 in Rhode Island (all prices are per megawatts-hour). These states are considered the high-price market. In the low-price market, in January 2009, the price of RECs was little over $1 in Texas, about $3 in the national voluntary, and about $7.50 in the west voluntary market. Overall, the prices have a high degree of variability.

In a significant fraction of the cases, a PPA defines a bundled price for the energy and the REC. In this case the buyer of energy owns the REC and when renewable attributes are claimed to meet the RPS then the buyer (through a REC tracking system) retires the RECs. In cases of unbundling of REC, there is a potential for "double counting" whereby both entities (buyer of REC and buyer of electricity) claim the renewable energy attributes.

Carbon credit is the other major mechanism for monetizing through tradable renewable energy credits. This is applicable in countries that have ratified the Kyoto protocol. One carbon credit is equivalent to 1 ton of CO_2 or equivalent gases. Kyoto protocol defines caps or quotas on greenhouse gas (GHG) emissions for a country. The country, in turn, assigns the quotas to utilities and businesses that produce GHG. If the business emits less than the assigned quota, then it can sell the credit; on the other hand, if a business emits more than the assigned quota, it then buys credits. This provides a monetary incentive to a utility or business to invest in renewable energy projects.

In developing countries, which have ratified the Kyoto protocol, renewable energy generators can create credits that can be monetized.

One of the mechanisms is the clean development mechanism (CDM) that creates certified emission reduction (CER), which are monetized by trading in an exchange.

Figure 13-1 contains data collected by Lawrence Berkeley National Laboratory (LBL)[2] on the sale price of electricity in the United States for projects with a PPA that has a bundled price for electricity and RECs. The LBL database contains 42% of wind capacity installed between 1998 and 2008. The average bundled sale price of energy was about $51.5/MWh for 1,769 MW of wind projects installed in 2008. Note these prices do not include federal or state incentives. In 2008, the production tax credit was $20/MWh.

Revenue Computations

After the price of electricity has been determined, the next step is to multiply it by the amount of energy produced in the time period. The time period for a preliminary financial model is a year, while it is a month for detailed financial model. Revenue is therefore:

$$R(i) = pr(i) \cdot en(i)$$

where i is the period, $R(i)$ is the revenue in period i, $pr(i)$ is the total price (including credits) in period i, and $en(i)$ is the energy production in period i.

Capital Costs

The capital cost of a wind project is called the total installed cost (TIC), which includes all the costs until commissioning of a project. TIC is expressed in terms of dollars (or Euros or any other currency) per kilowatt. According to LBL report[2] TIC in the United States were: $1,570/kW in 2006 and $1,915/kW in 2008 (see Fig. 13-2). The 2008 data is based on 61 onshore wind projects totaling 6,125 MW or 72% of wind power capacity installed.

WindPower Monthly reported the following average cost of a fully installed onshore wind farm across the world were:[5] €1,502/kW in 2008 averaged over 3,600 MW with a range of €1,300 to €1,700/kW, and €1,500/kW in 2009 averaged over 4,000 MW with a range of €1,200 to €1,800/kW. The price of turbine decreased from €1,100 in 2008 to €1,050/kW in 2009. (See Fig. 13-3 for a plot of costs in 2004 to 2009 from *WindPower Monthly*.)

The capital costs fall into two broad categories as listed in Table 13-1: Turbine costs and balance of plant (BOP) costs. The costs are from the NREL's jobs and economic development (JEDI) model.[6] This model contains state-specific default costs in the United States market. Although valuable for preliminary financial analysis, the default costs are approximate. It must be emphasized that capital costs are site- and size-specific, and that generic costs should be used with caution. In

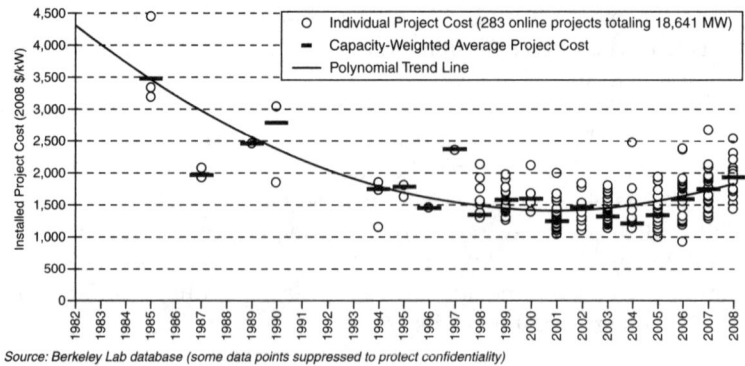

Source: Berkeley Lab database (some data points suppressed to protect confidentiality)

FIGURE 13-2   Total installed cost of wind projects in the United States.[2]

Table 13-1, total installed cost of $2,000/kW is used; the percentages are for the state of Colorado.

The cost of turbine is the most significant cost of a project. According to LBL[2] data, turbine costs hit a low point of $700/kW in 2000–2002; since then, the cost has doubled to $1,360/kW in 2008. The reasons are: Higher degree of sophistication of the machines with respect to higher capacity factor, higher degree of grid friendliness (LVRT, reactive power compensation, etc.) and higher prices of raw materials, like steel and copper. LBL[2] data suggests a significant

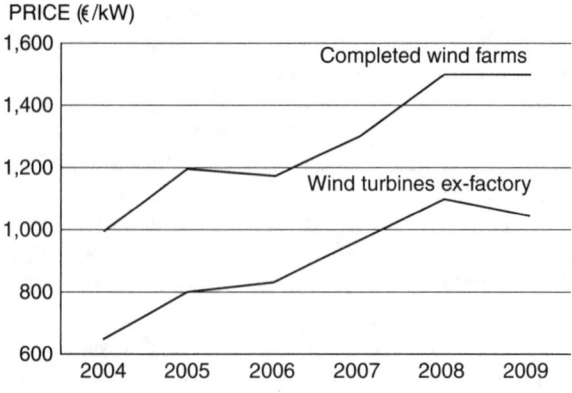

FIGURE 13-3   Average total installed cost of wind projects. The 2009 cost data is based on a sample of 4 GW of international projects. (From Milborrow, D. "Annual Power Costs Comparison: What a difference a year can make," *WindPower Monthly*, 2010, January.)

Capital Costs	% of TIC	$ per kW
Equipment costs		
Turbines (excluding blades and towers)	44.76	895
Blades	10.48	210
Towers	11.60	232
Transportation	8.01	$160
Equipment total	**74.85**	**1,497**
Balance of plant		
Materials		
Construction (concrete, rebar, equip, roads, and site prep)	10.82	216
Transformer	1.22	24
Electrical (drop cable, wire)	1.29	26
HV line extension	2.36	47
Materials subtotal	**15.68**	**314**
Labor		
Foundation	0.62	12
Erection	0.70	14
Electrical	1.02	20
Management/supervision	0.53	11
Misc.	3.80	76
Labor subtotal	**6.67**	**133**
Development/other costs		
HV Sub/Interconnection		
Materials	0.74	15
Labor	0.23	5
Engineering	1.01	20
Legal services	0.55	11
Land easements	0.00	0
Site certificate/permitting	0.26	5
Development/other subtotal	**2.79**	**$56**
Balance of plant total	**25.15**	**$503**
Total	**100.0**	**$2,000**

* For dollar cost, a total installed cost (TIC) of $2,000/kW is used.
   Source: Default costs in the NREL JEDI model. National Renewable Energy Laboratory. Jobs and Economic Development Impact (JEDI) Model, Release W1.09.03e. NREL, 2010. MS Excel-Based Model. Golden, CO.

TABLE 13-1   Detailed Categories of Capital Costs and the Approximate Breakdown in Percentage (%) and Dollars ($)*

improvement in capacity factors (CF) in the United States: 22% CF for projects installed before 1998, 30 to 33% CF for projects installed between 1998 and 2003, and 35 to 37% CF for projects installed between 2004 and 2007.

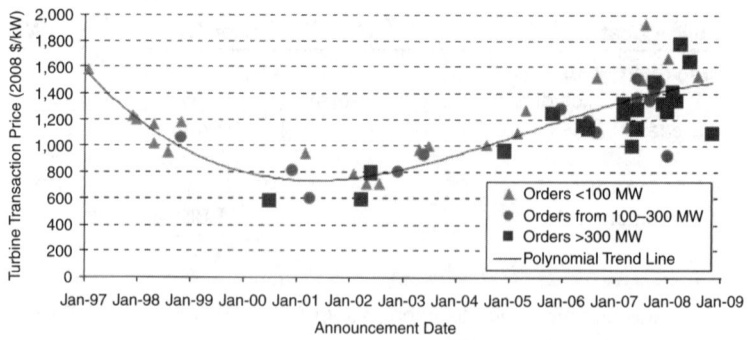

Figure 13-4   Turbine prices in the United States over the past 11 years. Data for 2008 is based on 58 wind projects totaling 21.1 GW of nameplate capacity.

Cost of Turbine

Turbine is the most significant cost of a wind project. It may be as high as 75%. The cost of turbines in international projects is seen in Fig. 13-3 and cost in United States projects is in Fig. 13-4. The cost of turbine is normally broken into various components: Wind turbine, tower, transportation to site, limited warranty (normally, 2 years), supply of manufacturer recommended spares and consumables, tool for use during installation, and supervision of installation and commissioning. Most manufacturers include the price of spares, consumable, and tools in the price of turbine. Towers are priced as a separate line item in the total price of turbine, because of the cost difference between different height towers. Turbine manufacturers provide personnel to supervise the installation and commissioning of turbines; note the developer performs the actual installation and commissioning.

The NREL JEDI model splits the BOP costs based on material, labor, and development/others. The details are in Table 13-1. An alternate method of splitting the BOP cost is along disciplines: civil works, electrical works, and development/others. The following section groups the costs by discipline to illustrate the types of activities performed in each discipline.

Cost of Foundation, Erection, Access Roads, and Other Civil Works

Primary categories of costs in this section include: Civil engineering, preparation of land, construction of foundation, concrete, steel, post-tensioning systems, and rental and operation of cranes for erection of tower and turbine. Other civil works included in this category are: Fencing, administration building, substation building, building of access roads, building of crane pads, improvement of public roads to transport blades, tower pieces, nacelle, and others.

Substation, Control System, Cables, Installation, and Others Related to Grid Connection

This cost component includes: Electrical engineering, transformers (LV to MV) for each turbine, transformer in substation (MV to HV), switch gear and protection equipment at each turbine and in substation, SCADA system, meters, underground cables, overhead cables, fiber-optic cables, FAA lights, and capacitor (if necessary). LV to MV transformer is usually not included in the turbine price because the step-up voltage varies by wind farm. Also included in this category are: Cost of labor to wire the turbine, pull cables inside the tower, laying of underground power and fiber-optic cables, interconnection to the grid, installation of lightning protection, commissioning of substation and turbines, communication systems to connect wind plant's SCADA system to that of the grid operator, and payment for grid upgrades beyond the point of common coupling.

Other Costs

There are two classes of costs: (i) Costs prior to start of construction and (ii) costs during construction until the end of commissioning. Prior to construction, the costs incurred are associated with prospecting, wind measurement, detailed wind study, environmental study, interconnection study, soil study, permits, civil and electrical engineering, cost of other consulting, cost of internal employees, financial modeling, and financing costs. The second class of costs include: Project management, building of onsite office, insurance against loss or damage of turbine and parts from port until installation, insurance against construction risks, interest on debt during construction, additional warranty beyond offered by manufacturer, and additional inventory of spares. Other miscellaneous costs include, cost of installing permanent met-towers in the wind farm for independently monitoring wind conditions.

Operating Costs

There are three categories:

- Finance charges—payment of principal and interest
- Maintenance charges—scheduled and unscheduled
- Operations charges—insurance, administration, and management of wind farm, transmission cost, land lease payments, taxes (property, sales)

Finance charges are the easiest to compute, it is a fixed payment every month to cover the cost of interest and return of principal.

Scheduled maintenance charges are also easy to compute in the warranty period. In 2009, most manufacturers provided warranty for 2–5 years. One notable exception is Enercon of Germany with a partner concept program that provides 12-year warranty for a yield-(kilowatt-hour produced) based fee. As the wind industry matures, manufacturers are increasingly providing warranties that include 95% availability and guaranteed performance for at least 5 years. For newer wind plants, maintenance cost is becoming predictable within the warranty period. In order to budget for unscheduled maintenance costs postwarranty, most operators establish a reserve or a sinking fund. A reserve fund contains money set aside each year for the explicit purpose of funding repairs and replacements of components.

Scheduled maintenance charges are in the range of $0.008 to $0.012/kWh during the warranty period. LBL[2] reports that the O&M cost (including land lease payments) for a project installed since 2000 has been $8/MWh. Postwarranty, the scheduled maintenance charges are difficult to predict given the limited experience with the new generation machines; however, $0.01 to $0.015/kWh would be a reasonable assumption. Annual contributions to a reserve fund of 0.75 to 1% of total installed cost is deemed appropriate to manage unscheduled maintenance postwarranty.

NREL JEDI model[6] provides a more detailed breakdown of the O&M costs into components, See Table 13-2. The default costs in the model are in terms of dollars per kilowatt, instead of dollars per kilowatt-hours used above.

Rest of the recurring costs is even more locale-specific compared to O&M costs, because these costs deal with insurance, taxes, land lease, and other related costs. Following are approximate cost ranges for US-based projects.

- Insurance cost for property and business interruption are 0.1–0.15% of the TIC

- Administration and management of wind farm include salaries of people employed in the wind farm onsite, administration, financial, and legal. This cost is in the range of 0.1 to 0.15%.

- Transmission cost, if applicable, is negotiated with the transmission provider; it may be in the range of $0.005/kWh.

- Land lease payments are negotiated with landowners. The cost is commonly negotiated as a percentage of revenue from energy production. The range for land lease payment is 2–5% of revenue. Another source of land lease payment data is from NREL JEDI model,[6] which uses $3,000 per megawatt as a default for US-based projects.

Operations and Maintenance Costs	% of TIC	$ per kW
Labor		
Personnel		
Field salaries	14.3	2.86
Administrative	2.3	0.46
Management	5.7	1.14
Labor/personnel subtotal	**22.3**	**4.47**
Materials and services		
Vehicles	2.2	0.44
Site maint/misc. services	0.9	0.17
Fees, permits, licenses	0.4	0.09
Utilities	1.7	0.35
Insurance	16.6	3.33
Fuel (motor vehicle gasoline)	0.9	0.17
Consumables/tools and misc. supplies	5.6	1.13
Replacement parts/equipment/spare parts Inventory	49.3	9.86
Materials and services subtotal	**77.7**	**15.53**
Total O&M cost	**100.0**	**20.00**

*For dollar cost, total O&M annual cost of $20/kW is used.

TABLE 13-2   Detailed Categories of O&M Costs and the Approximate Breakdown in Percentage (%) and Dollars($)*

- Taxes are country- or state-specific. The norm is to pay a sales or value-added tax in addition to property tax. With respect to property tax payments to local jurisdictions, in order to eliminate the variability of property tax payments (tax rate change or the assessment change) to local jurisdictions during the life of the project, a payment-in-lieu-of-taxes (PILOT) agreement is negotiated with the local taxing entities. PILOT allows a developer to pay a negotiated annual amount over a specified period. The amount is locale-specific; as a guideline,[7] $5,000–$6,000 per MW per year for 15-year period may be used for US-based projects.

- Other costs include transactions cost for selling energy in the futures market and selling carbon credits, depending on where the energy and credits are sold.

Depreciation and Taxes

In the United States, as in most countries, tax incentives play a big role in the financial modeling of a wind project. In the United States, there are three main incentives:

1. *Accelerated depreciation of assets.* In the United States, the Internal Revenue Service (IRS) allows for 5-year accelerated depreciation of renewable energy assets. In Mexico, a 1-year accelerated depreciation is allowed and the tax credits may be rolled over. The depreciation charges reduce taxable income.

2. *Production tax credit (PTC).* In the United States, this has become the primary incentive. In 2009, for every kilowatt-hour of renewable energy produced and delivered to the utility, the wind project owner gets a tax credit in the amount of 2.1c/kWh. This tax credit can be applied to offset tax on other income. The advantage of PTC is that it provides an incentive to produce energy, as opposed to an investment incentive in an energy project that may or may not produce.

3. *Investment tax credit.* In this case, the owner of a wind project is issued a tax credit that is based on the investment in the project. Note this tax credit is oblivious to amount of energy production. In 2009, as a part of the stimulus package, the United States offered a 30% ITC. A project owner can avail of either production tax credit or investment tax credit, but not both.

Financial Statements

Three types of financial statement are covered: Income statement, cash flow statement, and balance sheet. In wind projects, the income statement and cash flow may be combined. The combined statement conveys majority of the information.

Income Statement and Cash Flow for a Wind Project

Table 13-3 contains the income statement along with after-tax cash flow. A 20-year pro forma income statement will contain 20 columns containing values for each year. In addition to the above, a cash flow statement will normally include an accumulated liquidity line item, which aggregates all the cash flows for previous years. Some investment houses may request monthly pro forma income statements for the first few years.

Balance Sheet for a Wind Project

In a wind project, the balance sheet is not as interesting as the income statement. The elements of a balance sheet are in Table 13-4. In this table, dividend payments at the end of the life of the project are assumed. Annual dividend payments would involve appropriate adjustment to cash flow.

	Line Item in Income Statement	Values for Year *i*
	Annual revenue	
R1	Revenue from selling electricity	kWh produced multiplied by energy price in $/kWh
R2	Revenue from production tax credits	kWh produced multiplied by tax credit in $/kWh
R3	Revenue from renewable energy credits or carbon credits	kWh produced multiplied by price of REC or other credit in $/kWh
RA	**Total revenue**	**R1 + R2 + R3**
	Annual operating expenses	
01	Operations and maintenance	kWh produced multiplied by scheduled maintenance charges $/kWh + annual reserve fund payment based on TIC
02	Insurance	Annual insurance charges
03	Leaseholder payments	Payment usually based on percentage of revenue
04	Admin/financial/legal	General and administrative charges
05	Other expenses (transmission)	kWh produced multiplied by transmission charge (if applicable)
OE	**Total operating expenses**	**01 + 02 + 03 + 04 + 05**
	Annual depreciation and interest	
D1	Depreciation	TIC multiplied by an annual depreciation schedule
D2	Interest	Total outstanding loan multiplied by interest rate
DA	Total depreciation & interest	D1 + D2
TI	Annual taxable income	RA − (OE + DA)
TA	Taxes	TI* tax rate
NIA	**Net income**	**RA − (OE + DA + TA)**
CF	**After tax cash flow**	**NIA + D1-Principal payment on debt**

TABLE 13-3   Structure of an Income Statement for a Typical Wind Project

Financial Performance

Financial performance of a wind project is measured in terms of the following parameters: Levelized cost of energy, net present value, payback period, and internal rate of return.

	Balance Sheet Line Items	Values for Year *i*
	Current assets	
B1	Cash in bank	Year-end cash position
BA	**Total current assets**	**B1**
	Fixed assets	
FA1	Machinery and equipment	Total installed cost of the project
FA2	Less accumulated depreciation	Sum of depreciation claimed up to current year
FAA	Total fixed assets	FA1 − FA2
	Total assets	**FAA + B1**
	Liabilities and equity	
L1	Current liabilities	Short-term debt
L2	Long-term debt	Long-term debt
LA	**Total liabilities**	**L1 + L2**
	Owner's equity	
C1	Invested capital	Equity investment in project
C2	Retained earnings	From income statement
CA	Total owner's equity	C1 + C2
	Total liabilities + owner's equity	**LA + CA**

TABLE 13-4 Structure of a Balance Sheet for a Typical Wind Project

Levelized Cost of Energy (LCOE)

LCOE is a measure of the cost of energy produced by a wind farm in terms of dollars per kilowatts-hour. LCOE is an industry standard in the electricity generation for computing the lifecycle cost of electrical energy. It takes into account the following:

- Total installed cost (TIC)
- Annual recurring costs of operations, maintenance, and others
- Annual energy production for the life time of the project
- Cost of money

Levelized cost of energy, pr_{LCOE}, is the value of energy that yields zero NPV, that is:

$$\text{NPV} = 0 = \sum_{i=1}^{L} \left(pr_{LCOE}\, en\,(i) - rc\,(i)\right)/(1+r)^i - \text{TIC} \qquad (13\text{-}1)$$

where i is the year index, L is the life of the project in years, $en(i)$ is the amount of energy generated in year i, $rc(i)$ is the total recurring cost in year i, r is the discount rate, and TIC is the total installed cost.

$$\text{pr}_{\text{LCOE}} = \left(\sum_{i=1}^{L} rc(i)/(1+r)^i + \text{TIC} \right) \Big/ \sum_{i=1}^{L} en(i)/(1+r)^i \quad (13\text{-}2)$$

In the above formula, a simplification is utilized based on the assumption that TIC occurs in the first year. As expected, LCOE does not depend on the tariff, equity/debt structure of project, incentives, taxes, and other similar factors. When comparing LCOE of different sources of energy, the assumptions used in computing recurring cost and discount rate must be the same.

As an example, consider a 15 MW project with a production of 50,000 MWh of electrical energy annually. Assume the following:

- TIC = $27 million or $1,800/kW, discount rate = 8%, life of project = 20 years
- Operations and maintenance = $0.01/kWh, annual O&M cost = $500,000
- Annual reserve fund = 1% of TIC = $270,000
- Land lease + insurance and other administrative costs = $250,000 + $81,000 = $331,000

$$\sum_{i=1}^{L} en\,(i)/(1+r)^i = 490{,}907 \text{ MWh} \quad (13\text{-}3)$$

$$\sum_{i=1}^{L} rc(i)/(1+r)^i + \text{TIC} = 10{,}809{,}780 + 27{,}000{,}000$$

$$= \$37{,}809{,}780 \quad (13\text{-}4)$$

$$\text{pr}_{\text{LCOE}} = \$0.077/\text{kWh} \quad (13\text{-}5)$$

The interpretation and use of pr_{LCOE} is illustrated next. For the purposes of illustration, assume:

- Project is funded with 100% equity
- Project pays no taxes
- Project is in the United States with a fixed PPA
- Production tax credit (PTC) = $0.021/kWh for 10 years
- Average value of renewable energy credit (REC) = $0.005/kWh

This project would be profitable if the energy can be sold at an effective price that is greater than pr_{LCOE}. However, the effective price is usually not easy to compute because of the variability of the price of RECs and whether PTC are fully utilized. In this situation, a project with a fixed price PPA with a guaranteed power purchase price that is greater than or equal to pr_{LCOE} will be profitable; the revenue from PTC and REC will be the profit.

Often, a simpler method is used to compute LCOE:[8]

$$pr_{LCOE} = \frac{TIC}{en} fcr + rc \qquad (13\text{-}6)$$

where en is the annual average energy generation in kWh, fcr is the fixed charge rate, rc is the annual average recurring charge per kilowatt-hour. fcr is the fraction of TIC that must be allocated annually to account for capital cost, interest on debt, return on equity, and other fixed charges. In the following calculation, value of fcr is set to 11.58%, the value used in the NREL report.[8] Gipe[9] has a more detailed exposition of fcr for different investors and situations.

$$LCOE = \frac{\$27\,\text{million}}{50\,\text{million kWh}} 0.1158 + \frac{\$1.1\,\text{million}}{50\,\text{million kWh}} = \$0.077/\text{kWh}$$

Net Present Value (NPV)

NPV is computed using the net after-tax cash flow $(cf(i), i = 1 \text{ to } L)$ time series. It is the last row in Table 13-3.

$$NPV = \sum_{i=0}^{L} cf(i)/(1+r)^i \qquad (13\text{-}7)$$

where r is the discount rate and $i = 0$ is the year when investment is made and project is installed, $i = 1$ is the year when project is commissioned and starts producing energy for sale, and $cf(0) = TIC$. In a more rigorous model, multiyear investments may be modeled.

Payback Period

Simple payback period is a measure of the number of years it takes for a project to return the total investment. A concept of accumulated liquidity is used, which is the sum of net after-tax cash flow. For instance, accumulated liquidity after n years $(n \leq L)$ of a project is:

$$\text{Accumulated Liquidity } (n) = \sum_{i=0}^{n} cf(i) \qquad (13\text{-}8)$$

Years	0	1	2	3	...	9	10
NATCF	$(27)	$ 2.75	$ 2.75	$ 2.75	...	$ 2.75	$2.75
AL	$(27)	$(24.25)	$(21.50)	$(18.75)	...	$(2.25)	$0.50

*NATCF is the net after-tax cash flow and AL is the accumulated liquidity. All numbers are in millions.

TABLE 13-5   Illustration of the Computation of Simple Payback Period*

Simple payback period is the year when accumulated liquidity turns positive. Continuing with the example above and assuming total revenue of $0.077/kWh of production:

- TIC = $27 million
- Revenue = $77/MWh. Annual revenue = 55,000 MWh × $77/MWh = $3.851 million
- Total annual recurring costs = $1.101 million
- Net after-tax cash flow = $3.851 − $1.101 = $2.75 million

The accumulated liquidity computation is seen in Table 13-5. Simple payback period is 10 years. Note that simple payback period can be a misleading number because it does not take into account the discounting of future cash flow. As the name suggests, it is a simplistic measure of payback period. In fact, in this example, the real payback period with 8% discount rate is 20 years.

Internal Rate of Return (IRR)

The internal rate of return for the NATCF series is the interest rate received for an investment that yields regular cash flow. IRR is, therefore, the interest rate (same as discount factor) corresponding to zero NPV. Modified IRR is an IRR that takes into account the interest received on positive cash flow that is reinvested.

Impact of Tax Credits and Accelerated Depreciation on Financial Performance

The impact of production tax credit, investment tax credit, and accelerated depreciation is investigated in this section. First, examine the impact of 5-year accelerated depreciation allowed in the United States according to the following schedule (in %): 20, 32, 19.2, 11.52, 11.52, and 5.76. These percentages are applied to the total investment TIC in years 1 to 6 to compute depreciation. Taxable income is then computed by using:

$$\text{Taxable Income} = \text{Total Revenue} - \text{Total operating expenses} - \text{Depreciation}$$

The normal depreciation is called the "straight line" depreciation, which is 5% for each of the 20 years. Table 13-6 contains the detailed financial analysis with straight-line depreciation and 35% income tax rate. Here, earnings before-interest, taxes, and depreciation (EBITDA) = revenue – total annual operating expenses. Taxable income is computed by subtracting depreciation from EBITDA. In this scenario, the project will pay $0.49 million in taxes each year.

The 5-year accelerated depreciation case is presented in Table 13-7. In years 1 to 3, the taxable income is negative. Assuming tax loss carry forward is not allowed, then the tax benefit cannot be utilized and, therefore, the cash flow is taxable income + depreciation. A tax benefit of $0.64, $1.61, and $0.58 (in millions) cannot be utilized. However, if this wind farm were owned by an entity that has appropriate amount of tax liability (say, $2.06 million each year), then it could lower its tax liability by the "tax benefit" amount. This scenario is in Table 13-8, where the tax benefit has been added to compute the cash flow, which is equal to taxable income + depreciation + tax benefit.

In the next scenario, consider accelerated depreciation plus production tax credit (PTC). In 2009, PTC was $0.021/kWh. For every kilowatt-hour of electricity production the company is entitled to a tax credit. Obviously, if the company has no tax liability, like in years 1 to 3, then the tax credit cannot be used. This project generates a tax credit of $1.05 million each year. As illustrated in Table 13-9, if the wind farm owner has sufficient amount of tax liability (from some other business), then it would lower the tax liability by $1.05 + $0.56, $1.05 + $1.69, $1.05 + $0.48 in years 1 to 3.

The final scenario is an illustration of investment tax credit (ITC). In 2008, the United States added a new incentive that is an investment tax credit of 30%, which may be taken in the first year of production. ITC may be used in lieu of PTC. As shown in Table 13-10, the amount of ITC is added to the tax benefit, yielding a net tax benefit of $8.74 million. With ITC, the depreciable value of the asset must be reduced by 15%.

The financial performance of the various tax-related scenarios is in Table 13-11. Tax incentives play a significant role in improving the performance of wind projects. As illustrated, most wind projects do not produce enough income to avail of PTC or ITC. In some case, PTC or ITC can be rolled over and applied to subsequent years of the project. Even when roll over is allowed, most wind projects cannot effectively use all the tax credits. This is the reason for a proliferation

Years	0	1	2	3	4	5	6	11	12	20
Electricity Sales Revenue per PPA		3.85	3.85	3.85	3.85	3.85	3.85	3.85	3.85	3.85
Green Tag Rate ($/kWh)		0	0	0	0	0	0	0	0	0
PTC		0	0	0	0	0	0	0	0	0
Revenue		**3.85**	**3.85**	**3.85**	**3.85**	**3.85**	**3.85**	**3.85**	**3.85**	**3.85**
Operations & Maintenance		0.50	0.50	0.50	0.50	0.50	0.50	0.50	0.50	0.50
O&M Contingency Fund		0.27	0.27	0.27	0.27	0.27	0.27	0.27	0.27	0.27
Other Expense		0.33	0.33	0.33	0.33	0.33	0.33	0.33	0.33	0.33
Total Annual Operating Expenses		**1.10**	**1.10**	**1.10**	**1.10**	**1.10**	**1.10**	**1.10**	**1.10**	**1.10**
EBITDA		**2.75**	**2.75**	**2.75**	**2.75**	**2.75**	**2.75**	**2.75**	**2.75**	**2.75**
Depreciation		1.35	1.35	1.35	1.35	1.35	1.35	1.35	1.35	1.35
Debt Interest Payment		0	0	0	0	0	0	0	0	0
Taxable Income		**1.40**	**1.40**	**1.40**	**1.40**	**1.40**	**1.40**	**1.40**	**1.40**	**1.40**
Total Tax Benefit/(Liability)		(0.49)	(0.49)	(0.49)	(0.49)	(0.49)	(0.49)	(0.49)	(0.49)	(0.49)
Net After-Tax Project Cash Flow	**(27.00)**	**2.26**	**2.26**	**2.26**	**2.26**	**2.26**	**2.26**	**2.26**	**2.26**	**2.26**
Accumulated Liquidity	**$(27.00)**	$(24.74)	$(22.48)	$(20.22)	$(17.96)	$(15.70)	$(13.44)	$(2.15)	$0.11	$18.19

*Some years are skipped for brevity.
†All numbers are in millions.

TABLE 13-6 Income Statement with Linear Depreciation*,†

Years	0	1	2	3	4	5	6	12	13	20
Revenue		3.85	3.85	3.85	3.85	3.85	3.85	3.85	3.85	3.85
Total Annual Operating Expenses		1.10	1.10	1.10	1.10	1.10	1.10	1.10	1.10	1.10
EBITDA		2.75	2.75	2.75	2.75	2.75	2.75	2.75	2.75	2.75
Depreciation		5.40	8.64	5.18	3.11	3.11	1.56	0.00	0.00	0.00
Taxable Income		(2.65)	(5.89)	(2.44)	(0.36)	(0.36)	1.19	2.75	2.75	2.75
Total Tax Benefit/(Liability)		0.93	2.06	0.85	0.13	0.13	(0.42)	(0.96)	(0.96)	(0.96)
Net After-Tax Project Cash Flow	(27.00)	2.75	2.75	2.75	2.75	2.75	2.33	1.79	1.79	1.79
Accumulated Liquidity	$(27.00)	$(24.25)	$(21.50)	$(18.75)	$(16.00)	$(13.26)	$(10.92)	$(0.20)	$1.58	$14.09

*Tax benefit is not monetized.
†All numbers are in millions.

TABLE 13-7 Income Statement with Accelerated Depreciation*†

Years	0	1	2	3	4	5	6	9	10	20
Revenue		3.85	3.85	3.85	3.85	3.85	3.85	3.85	3.85	3.85
Total Annual Operating Expenses		1.10	1.10	1.10	1.10	1.10	1.10	1.10	1.10	1.10
EBITDA		2.75	2.75	2.75	2.75	2.75	2.75	2.75	2.75	2.75
Depreciation		5.40	8.64	5.18	3.11	3.11	1.56	0.00	0.00	0.00
Taxable Income		(2.65)	(5.89)	(2.44)	(0.36)	(0.36)	1.19	2.75	2.75	2.75
Total Tax Benefit/(Liability)		0.93	2.06	0.85	0.13	0.13	(0.42)	(0.96)	(0.96)	(0.96)
Net After-Tax Project Cash Flow	(27.00)	3.68	4.81	3.60	2.88	2.88	2.33	1.79	1.79	1.79
Accumulated Liquidity	$(27.00)	$(23.32)	$(18.51)	$(14.91)	$(12.04)	$(9.16)	$(6.83)	$(1.47)	$0.32	$18.19

*Details that are the same as the previous table are not shown in this table.
†All numbers are in millions.

TABLE 13-8 Same Case as in Table 13-6 Except Tax Benefit Is Monetized*,†

Years	0	1	2	3	4	5	6	7	8	20
Electricity Sales Revenue per PPA		3.85	3.85	3.85	3.85	3.85	3.85	3.85	3.85	3.85
Green Tag Rate ($/kWh)		0	0	0	0	0	0	0	0	0
PTC		1.05	1.05	1.05	1.05	1.05	1.05	1.05	1.05	0.00
Revenue										
Operations & Maintenance		4.90	4.90	4.90	4.90	4.90	4.90	4.90	4.90	3.85
O&M Contingency Fund		0.50	0.50	0.50	0.50	0.50	0.50	0.50	0.50	0.50
Other Expense		0.27	0.27	0.27	0.27	0.27	0.27	0.27	0.27	0.27
		0.33	0.33	0.33	0.33	0.33	0.33	0.33	0.33	0.33
Total Annual Operating Expenses		1.10	1.10	1.10	1.10	1.10	1.10	1.10	1.10	1.10
EBITDA		3.80	3.80	3.80	3.80	3.80	3.80	3.80	3.80	2.75
Depreciation		5.40	8.64	5.18	3.11	3.11	1.56	0.00	0.00	0.00
Debt Interest Payment		0	0	0	0	0	0	0	0	0
Taxable Income		(1.60)	(4.84)	(1.39)	0.69	0.69	2.24	3.80	3.80	2.75
Total Tax Benefit/(Liability)		0.56	1.69	0.48	(0.24)	(0.24)	(0.79)	(1.33)	(1.33)	(0.96)
Net After-Tax Project Cash Flow	(27.00)	4.36	5.49	4.28	3.56	3.56	3.01	2.47	2.47	1.79
Accumulated Liquidity	$(27.00)	$(22.64)	$(17.15)	$(12.86)	$(9.31)	$(5.75)	$(2.73)	$(0.26)	$2.20	$25.01

*All numbers are in millions.

TABLE 13-9 With PTC and Tax Benefit Monetized*

Years	0	1	2	3	4	5	6	7	8	20
Electricity Sales Revenue per PPA		3.85	3.85	3.85	3.85	3.85	3.85	3.85	3.85	3.85
Green Tag Rate ($/kWh)		0	0	0	0	0	0	0	0	0
PTC		0.00	0.00	0.00	0.00	0.00	0.00	0.00	0.00	0.00
Revenue		3.85	3.85	3.85	3.85	3.85	3.85	3.85	3.85	3.85
Operations & Maintenance		0.50	0.50	0.50	0.50	0.50	0.50	0.50	0.50	0.50
O&M Contingency Fund		0.27	0.27	0.27	0.27	0.27	0.27	0.27	0.27	0.27
Other Expense		0.33	0.33	0.33	0.33	0.33	0.33	0.33	0.33	0.33
Total Annual Operating Expenses		1.10	1.10	1.10	1.10	1.10	1.10	1.10	1.10	1.10
EBITDA		2.75	2.75	2.75	2.75	2.75	2.75	2.75	2.75	2.75
Depreciation		4.59	7.34	4.41	2.64	2.64	1.32	0.00	0.00	0.00
Debt Interest Payment		0	0	0	0	0	0	0	0	0
Taxable Income		(1.84)	(4.60)	(1.66)	0.11	0.11	1.43	2.75	2.75	2.75
Investment Tax Credit (30% of TIC)		8.10								
Total Tax Benefit/(Liability)		8.74	1.61	0.58	(0.04)	(0.04)	(0.50)	(0.96)	(0.96)	(0.96)
Net After-Tax Project Cash Flow	(27.00)	11.49	4.36	3.33	2.71	2.71	2.25	1.79	1.79	1.79
Accumulated Liquidity	$(27.00)	$(15.51)	$(11.15)	$(7.82)	$(5.11)	$(2.40)	$(0.15)	$1.64	$3.43	$24.87

*All numbers are in millions.

TABLE 13-10 With ITC and Tax Benefit Monetized*

Tax Scenarios	NPV	IRR	Simple Payback
Linear depreciation and no tax incentives	$(4.46)	5.5%	12
5-year accelerated depreciation with no carry over of tax benefits	$(4.88)	4.9%	13
5-year accelerated depreciation with tax benefits utilized	$(1.66)	6.9%	10
5-year accelerated depreciation + PTC with tax benefits utilized	$2.58	9.8%	8
5-year accelerated depreciation + ITC with tax benefits utilized	$4.22	11.5%	7

* NPV is in millions.

TABLE 13-11   Financial Performance Parameters

of tax-equity partnerships in which an entity that generates sufficient taxable income in an unrelated area (like banking, insurance) owns a majority stake in a wind project. Such a tax-equity partnership is able to "monetize" the tax incentives. After most of the tax benefits have been realized, then the majority ownership of the project flips to the project developer. Through such a scheme, PTC or ITC are monetized in the project.

Financing and Structure of Wind Projects

There are numerous methods of financing and structuring a wind project. The methods may be generalized into two broad categories:[10]

1. *Corporate structure.* Large utility companies, energy companies, and large wind development companies from Europe are examples of this structure. Financing is done primarily through internal funds and the tax benefits are used internally to offset income in the project portfolio.

2. *Tax equity and flip investors.* In this corporate structure, tax investors provide a fraction of the capital and, in return, get tax benefits. After much of the tax benefits have been realized and a predetermined IRR has been achieved for the tax investor, the ownership of the project flips from the tax investor to the project developer. The reason for such a structure is to bring in capital to the project from investors that are able to utilize the tax benefits. This structure has numerous variations:

a. *Strategic tax investor.* This is favored by developers that have the expertise but do not have the capital to fund a wind project. Strategic tax investors are brought in to fund a very large fraction (as high as 99%) of the project. In return, the investors gets 99% of the cash flow and tax benefits. This continues until the strategic investor realizes an expected level of IRR. At which point, the ownership of the project flips to the developer, who buys the majority stake in the project at market value from the strategic investor. For tax considerations, the strategic investor has to own at least 5% of the project after the flip.

b. *Institutional tax investor.* This is similar to strategic tax investor, except that the developer contributes some capital (as high as 30–40%). The cash and tax benefits are not split according to the equity ownership. Instead, the institutional investor takes 100% of the tax benefit, while the developer gets 100% of the cash benefit. When the developer has received all the money that it had invested, then all the cash flows to the institutional investor. After the institutional investor has reached an expected level of IRR, the project ownership then flips. At this point, most of the cash and tax benefits are transferred to the developer. This is the most commonly used flip structure.

c. *Pay-as-you-go.* This is similar to the previous two tax equity investments, except that the tax investor pays into the project to increase equity over time as the PTC tax benefits are realized. In addition to the initial investment, the tax equity investor pays into the project as energy is produced and as the PTCs are monetized by the tax equity partner. For each dollar of PTC tax benefit that is realized by the tax investor, the tax investor pays 80–90 cents as equity contribution into the company. The exact mechanics is case-specific; in most cases, the tax equity investor pays the developer, while some amount of cash flow goes to the tax investor.

d. *Cash leveraged.* This structure is a variation on the institutional investor. The variation is addition of debt financing. In most cases, the amount of debt is 40–60% of the total investment and is secured by the assets.

e. *Back leveraged.* This structure is similar to the cash leveraged case, except that the developer takes on debt as opposed to debt taken by the wind company that is jointly owned by the developer and tax equity investor.

Table 13-12 contains results of LBL calculation for the different project structures. Although the numbers illustrate the norms in the industry,

	Corporate	Strategic Flip	Institutional Flip	-As-You-Go	Cash Leveraged	Backleveraged
Cost assumptions						
Hard cost	$1,600	1,600	1,600	1,600	1,600	1,600
Soft cost	$125	$183	$183	$183	$215	$183
Total installed cost	$1,725	$1,783	$1,783	$1,783	$1,815	$1,783
Tax investor after tax return						
10-year Target IRR	n/a	6.5%	6.5%	6.5%	9%	6.5%
20-year IRR, Computed	n/a	7.02%	7.12%	7.02%	9.29%	7.12%
Debt assumptions						
Interest rate	n/a	n/a	n/a	n/a	6.7%	6.7%
Maturity					15 years	15 years
Developer after tax return						
10-year IRR	6.64%	6.5%	0	5.75%	9%	−10.08%
20-year IRR	10%	37.44%	10.44%	11.52%	30.58%	11.91%
20-year LCOE per kWh	$0.063	$0.061	$0.053	$0.059	$0.05	$0.053

Source: LBL. Harper, J. P. Karcher, M. D., Bolinger, M. *Wind Project Financing Structures: A Review & Comparative Analysis,* Ernest Orlando Lawrence Berkeley National Laboratory, Berkeley, CA, 2007. LBNL-63434.

TABLE 13-12   Assumptions and Financial Performance of Various Financing Structures Used in Wind Projects

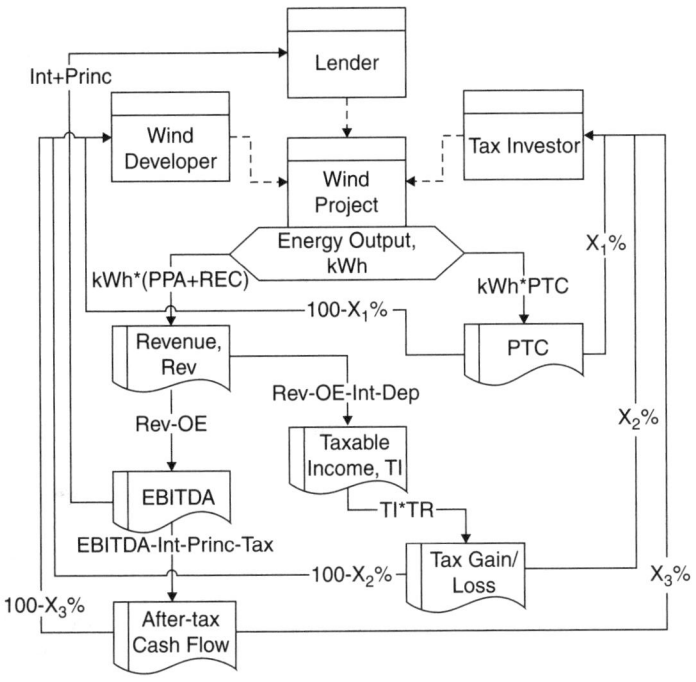

Figure 13-5   Schematic of corporate structure and flow of money. OE, Operating expense; Int, Interest payment; Dep, Depreciation; Princ, Principal; TR, Tax rate.

all the numbers are sensitive to modeling assumptions. For details, see LBL report.[10] The cash-leveraged case is the best performing model because debt is used, which is a cheaper form of capital compared to equity.

A schematic of the corporate structure and the cash flow is shown in Figure 13-5. Wind developer, tax investor, and lender are the three entities involved in the wind company. The revenue is broken into tax benefit-related cash flow and nontax benefit-related cash flow. There are two streams of tax–benefit-related cash flow: Production tax credit and tax gain/loss. There is single stream for nontax benefit-related cash flow; loan interest and principal are paid out of this. The three cash flow streams are split between the tax investor and the wind developer as $x_1\%$, $x_2\%$, $x_3\%$. The percentages depend on the structure and the percentage change at the point of flip or some other defined event.

Financial Evaluation of Alternatives

In this section, a few examples are presented to illustrate the large number of choices faced by a wind developer and why the intuitive selections are not necessarily the best. The choices may lead to a large number of permutations and combinations and a spreadsheet-based sensitivity analysis combined with wind analysis software can help to make appropriate decisions.

- *Wind speed.* When making decisions about selecting among competing sites, small differences (10% or less) in average annual wind speed should not lead to automatic decision of selecting site with higher wind speed. An extreme example is used to illustrate this: Site A has a steady wind speed of 6 m/s. Site B has an average wind speed of 5.5 m/s with a step profile: 3.0 m/s 50% of the time and 8 m/s 50% of the time. Power curves of most turbines will yield more than double the energy at 8 compared to 6 m/s. As this illustrates the same amount of energy is produced annually, even though the average wind speed is different; therefore, annual energy production should be computed and used for comparing sites.

- *Higher elevation.* Sites at higher elevation tend to have higher wind speed, for example, a mountaintop. Sites with higher elevation will have lower air density and, in most cases, higher turbulence because of mountainous terrain. Both these factors lead to lower energy production. In addition, the total installed cost is also likely to be higher because of lack of roads and infrastructure, which will lead to higher transportation, turbine installation, and utility interconnection costs. An energy production model that takes into account air density and turbulence coupled with a realistic cost estimation model is required to compare the scenarios.

- *Tower height.* Taller towers experience stronger winds (function of shear) and, therefore, higher energy production. Taller towers also result in higher project cost because: Cost of tower itself, cost of cranes to install, and cost of foundation to support the heavier tower. Project specific analysis can provide the appropriate tradeoff. As an example, consider a comparison of 80- versus 100-m tower.
  - From a cost standpoint: (a) Tower cost is approximately 50% higher, (b) other costs like transportation cost, crane cost, and foundation costs are likely to be 1.5–2 times. From Table 13-1, the cost of tower is 11.6% of total installed cost; total transportation cost is 8%, assume 25% is related to tower transportation; construction cost is about 11%, assume 25% is related to tower. Therefore, the total cost impacted is about 16%. For a 2.5 MW turbine at $2,000/KW the impacted cost is $800,000. If 80-m tower is replaced with 100-m tower, the cost is likely to increase by $360,000.

- From a revenue standpoint, a 2.5 MW turbine with 7 m/s average wind speed and shear of 0.24 will produce 7.26 million kWh per year with 80-m tower. The same turbine with a 100-m tower will produce about 0.7 million kWh of additional energy. Assuming revenue is 7 cents/kWh, this will produce $49,000 per year. In an alternate scenario, if the shear at site is 0.15, then the additional energy is 0.44 million kWh per year and additional annual revenue is $30,800 with a 100-m tower.

- The NPV with 8% discount rate of additional revenue with shear of 0.24 is $481,090. For shear of 0.15, the NPV is $302,400. When compared with the additional cost of $360,000, the choice is clear.

- *Size and number of turbines.* Should there be a 75 MW wind farm with fifty 1.5 MW turbines or thirty 2.5 MW turbines? This decision requires detailed computation of wind farm layout, average annual energy production, wake losses, and related analysis. In addition, it requires a detailed cost model with sufficient degree of certainty that allows a decision maker to make a choice.

The most difficult alternatives to analyze are ones that are difficult to quantify. Examples are, buying 20-year warranty from a new manufacturer with weak balance sheet versus buying from an established manufacturer that offers only a 5-year warranty.

As the above examples illustrate, the process of choosing among dozens of alternatives requires rigorous analysis in order to make optimal selections.

References

1. DSIRE-Database of State Incentives for Renewables & Efficiency. Illinois-Incentives/Policies for Renewables & Efficiency. *DSIRE* [Online] 2010. http://www.dsireusa.org/incentives/incentive.cfm?Incentive_Code=IL04R&re=1&ee=1.
2. Wiser, R., and Bolinger, M. *2008 Wind Technologies Market Report,* Berkeley, CA: Lawrence Berkeley National Laboratory, Berkeley, CA, 2009.
3. Porter, K. *Feed-In Tariffs,* California Energy Commission, IEPR Workshop. Sacramento, CA, 2006. http://www.energy.ca.gov/2007_energypolicy/documents/2006-08-22_workshop/presentations/4-FEED-IN_TARIFFS-K-PORTER.PDF.
4. Holt, E. A., Wiser, R., and Bolinger, M. *Who Owns Renewable Energy Certificates? An Exploration of Policy Options and Practice,* Ernest Orlando Lawrence Berkeley National Laboratory, Berkeley, CA , 2006. LBNL-59965.
5. Milborrow, D. "Annual Power Costs Comparison: What a difference a year can make." *WindPower Monthly,* 2010, January.

6. National Renewable Energy Laboratory. Jobs and Economic Development Impact (JEDI) Model, Release W1.09.03e. NREL, Golden, CO, 2010. MS Excel-based model.

7. Ohio Legislative Service Commission: Fiscal Note & Local Impact Statement [Online] March 8, 2010. http://www.lbo.state.oh.us/fiscal/fiscalnotes/128ga/SB0232IN.htm.

8. Thresher, R., and Laxson, A. *Advanced Wind Technology: New Challenges for a New Century*, National Renewable Energy Laboratory, Golden, CO, 2006. NREL/CP-500-39537.

9. Gipe, P. *Wind Energy Comes of Age*, Wiley, New York, 1995.

10. Harper, J. P., Karcher, M. D., and Bolinger, M. *Wind Project Financing Structures: A Review & Comparative Analysis.* Ernest Orlando Lawrence Berkeley National Laboratory, Berkeley, CA, 2007. LBNL-63434.

CHAPTER 14

Planning and Execution of Wind Projects

By failing to prepare, you are preparing to fail.
—Benjamin Franklin

Introduction

Similar to other energy projects, wind projects are long and complex. Wind resource assessment is the longest duration task in the process, while installation and commissioning happen in short duration. Therefore, installation and commissioning require sophisticated management of the supply chain of the project.

This chapter starts with the high-level components of a wind project. The first component of a wind project is the development phase. The five phases of development are described next. The following section covers construction, installation, and commissioning. Here, construction of infrastructure, foundation, turbine erection, collection system, substation, and commissioning are described. The final section deals with operations of a wind project.

High-Level Project Plan and Timeline

A wind project can be broken into three distinct phases as described in Fig. 14-1:

1. *Development.* The development phase of a wind project involves all the activities starting from prospecting, wind

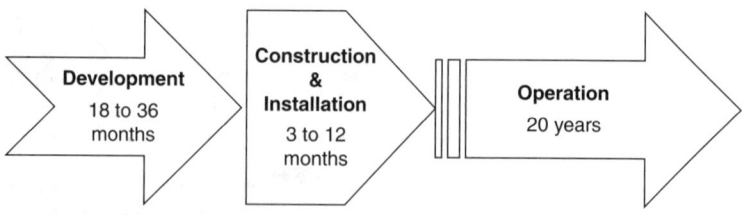

Figure 14-1 Three phases of wind project.

measurement, wind assessment, project siting, permitting, engineering, financing, engineering procurement, and construction (EPC) contracting, up to preconstruction. The duration of this phase is anywhere from 18 to 36 months. The longest lead-time item is wind measurement; most financiers require at least 2 years of onsite measurement for large projects.

2. *Construction and installation.* In this phase, all the physical onsite activities occur starting from site preparation, turbine transportation, foundation construction, turbine erection, collection system and substation installation, and project commissioning. The duration of this phase is 3–12 months, depending on the size of the project.

3. *Operation.* In this phase, energy is produced by the wind project. Activities in this phase include, wind plant monitoring, and scheduled and unscheduled maintenance.

Development

The development phase may be decomposed into the following five stages:

- Prospecting
- Wind measurement and detailed wind assessment
- Project siting, PPA, and interconnection
- Project engineering and procurement
- Project financing

As shown in Fig. 14-2, the stages may have significant time overlap.

Task Name	Year 1				Year 2				Year 3			
	Qtr 1	Qtr 2	Qtr 3	Qtr 4	Qtr 1	Qtr 2	Qtr 3	Qtr 4	Qtr 1	Qtr 2	Qtr 3	Qtr 4
1 Prospecting												
2 Measurement & Detailed Assessment												
3 Siting, PPA, Interconnection												
4 Engineering & Procurement												
5 Financing												

Figure 14-2 Typical timeline for the development phase.

Prospecting

This is the first step in a wind project with activities that involve evaluating various sites. The following steps are a guide to prospecting; experienced wind project developers may not follow all the steps described below.

1. Identify regions with good wind resources, good transmission, good sale price for energy, good incentives, and good access. Prospecting is a subjective process that involves substantial judgment. Experienced wind developers working on a large wind project will have cutoff criteria like, at least 7 m/s average wind speed, at least 115 kV transmission line within a mile, at least $0.055/kWh of PPA, at least $0.02/kWh in incentives, and site with average construction costs. This step is not necessary if the wind project under consideration is a medium- (< 5 MW) or small-wind project in a predetermined area (ranch, farm, house).

2. *Preliminary wind assessment.* In this step, publicly available wind data or previously measured wind data is used to perform preliminary assessment of wind conditions. The outcome of this stage is: Average wind speed, average annual energy production, and average capacity factors. In addition, preliminary financial assessment is performed to check if a site meets financial performance criteria.

3. *Site visit to evaluate suitability of site for wind development.* Although desktop-based tools provide detailed Geographical Information Systems (GIS) data and 3-D images of a site, there is no substitute to a site visit. During the site visit, suitability is determined with respect to key development issues: Transmission, zoning, site access, land ownership, easements, setbacks, natural resources, compatibility with existing land use, and special site conditions. Site visit should include detailed documentation of landscape, roads, transmission, wildlife, and vegetation.

The outcome of the prospecting step is selection of sites for wind measurement.

Wind Measurement and Detailed Wind Assessment

In this step, onsite wind measurement is performed and the data is used to perform a detailed wind assessment. The tasks involved in this stage are:

- *Wind measurement locations and methods.* Based on preliminary wind assessment, locations for wind measurement are identified; this is described in more detail in Chapter 6. In addition, methods of measurement (met-towers, SODAR, and LIDAR), type of equipment and sensors, and the mode of measurement are determined. If SODAR and LIDAR are used, then the option of relocating to measure at multiple sites is available. If this option is availed, then an itinerary of locations must be determined. The cost of traditional 60-m met-tower with standard set of instruments, data-logger and communication equipment, installation, and decommissioning is about $35,000. Other approximate costs in 2010 are: An 80-m tower is an additional $5,000; higher quality instruments will cost an additional $3,000; SODAR unit is about $45,000 to $50,000 for the basic unit, and additional $10,000 for trailer; LIDAR unit is $120,000–$150,000. In situations where the high quality of wind resource is known and planned wind project is large, then a few permanent met-towers may be installed.

- *Land lease and permitting.* Land lease during wind measurement is typically $2000–$5000 per year. Two types of agreements are signed with the landowner: Wind measurement and options. In most cases, the developer seeks an exclusive option for a period of 2 to 7 years to measure wind speed, perform geotechnical tests, survey land, and other related activities. The exclusivity clause will prevent the landowner from leasing land to other developers and may place restrictions on the nature of information that can be divulged to others. Projects with towers higher than 60 m are required to file a request with the FAA (see Chapter 11). In addition, local permits may be required to erect temporary structures, like met-towers.

- *Met-tower installation and commissioning.* This activity may take 2–3 days and a four to five-member crew. Since met-towers cannot be erected during conditions of high wind and gusts, wind forecasts must be checked before scheduling a met-tower installation. Under perfect conditions and very experienced crew, a met-tower may be installed in a day. After the installation is complete, commissioning involves checking if the wind measurement data is being logged, transmitted, and received. The first few hours of data must then be thoroughly reviewed and validated against concurrent data from neighboring airports or weather stations.

- *Wind data management and reporting.* A standard practice is to receive daily wind data by email, which is automatically inserted into a database with a variety of checks and filters. Web-based reporting is then available to view data in a variety of reports.

- *Decommissioning of Met-towers.* After the wind measurement campaign, the land is returned to its previous state. Since most towers are tilt-up with no foundation, decommissioning involves disassembly and transportation of the met-tower.

- *Detailed wind assessment.* Onsite wind measurement data is processed in software programs like WindPRO and Wind-Farmer, as described in Chapter 6. Methods like measure-correlate-predict are used to determine long-term wind speed average. This process may be done once a quarter to check correlations with one or more long-term reference wind speed data series. In some situations, quarterly reports may provide the basis for a go/no-go decision; if the wind conditions are not up to par, the project may then be aborted. In addition, measurements from multiple locations may be used to develop a wind flow model that interpolates and extrapolates wind speed and direction over the entire area of interest. At the end of a measurement campaign, detailed data about: Wind speed, wind shear, energy density, energy production, wind direction, turbulence, air density, and others are computed. In addition, estimates of uncertainties are computed. Since turbine selection may not be completed, the above computations are made with several turbine power curves. The duration of this task is about 2 to 4 weeks.

- *Detailed financial assessment.* Pro forma financial assessment is generated to determine the financial performance of the project. At this stage, most of the revenue and cost numbers used in this assessment may be estimates; however, the estimates should be site-specific and not generic industry averages. (See Chapter 13 for details.) Duration of this task is about 2 weeks.

- *Detailed project plan.* No business plan is complete without a detailed project plan, that consists of tasks, timeframes, milestones, deliverables, and decision points. The project plan should be updated regularly to reflect new realities and new information.

The outcome of the wind measurement and detailed wind assessment stage is a business plan that is taken to investors or investment committees to fund the subsequent phases of the project.

Project Siting, Interconnection, and PPA

In this step, a broad range of tasks are performed to clear regulatory and environmental hurdles related to siting of a wind project. This step is very locale-specific starting with the environmental features at the proposed site, to regulatory and compliance requirements of local, state, and federal governments. Therefore, the effort and timeframe for this step is highly variable. The tasks involved in this step include:

Critical Issues Analysis

One of the first tasks in project siting is to perform a critical issues analysis, which is a desktop-based preliminary analysis of all the environmental, regulatory, and compliance issues. American Wind Energy Association (AWEA)'s siting handbook contains a list of considerations:[1]

- Required permits, licenses, and regulatory approvals
- Threatened or endangered species or habitats
- Avian and bat species or habitat
- Wetlands and protected areas
- Community facilities and services
- Land development constraints
- Telecommunications interference
- Aviation considerations
- Visual/aesthetic considerations
- Noise and shadow flicker
- Locations of known archaeological and historical resources
- Other locale-specific considerations

This task may take about 2 months to complete. The analysis will create a list of locale-specific environmental, regulatory, and compliance issues that should be tackled in the subsequent steps. The output of this activity should be a work plan and estimate for the subsequent steps.

Environmental, Regulatory, and Compliance Tasks

The detailed activities associated with this broad category are locale-specific, as are timeframe and costs associated with these activities. Details of the activities are described in Chapter 12.

Utility Interconnection Study

For large wind farms, the duration of this step is at least 9 months and, in most cases, it takes about 18 months. Since it is a long lead-time activity, it should be started as soon as detailed wind assessment is complete and a decision has been made to proceed with the project. Details of the interconnection study are described in Chapter 11.

Power Purchase Agreement (PPA)

PPA is a legal contract between the wind project and the utility. If the utility that is buying the energy has experience with wind projects, then it may propose a standard contract. The duration of this task is at least 2 months for a simple project and may take 6 months or more. The reason for a high degree of variance is that it requires negotiations related to: (a) Pricing of energy, (b) sharing of renewable energy credits and other potential incentives, (c) capacity and delivery of energy, (d) incentives, penalties, and exceptions related to planned versus actual date of commissioning, and (e) allocation of risk related to interconnection, siting, and permitting. This activity should be started as soon as detailed wind assessment is complete.

Land Lease Agreement

Long-term land lease agreement is a legal contract between the project and the landowners. Often, land leases are signed not just for the property that contains the wind farm, but also for adjacent lands that are in the prevailing direction of wind. This protects the wind farm from other wind development that may infringe upon the wind resource.

Community Involvement

This ongoing activity must be performed from the outset by the project developer to build support for the project in the local community. An effective public outreach program must provide a venue to listen and address the concerns of neighbors. The most dominant concerns are likely to be environmental, viewshed, noise level, and property value.

Project Engineering and Procurement

Project engineering of a wind energy project includes a variety of tasks, some that are performed by the developer and others that are the domain of a contractor.

Turbine Selection

Evaluation of turbines for the project at hand should start soon after 1 year of met-tower data. Most turbine manufacturers will not provide a quote until they have examined 1 year of onsite wind data. For projects on a tighter schedule, turbine selection may be commenced with 6 months of wind data and MCP analysis, at which point the turbine class can be determined with some certainty. A contract with the turbine manufacturer is usually not signed until the financing entity is chosen and the financing entity has approved the pricing, the terms, and conditions for delivery, supervision, warranty, and others. In addition, a significant down payment, in the range of 30%, is required at the time of contract signing with the turbine manufacturer.

Turbine selection allows subsequent engineering to begin. Inputs required for engineering are: Turbine-rated capacity, blade size, tower size, generator type, weight and dimensions of components, power curve, noise data, and variety of other inputs.

Project Layout and Civil Engineering of Infrastructure

Project layout is a task that begins with sizing of a wind project and layout of turbines using software like WindPRO or WindFarmer. This is a first-pass layout of turbines that meets all the known constraints:

- Setbacks from public roads, water bodies, inhabited structures, transmission lines, and property boundaries
- Setbacks from exclusion areas like endangered species habitat, wetlands, and others
- Setbacks from microwave Fresnel zones, airports, radars, and other telecommunications links

With above preliminary micrositing of turbines, the process of infrastructure civil engineering can begin. The steps include:

- Land survey of the property of interest to delineate property boundary accurately, specify elevation contour lines to a higher degree of precision, and identify features on the property like transmission lines, roads, water bodies, vegetation, and others. With this information, areas of higher slopes are either marked as exclusion areas or turbines placed in these areas are moved to alternate locations.

- Site visit by the civil engineering team, wind energy modeling team, and landowners. The intent is to assess the suitability of each proposed wind turbine site with respect to access roads, drainage, turbine foundation, crane walk and crane pad, and proximity to exclusion areas. In addition, collection system (transmission inside the wind farm) and substation layout, entrance from public roads into the wind farm, fencing and security, and other related aspects are assessed. Landowner's preferences related to agriculture and other needs are assessed.

- Transportation planning involves evaluation of public transportation infrastructure and its ability to support transportation of equipment for the wind project. In most locations, the public road infrastructure is unable to support the weight and width of the loads, and turn radius. A site visit can provide data to design the necessary changes.

- Geotechnical study for testing of subsurface soil conditions at each turbine location. Soil testing involves boring holes to collect samples. Foundation design typically requires sampling at four boreholes that are 30–50 ft deep, depending on the turbine and type of foundation. Three holes are bored at vertices of an equilateral triangle and one hole is bored at the center. The samples are analyzed for soil characteristics.

- Design of all the infrastructure components. In this step, all the access roads, drainage, turbine erection crane pads, and crane walk areas are designed. The following is an illustrative example of the type of infrastructure required for a wind farm:[2]

- Access roads are permanent infrastructure that must be between 16 and 20 ft wide, 150-ft inside turn radius with load-carrying capacity of 250,000 lb or more, and single axle load of 15 tons per axle for concrete trucks and 10 tons per axle for transporters.

- The turbine erection pads must be able to withstand loads of 6000 pounds per square foot during lifting of nacelle, rotor, and tower sections. In most large projects, a 500- to 600-ton crane is deployed.

- Crane walk area is a temporary path adjacent to the access road for the crane to move between two turbine locations. This path must be 32–36 ft wide with minimal slope and minimal dip/crest to ensure that a 300-ft crane with high center of gravity can use to travel between sites.

Foundation Design

Foundation design is described in Chapter 9. This task is typically performed by a foundation design consultant.

Electrical Design

Electrical design involves design of the collection systems, substation, and interconnection to the grid. Design of collection systems require soil thermal resistivity testing along the path of the buried cables. With this information, the correct size cables are chosen. Other items in the collection system are the protection system and grounding system. The protection system is a collection of switchgear (fuses and circuit breakers) that protect equipment from fault current (sharp rise in current because of short circuit) by isolating the area with the fault. It also allows isolation of parts of a circuit for the purpose of conducting repair and maintenance. Utility will review and approve the protection system and substation design before allowing interconnection. Substation design was described in Chapter 11.

Permits

The critical issues analysis step identified the list of permits required and the agencies that issue them. In this task, activities related to obtaining the permits are performed. In most locations, all levels of the government get involved: Federal, state, and local. In an ideal situation, a one-stop comprehensive permit issued by single agency would eliminate the uncertainty around the permitting process. Some states have a single agency that manages most of the permitting, except those that the federal agency has not delegated to the state agency. Following is a list of the most common permits:

1. *Transportation permit.* Turbine parts, construction equipment, and cranes are large and heavy, requiring transportation

permit from several departments of transportation that have jurisdictions over different roads on the route to the site. For example, the state and local departments of transportation must approve transportation, improvements to existing roads, and building of access roads.

2. *Wildlife permit.* This is a general category that may contain a variety of specific permits depending on jurisdictions. In the United States, migratory birds and endangered species-related permits are obtained from the federal agency, US Fish and Wildlife; state-specific wildlife resources are managed by the Department of Natural Resources of individual states.

3. *Aviation permit.* Since turbines are tall structures, aviation authorities regulate construction to ensure that these structures are not an obstruction to airspace and have no adverse impact on the radar systems that monitor airspace.

4. *Radar and electromagnetic interference.* Agencies that operate radars include Weather Service, Homeland Security, and Department of Defense. Consulting with these agencies to obtain clearances (not necessarily permits) will ensure that there are no objections.

5. *Cultural resources.* If the site has archeological, architectural, and traditional cultural resources, then a permit will be required. In the United States, the Advisory Council of Historical Preservation and the State Historic Preservation Offices review the impact on cultural resources. Consultation with these agencies is required to determine the existence of cultural resources.

6. *Land use permit.* In the United States, if the wind project is on government-owned land, accesses federally owned transmission lines, or receives federal grants, then one of the impacted federal agencies must perform an environmental impact study. The outcome is either: Categorical exclusion, environmental assessment, or environmental impact statement. Some states have developed model zoning ordinances for wind projects. These ordinances are used as-is or as guidelines by the local county or district. At the local county level, a conditional-use or special-use permit will be required to build a wind project.

7. *Construction permit.* Building permits are often required at the local county level to demonstrate that construction will adhere to building and engineering codes and standards.

8. *Water and wetlands.* Permits will be required when constructing in: (i) Areas that are in proximity to water bodies that are used for drinking water and navigation; (ii) designated

wetlands. Stormwater discharge permit will also be required to discharge stormwater from construction activities. At the federal level, the US Army Corp of Engineers and the Environmental Protection Agency regulate construction activity that may harm the water resources; at the state level, the state environmental quality agency administers the federal programs, in addition to regulating state-specific resources.

Logistics Planning

Wind projects involve complex coordination and scheduling of tasks. This is especially true during the construction phase, which is compressed into a very tight timeframe. Poor logistics planning can be costly, because of the high cost of crane rentals, operators and setup, and, in general, high cost of capital. A wind project, like other projects, has complex relationships between different parts of the supply chain. Wind projects work with just-in-time (JIT) manufacturing, JIT transportation with no inventories or storage, and JIT erection. The schedules must be coordinated between the civil contractor, turbine supplier, crane supplier, transportation contractors (with all three modes: water, rail, and road), electrical contractor, and others. However, unlike other energy projects, weather can play a big role. Rain and water logging can delay excavation and foundation work; rain, thunderstorms, and high wind can delay erection of turbines. In some locations, winds are so strong that erection cannot be done for 6 months. Therefore, if the erection window is missed, the project may be delayed for 6 months.

Logistics planning is further complicated because of shortage of most of the critical equipment and labor:

- Turbine manufacturing has a long lead time
- Trucks and railcars to transport oversized components are in short supply and, therefore, difficult to schedule and reschedule. For a large utility-scale turbine, it takes 9 tractor trailers to transport one turbine. Even for a single turbine, sequencing of deliveries is critical to avoid logjam at a site that may not have storage space. For instance, blades are scheduled to arrive after nacelle, which is scheduled to arrive after the towers.
- Site supervision by turbine manufacturer is a specialized skill that must be scheduled
- Large 500- to 600-ton cranes and operators are in short supply and the cranes are difficult to transport

A general contractor must, therefore, develop a detailed comprehensive schedule, obtain regular updates from all task owners, update the schedule regularly based on new data, and communicate to the entire team.

Bidding and Contracting: Engineering Procurement and Construction (EPC) and Other Forms of Contract

EPC or turnkey contracts are the most common form of contract between the wind project owner and the contractor. As the name implies, the EPC contractor delivers to the project owner a completed project at a prespecified cost, in a prespecified timeframe, and that performs to a prespecified level of production and quality. An alternative to EPC contract is individual contracts with numerous vendors: Design contract, construction contract, electrical contract, erection contract, project management contract, etc. An ideal EPC contract is one that is fixed cost with fixed delivery date, a security deposit to guarantee performance, liquidated damages for delay in completion and substandard performance, and large cap on liability. A security deposit or a performance bond is part of an EPC contract to protect the project owner in case the contractor does not fulfill its obligations. Normally, the security amount is 5–10% of the total contract price and it is held at a bank. The advantage of EPC contract is the assignment of single point of responsibility. The disadvantage is that the EPC contract is more expensive because of built-in contingencies and, in a tight market, contractors are unwilling to bid on such contracts.

In this activity all the tasks related to EPC contracting are performed including (a) preparing requirements document that describes in detail all the particulars of the project, (b) sending the requirements documents to potential bidders, (c) receiving bids, (d) evaluating bids, and (e) choosing a contractor.

Project Financing

Presentation to investors and investment committees begins after the detailed wind assessment and detailed financial assessment are complete. This is a negotiation-intensive process and it is advisable to work with a financing partner during detailed wind and financial assessments. The reason is a financing entity has its own checklist and criteria for evaluating a project. It is, therefore, crucial to involve the financing entity in a project as early as possible to obviate the need for rework. Examples of rework include, longer wind data collection, higher uncertainty factors in assessment, and, therefore, lower P84, P90, and P99 estimates, changes to PPA, changes to construction contracts, and others. In addition, a developer, early in the process, should ask the financing entity to review and verify the assumptions, soundness of the measurement approach, and subsequent technical analysis.

In any renewable energy project, a financing entity is attempting to manage the following three types of risks; understanding of these risks from the perspective of the financier will assist a developer to avoid rejections or significant rework:

- *Revenue risk*. Since PPA is the main source of cash flow, it is closely scrutinized. Details of PPA are presented in Chapter 13. Some of the items in the PPA that are of concern are: Curtailment clauses, penalties related to inaccuracy of day-ahead forecasts, cost of providing reactive power, and penalties related to delay in commissioning. Note that all these may not apply to a project at hand.

- *On time completion risk*. This risk pertains to ensuring that the project is completed on time or worse, not abandoned midstream because of insurmountable hurdles. A financier will ensure that all the permits have been obtained and there are no serious hurdles to overcome. In addition, a financier will scrutinize the EPC contract with the goal of eliminating ambiguities in scope, roles, and responsibilities, and link payments to milestones or completed work.

- *Operational risk*. This risk pertains to ability to generate the projected amount of energy in order to meet the cash flow needs. This is typically accomplished through performance warranties, and operations and maintenance contracts; the financing entity will therefore scrutinize these contracts to ensure that high availability is built into the contract.

Construction, Installation, and Commissioning

In this phase of the project, construction, erection and commissioning of the wind project are performed. For convenience, this phase is divided into four stages, as illustrated in Fig. 14-3:

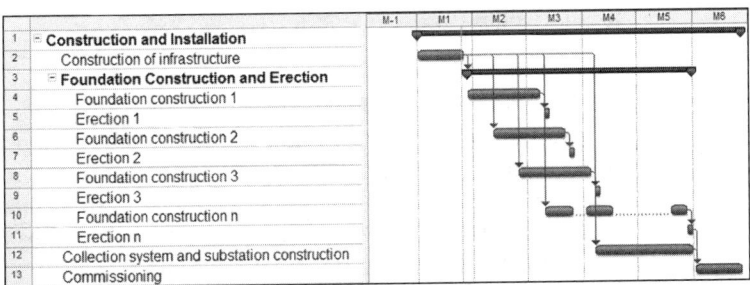

FIGURE 14-3 Typical timeline for construction and installation. Months are indicated as M1, M2, ... The total duration depends on number of turbines.

1. Construction of infrastructure
2. Foundation construction and turbine erection
3. Collection system and substation construction
4. Commissioning

Construction of Infrastructure

In this stage, all the civil and other infrastructure work prior to turbine specific work is performed. Since most of the activities were described in some detail in the planning phase, descriptions in this section will be brief.

Site Preparation

In this activity, all tasks related to preparing land and setting up temporary office are performed. Tasks include:

- *Setup of a temporary office onsite.* The onsite office is usually two trailers with electric and internet connections (if available). This office serves as an administrative office.
- *Upgrade of public roads.* This involves working with the local department of transportation to enhance the public roads in order to support the transportation of construction equipment and turbine components.
- *Wind farm land preparation.* This involves clearing of brush and trees, leveling land, construction of access road, and other tasks to make the entire wind farm easily accessible to earth-moving equipment and cranes. Depending on the site, there may be significant amount of earthmoving involved.
- *Wind turbine land preparation.* At each turbine location, the following areas are constructed:
  - Crane pads for both main and tail crane
  - Tower, nacelle, and blades staging area
  - Rotor assembly area
  - Storm water drainage
  - Foundation excavation and compaction
- *Temporary storage area and other areas.* Within the wind farm, a temporary storage area is created to store items like cables, rebar, and other material. In addition, an area may be required for concrete preparation. Most of the heavy items are moved directly to the turbine site.

In all the construction activity, a storm water pollution prevention plan is prepared that minimizes erosion and sedimentation. All the constructed areas are graded and areas disturbed because of construction are restored to the original state by reseeding with native vegetation.

Foundation Construction and Turbine Erection

Foundation Construction

The foundation design provides the specifications and instructions for foundation construction. From a project-planning perspective, the generic steps are:

- Hundreds of cubic meters of soil is excavated or drilled and blasted, depending on the type of soil.
- An outer form is placed, followed by rebar, and then by a bolt cage assembly along with conduits for cables and lightning protection.
- Concrete is then poured and cured. Curing time may be as much as 1 month, depending on the size of the foundation.
- The outer form is then removed and the area around the foundation is backfilled and compacted.
- In the middle of the foundation is an embed ring with bolts sticking out. The tower is bolted to this embed ring.

A picture of partially complete foundation is in Fig. 14-4.

FIGURE 14-4   Spread foundation after concrete has been poured in the footing of the foundation. (Courtesy Vensys Energy AG.)

Erection

Turbine manufacturer provides an erection plan along with crane requirements for erection. Normally, two cranes are used: A main crane with a rated capacity of 500– to 650 metric tons (for utility-scale turbines), and an auxiliary (or tail) crane with a rated capacity of 90 metric tons. Typically, the main crane is a mobile crawler crane, while the auxiliary crane is a mobile hydraulic crane.

Turbine erection is a 2- or 3-day process per turbine. The first step in the process is to assemble the tower, which is in three or more sections. Figure 14-5 is a photograph showing lifting of a tower section. Two cranes are used to lift the sections and stack them. After the tower sections have been stacked and bolted, the nacelle is lifted. Depending on the turbine manufacturer, there are two options for lifting of nacelle: (i) Single lift of nacelle with generator or (ii) two lifts: first, the nacelle is raised without a generator, then the generator is lifted and placed in the nacelle. The final assembly to be lifted is the rotor with blades. In order to prevent the blades from swinging and hitting the tower during the lift, various strategies are used including using two cranes working in tandem.

Figure 14-5 Lifting of tower sections by the main crane (right). Auxiliary crane (left) assists with the lift. (Courtesy Vensys Energy AG.)

Figure 14-6 Main 500-t crane lifting the rotor assembly of 1.5-MW Vensys 77, 100-m hub in Nuekirchen wind farm near city of Eisenach in Thuringen Germany. Ropes are tied to the two top blades to ensure that it does not hit the tower. (Courtesy of Vensys Energy AG).

The blade assembly starts in the horizontal plane. The main crane attaches a sling, as shown in Fig. 14-6, and the auxiliary crane lifts the bottom blade. After both cranes have lifted the entire assembly to sufficient height, the auxiliary crane lowers the bottom blade while the main crane continues to lift the rotor. This continues until the entire assembly is vertical, as in Fig. 14-6.

Most of the joints in the turbine are bolt joints—from foundation to tower joint, between towers, blades to hub, hub to generator, generator to nacelle, and myriad of others. Insufficient tightening of bolts has been a significant cause of failures. Torque-based methods for tightening of bolts have been a source of problem. Reliable functioning of bolts requires that the bolts be subject to adequate tension. Correctly tensioned bolts are subjected to a small change in tension as external loads are applied, which leads to high fatigue life. Torque is not considered an accurate measure of tension, because friction between the bolt and nut can vary. Hydraulic tensioning of bolts is an alternate method of tightening, in which the bolts are tensioned to an appropriate level (desired tension + load transfer relaxation) and then the nut is turned down. This method is more commonly used. Other methods include use of direct tension indicating (DTI) compressible washers that squirt out a colored silicone when the correct bolt tension is applied.

Collection System and Substation Construction

This section pertains to construction and installation of the electrical infrastructure. A crew that specializes in electrical systems performs does these tasks.

- *Collection system.* Most projects use underground power and communications cables from the turbines to the substation. Trenches are dug 4–6 ft deep and 1 ft wide. Power cables are placed first and backfilled with soil that has acceptable thermal resistivity. The communications cables are placed over the power cables with several inches of backfill in the middle. Specialized trucks are used for laying cables. If the backfill soil does not have acceptable conduction properties to move the heat, then the conductors can fail leading to expensive repairs. Therefore, heat conductivity of soil must be tested and if backfill soil is inappropriate, then alternate backfill material is used. In addition, the cables must be inspected to ensure there is no pinching of cables before the trenches are compacted.

- *Substation and maintenance building construction.* Substation normally contains pad-mounted transformer, metal-enclosed switchgear, and other components like capacitors, if necessary. In any electrical installation, the following are key to safe operations: Proper grounding; testing of relay settings; tight connections between conductors and equipment.

- *SCADA Systems.* Installation and testing of SCADA system is done before commissioning can start. This involves installation of all the telecommunications equipment, SCADA server, SCADA software, and configuration. Testing of the SCADA system involves testing if each device that is connected to the SCADA system is able to:
  - Transmit status data when requested
  - Store operating data received from a device
  - Control of wind turbines based on stored logic
  - Communicate with external systems like grid operator, turbine manufacturer, and others
  - Provide remote access
  - Generate reports

Commissioning

Commissioning of a wind farm is usually an elaborate handover process of the project from the contractor and turbine manufacturer to the project owner. The following entities are usually involved:

- Turbine manufacturer, whose commissioning personnel perform the commissioning exercise
- Contractor, whose civil and electrical personnel assist the turbine manufacturer's representatives with testing, monitoring, and fixing
- Project owner, who takes over and owns the project at the end of commissioning
- Wind farm operator, who operates and maintains the facility for the project owner
- Local utility, who will buy the energy produced by wind farm
- Third-party independent expert, who works on behalf of the project owner to oversee the commissioning process

To avoid problems, the types of tests and outcomes during commissioning should be documented in the contract with the turbine manufacturer and the contractor. During contract negotiation, the turbine manufacturer's or contractor's preexisting checklist should be reviewed, and this list, along with any modifications, should be added to the contract.

The objective of the commissioning phase is to ensure that the complete wind plant is safe to operate, is producing energy in a reliable manner and of acceptable quality, and any outstanding defects have been identified. In addition, a plan of action has been agreed to by turbine manufacturer, contractor, and project owner to resolve outstanding issues. After the commissioning process, the ownership of the wind project turns over to the project owner. Typical criteria for passing the commissioning process are:

1. *All the utility interconnection criteria are met.* Prior to commissioning, the local utility or the regional transmission operator provides checklists and the operations engineers from the utility/RTO inspect the plant. As an example, consider a two-step procedure employed by Electric Reliability Council of Texas (ERCOT)[3]

    a. *Request to commission station.* ERCOT provides a "request to commission station" checklist. After conditions in the checklist are met, it is submitted to the utility/RTO by the operator. An approval and date of initial energize is issued for the interconnection of the wind farm after the following three conditions are met:
    - Sign off on the checklist by the utility/RTO
    - Resolution of all pending issues with utility's operations engineering
    - SCADA system is ready to monitor energy production

b. *Request for initial synchronization.* ERCOT provides a "request to initial synchronization" checklist. After this checklist is submitted, ERCOT then issues an approval to proceed with initial synchronization. One day ahead of synchronization, the shift supervisor is informed of the startup of test device; on the day of the startup, the project reconfirms start up of the test unit with the shift supervisor. After successful synchronization, any subsequent testing or operations is communicated by means of resource schedule.

2. 95% availability during 250 hours of continuous operation

3. Proper functioning of startup, normal shutdown, and emergency shutdown procedures

4. Proper functioning of switch gear in response to variety of fault conditions

5. Communication of data to SCADA system

Operations

After the wind farm is commissioned and is operational, the goals shift to:

- Maximizing energy production for the remaining life of project. This goal is achieved through maximizing availability and yield of each turbine.

- Minimizing operations and maintenance costs for the remaining life of project

- Managing day-to-day tasks, like providing day-ahead forecasts, operating the wind farm in a safe manner, protecting assets, and being a good neighbor

The difficult challenge for any operations group is to achieve these competing goals not on a day-to-day basis, but for the entire life of the project.

There are three organizational models for O&M: Project owner manages O&M, third-party manages O&M, and turbine manufacturer manages O&M for an extended period (10–12 years). The trend is toward the latter two models. Third-party O&M contracts with performance-based incentives is good model for managing a wind farm. Incentives can take the form: Profits from availability in excess of 97% are shared. Here the risk shifts from the project owner to a professional O&M company. The emerging model is the offer by turbine manufacturers of extended warranties of 10+plus years with a bundled contract for operations and maintenance. Both contracts are at a cost premium compared to owner-operated wind farm; however, the two models lower the risk to financiers.

References

1. American Wind Energy Association. *Wind Energy Siting Handbook,* AWEA, Washington, DC, 2008.
2. Sedgwick, B. "Wind farm infrastructure: A primer," *North American Wind Power.* 2007, July.
3. ERCOT, Operations Support Engineering, New Generator Commissioning Checklist [Online] April 2009. http://www.ercot.com/services/rq/re/reg/New%20Generation%20InterConnection%20-%20QSE%20checklist%20-%20v1%205.doc.

Index

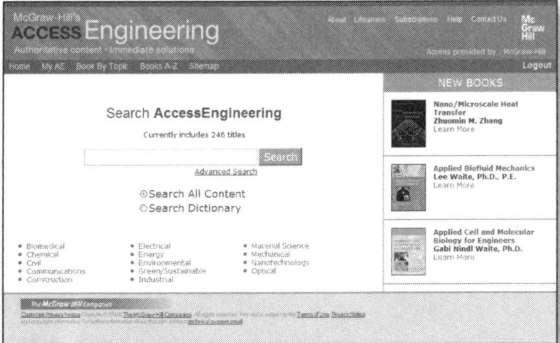
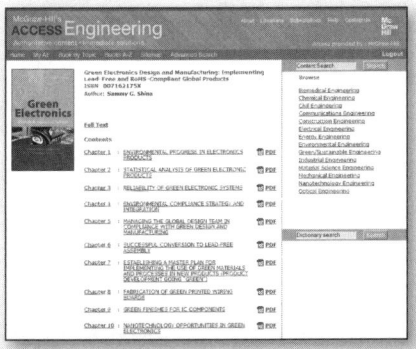 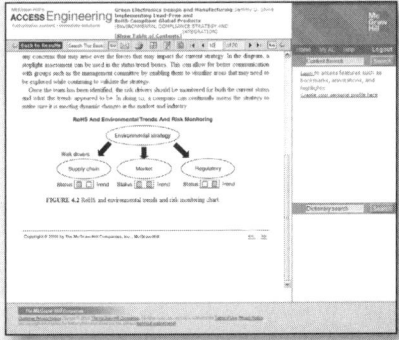